Laboratory Experiments in Microbiology

The Benjamin/Cummings Series in the Life Sciences

F. J. Ayala
Population and Evolutionary Genetics: A Primer (1982)

F. J. Ayala and J. A. Kiger, Jr.
Modern Genetics, second edition (1984)

F. J. Ayala and J. W. Valentine
Evolving: The Theory and Processes of Organic Evolution (1979)

C. L. Case and T. R. Johnson
Laboratory Experiments in Microbiology (1984)

R. E. Dickerson and I. Geis
Hemoglobin (1983)

P. B. Hackett, J. A. Fuchs, and J. W. Messing
An Introduction to Recombinant DNA Techniques: Basic Experiments in Gene Manipulation (1984)

L. E. Hood, I. L. Weissman, W. B. Wood, and J. H. Wilson
Immunology, second edition (1984)

J. B. Jenkins
Human Genetics (1983)

K. D. Johnson, D. L. Rayle, and H. L. Wedberg
Biology: An Introduction (1984)

R. J. Lederer
Ecology and Field Biology (1984)

A. L. Lehninger
Bioenergetics: The Molecular Basis of Biological Energy Transformations, second edition (1971)

S. E. Luria, S. J. Gould, and S. Singer
A View of Life (1981)

E. N. Marieb
Human Anatomy and Physiology Lab Manual: Brief Edition (1983)

E. N. Marieb
Human Anatomy and Physiology Lab Manual: Cat and Fetal Pig Versions (1981)

E. B. Mason
Human Physiology (1983)

A. P. Spence
Basic Human Anatomy (1982)

A. P. Spence and E. B. Mason
Human Anatomy and Physiology, second edition (1983)

G. J. Tortora, B. R. Funke, and C. L. Case
Microbiology: An Introduction (1982)

J. D. Watson, N. Hopkins, J. Roberts, and J. Steitz
Molecular Biology of the Gene, fourth edition (1984)

W. B. Wood, J. H. Wilson, R. M. Benbow, and L. E. Hood
Biochemistry: A Problems Approach, second edition (1981)

Laboratory Experiments in Microbiology

Christine L. Case
Skyline College

Ted R. Johnson
St. Olaf College

The Benjamin/Cummings Publishing Company, Inc.
Menlo Park, California * Reading, Massachusetts
London * Amsterdam * Don Mills, Ontario * Sydney

Sponsoring Editor: *Jane R. Gillen*
Production Coordinators: *Pat Franklin Waldo, Deborah Gale*
Copy Editor: *Carol Dondrea*
Interior Designer: *John Edeen*
Cover Designer: *Michael Rogondino*
Illustrators: *Cathleen Jackson-Miller with Kenneth R. Miller*
Adapted Art: *Barbara Haynes*

Sources for historical quotes used in Unit Introductions

Part 1 Fred, E. B. 1933. "Antoni van Leeuwenhoek." *J. Bacteriol.* 25:1

Part 2 Lechevalier, H. A. and M. Solotorovsky. 1974. *Three Centuries of Microbiology.* N.Y.: Dover Publications, Inc., p. 79.

Part 3 Society of American Bacteriologists (D. H. Bergey, Chairperson). 1923. *Bergey's Manual of Determinative Bacteriology,* 1st ed. Baltimore: Williams and Wilkins Co., p. 1.

Part 4 Lechevalier, H. A. and M. Solotorovsky, *ibid.,* p. 26.

Part 5 Stanley, W. M. and E. G. Valens. 1961. *Viruses and the Nature of Life.* New York: E. P. Dutton and Co., Inc. p. 8.

Part 6 Lechevalier, H. A. and M. Solotorovsky, *ibid.,* p. 502.

Part 7 Lechevalier, H. A. and M. Solotorovsky, *ibid.,* p. 25.

Part 8 Lechevalier, H. A. and M. Solotorovsky, *ibid.,* p. 45.

Part 9 Bulloch, W. 1938. *The History of Bacteriology.* N.Y.: Dover Publications, Inc., p. 46.

Part 10 Lechevalier, H. A. and M. Solotorovsky, *ibid.,* p. 263.

Part 11 Coetsch, R. N., ed. 1960. *Microbiology: Historical Contributions from 1776–1908.* New Brunswick, N.J.: Rutgers University Press, p. 130.

Part 12 DeKruif, P. 1953. *Microbe Hunters.* N.Y.: Harcourt, Brace, and World Inc., p. 195.

Library of Congress Cataloging in Publication Data

Case, Christine L., 1948-

Laboratory experiments in microbiology.

Includes index.
1. Microbiology—Laboratory manuals. I. Johnson, Ted R., 1946- II. Title.
QR63.C37 1984 576′.028 83-15743

ISBN: 0-8053-5040-3
12 13 14 15 -AL- 95 94 93 92 91

The Benjamin/Cummings Publishing Company, Inc.
2727 Sand Hill Road
Menlo Park, California 94025

Preface

Laboratory Experiments in Microbiology is a comprehensive, modern manual for undergraduate students in diverse areas, including the biologic sciences, the allied health sciences, agriculture, environmental science, nutrition, pharmacy, and various preprofessional programs. It contains more than 75 thoroughly class-tested exercises covering every area of microbiology and ranging from basic techniques to more advanced skills and applications. Most exercises require about an hour of laboratory time. By selecting an appropriate combination of exercises, the instructor can use this manual for almost any microbiology course, with any accompanying textbook.

Our Approach

Our goal in writing this manual has been twofold—to teach microbiologic techniques and to show students the importance of microbes in our daily lives and their central roles in nature. The exercises in this manual provide a solid base of fundamental microbiologic techniques and progress to more advanced skills and applications. However, we have not segregated applications in later exercises but have been careful to include them wherever feasible, right from the start. Almost every exercise includes an actual experiment. We hope in this way to make laboratory sessions *interesting* for students as well as to provide a variety of opportunities for *reinforcement* of the technical skills they have learned. To demonstrate further some practical uses of microbiology, frequently we have included material with direct application to procedures performed in clinical and commercial laboratories.

Scope and Sequence

This manual is divided into 12 parts. The part introduction explains the unifying theme for that part. Exercises in the first three parts provide sequential development of fundamental techniques. The remaining exercises are as independent as possible to allow the instructor to select the most desirable sequence. A *Techniques Required* section preceding each experiment lists prerequisites from earlier exercises.

Part 1, *Microscopy,* emphasizes observation through the microscope. Practice in use and care of the microscope is followed by observations of microbes, to familiarize students with their sizes.

Part 2, *Staining Methods,* begins with an explanation of how to handle bacterial cultures and teaches the use of the most common stains, including preparation of stained samples and their examination. Exercise 10, "Morphologic Unknown," is a unique exercise introducing the concept of unknowns, and can be used to test knowledge of staining methods.

Part 3, *Cultivation of Bacteria,* stresses aseptic technique and covers the isolation of bacterial strains and the maintenance of bacterial cultures. Exercises dealing with special media and bacterial nutrition prepare the student for the next part.

Part 4, *Metabolic Activities of Bacteria,* includes seven exercises on bacterial metabolism that provide the tools for Exercise 24, "Unknown Identification." These exercises are also useful for learning biochemistry. Exercise 25, "Rapid Identification Method," demonstrates computerization of laboratory work.

Part 5, *Bacterial Genetics,* is of particular interest because of recent notable advances in this field. It begins with isolation of a bacterial mutant. The other four exercises in this part show how mutants are used in genetic research. The last exercise is the testing of suspected chemical carcinogens by the Ames Test.

Part 6, *Fungi, Protozoans, and Algae,* examines the diversity and ecologic niches of eucaryotic organisms studied in microbiology. Free-living representatives are included in this unit; parasitic forms are the subject of exercises in Part 12, Medical Microbiology.

Part 7, *Viruses,* provides opportunities to isolate and cultivate viruses. Techniques involving bacteriophages are covered in three exercises; a plant virus is cultured in the last exercise.

Part 8, *Growth of Microorganisms,* deals with the effects on growth of environmental conditions such as temperature and the presence of oxygen.

Part 9, *Control of Microbial Growth,* provides practical applications of concepts of microbial growth. Methods of controlling unwanted microbes in food or a clinical environment are examined in this part.

Part 10, *Microbes in the Environment,* includes standard methods for the examination of food and water for microbiologic quality. Biogeochemical cycles and soil microbiology are the subjects of Exercises 51, 52, and 53. Exercise 55, "Microbes Used in the Production of Foods," is an opportunity to study food microbiology.

Part 11, *Medical Microbiology,* emphasizes procedures employed in the clinical laboratory. Exercises on normal and pathogenic bacteria in the human body are followed by identification of an unknown from a simulated clinical sample. Animal viruses are difficult to culture in the laboratory; so, in a novel approach, paper models are used in "Animal Viruses," Exercise 64. This part concludes with exercises on parasitic fungi, protozoans, and helminths.

Part 12, *Immunology,* covers the host's response to infectious disease with exercises on nonspecific resistance, inflammation, and phagocytosis. Serologic tests, including precipitation and agglutination reactions, conclude this part.

Several appendices at the end of the manual provide a convenient reference to techniques required in several exercises.

To meet the specialized needs and interests of various student populations, both "advanced" and "enrichment" exercises are interspersed throughout the book. The advanced exercises are so designated because they depart from the usual sequence in general microbiology by introducing new techniques or specialized information. The enrichment exercises also depart from the usual sequence but do *not* involve new techniques or specialized information. The *Instructor's Guide* (discussed in a later section) identifies the advanced and enrichment exercises.

Organization of Each Exercise

The exercises in this manual may involve mastering a skill or procedure or understanding a particular concept. Most of the exercises are investigative by design, and the student is asked to analyze the experimental results and draw conclusions.

- Each exercise begins with a section called **Objectives,** which lists skills or concepts to be mastered in that exercise. The objectives can be used to test mastery of the new material after completing the exercise.
- The **Background** provides definitions and explanations for each exercise. The student is expected to refer to a textbook for more detailed explanations of the concepts introduced in the laboratory exercises.
- **Materials** lists include supplies and cultures needed for the exercise.
- **Techniques Required** gives any necessary cross-references to earlier exercises.
- **Procedures** are step-by-step instructions, stated as simply as possible, and frequently supplemented with diagrams. Questions are occasionally asked in the Procedures section to remind the student of the rationale for a step.
- The **Laboratory Report** at the back of the book provides a space to record results. Questions in the Laboratory Report ask for interpretation of results. In most instances, the results for each student team will be unique; they can be compared to the information given in the Background and other references but will not be identical to these references. The questions are designed to lead the student from a collection of data or observations to a conclusion. The range of questions in each exercise requires students to think about their results, recall facts, and then *use* this information to answer questions.

Illustrations

This book is generously illustrated with diagrams, drawings, and photographs, including 8 pages of photographs in full color. In all cases, the illustrations have been designed as learning aids to be *used* by the student.

Instructor's Guide

The comprehensive *Instructor's Guide* provides the instructor with all the information needed to set up and teach a laboratory course with this manual. It includes:

- Suggested laboratory class schedules
- General instructions for setting up the lab
- Information on obtaining and preparing cultures, media, and reagents
- A master table showing the techniques required for each exercise
- A table giving cross-references for each exercise to specific pages in each of the leading textbooks
- For each exercise: helpful suggestions, detailed lists of materials needed, and answers to all the questions in the student manual

To make *Laboratory Experiments in Microbiology* easy to use, the experiments have been carefully designed to use inexpensive, readily available, nonhazardous materials insofar as is possible. The exercises have been thoroughly tested in our classes in California and Minnesota by students with a wide variety of talents and interests. Our students have enjoyed their microbiology laboratory experiences; we hope yours will too!

Acknowledgements

We gratefully acknowledge the contribution of the material in Exercise 37 on phage typing by Elaine Correia and Sandra Burr of Crafton Hills College, and the material on replica plating used in Exercise 26 and Appendix B by C. W. Brady of the University of Wisconsin, Whitewater.

We are indebted to our students who participated in the development of these exercises in their lab work and evaluations and to the students who class tested this manual.

Lorraine Caraway, who patiently and skillfully typed the manuscript, also deserves our thanks.

We would like to commend the staff at Benjamin/Cummings for their support. In particular, we thank Pat Waldo, Jane Gillen, and Deborah Gale.

Last, but not least, our gratitude goes to Don Biederman, who provided timely encouragement and unfailing support, and Michelle Johnson, who gave her professional insights and was a sustaining presence.

Christine L. Case
Ted R. Johnson

Reviewers

James P. Amon, Wright State University
Larry H. Knipp, North Park College
Allan Konopka, Purdue University
Jeffrey M. Libby, Miami University
Peter P. Ludovici, University of Arizona
Kathi W. Moody, University of Florida
William L. Tidwell, San Jose State University

Contents

Introduction

Life would not long remain possible in the absence of microbes.

LOUIS PASTEUR

Welcome to microbiology! Microorganisms are all around us, and, as Pasteur pointed out a century ago, they play vital roles in the ecology of life on earth. In addition, some microorganisms provide important commercial benefits through their use in the production of chemicals (including antibiotics) and certain foods. Microorganisms are also major tools in basic research in the biologic sciences. Finally, as we all know, some microorganisms cause disease—in humans, other animals, and plants.

In this course, you will have first-hand experience with a variety of microorganisms. You will learn the techniques required to identify, study, and work with them. Before getting started, you will find it helpful to read through the suggestions on the next few pages.

Suggestions to Help You Begin

1. Science has a vocabulary of its own. New terms will be introduced in **boldface** throughout this manual. To develop a working vocabulary using these terms, make a list of these new terms and their definitions.
2. Because microbes are not visible without a microscope, common names have not developed for them. The word *microbe,* now in common use, was introduced in 1878 by Sedillot. The microbes used in these exercises are referred to by their *scientific names.* The names will be unfamiliar at first, but do not let that deter you. Practice saying them aloud. Most scientific names are taken from Latin and Greek roots. If you become familiar with these roots, the names will be easier to remember.
3. Microbiology usually provides the first opportunity undergraduate students get to work with *living organisms.* Microbes are relatively easy to grow and lend themselves to experimentation. Since there is variability in any population of living organisms, not all the experiments will "work" as the textbook says. The following exercise will illustrate what we mean:

   Write a description of *Homo sapiens* for a visitor from another planet: ______________________

   ____

   After you have finished, look around you. Do all of your classmates fit the description exactly? Probably not. Moreover, the more detailed you make your description the less conformity you will observe. During lab, you will make very detailed descriptions of an organism and probably find that this description does not match your reference exactly.
4. Microorganisms must be cultured or grown to complete most of the exercises in this manual. Accurate record keeping is therefore essential. Cultures will be set up during one laboratory period and will be examined for growth at the next laboratory period. You should mark the steps in each exercise with a bright marking pen or bookmark so that you can return to complete your Laboratory Report on that exercise. *Accurate records* and *good organization* of laboratory work will enhance your enjoyment and facilitate your learning.
5. *Observing* and *recording* your results carefully are the most important part of each exercise. Ask yourself the following questions for each experiment:

   What did the results indicate?

   Are they what I expected? If not, what happened?
6. If you do not master a technique, try again. In most instances, you will need to utilize the technique again later in the course.
7. Be sure you can answer the questions that are asked in the Procedure for each exercise. These questions are included to reinforce important points that will ensure a successful experiment.
8. Finally, carefully study the general procedures and safety precautions that follow.

General Procedures in Microbiology

In many ways, working in a microbiology laboratory is like working in the kitchen. As some famous chefs have said:

*Our years of teaching cookery have impressed upon us the fact that all too often a debutant cook will start in enthusiastically on a new dish without ever reading the recipe first. Suddenly an ingredient, or a process, or a time sequence will turn up, and there is astonishment, frustration, and even disaster. We therefore urge you, however much you have cooked, always to read the recipe first, even if the dish is familiar to you.... We have not given estimates for the time of preparation, as some people take half an hour to slice three pounds of mushrooms, while others take five minutes.**

1. Read the laboratory exercises *before* coming to class.
2. *Plan* your work so that all experiments will be completed during the assigned laboratory period. A good laboratory student, like a good cook, is one who can do more than one procedure at a time—that is, one who is efficient.
3. Use only the *required* amounts of materials so that everyone can do the experiment.
4. *Label* all of your experiments with your name, date, and lab section.
5. Even though you will do most exercises with another student, you must become familiar with all parts of each exercise.
6. Keep *accurate* notes and records of your procedures and results so you can refer to them for future work and tests. Many experiments are set up during one laboratory period and observed for results at the next laboratory period. To be sure you perform all the necessary steps and observations, your notes are essential.
7. *Demonstrations* will be included in some of the exercises. Each student should study the demonstrations and learn the content.
8. Let your instructor know if you are color blind; many techniques require discrimination of colors.
9. *Clean* up your work area when you are finished. Leave the laboratory clean and organized for the next student. Remember:

   Stain and reagent bottles should be returned to their original location.

   Slides should be washed and put back into the box clean.

   All markings on glassware (e.g., Petri plates and test tubes) should be removed before putting glassware into the marked discard trays.

   Glass Petri plates should be placed agar-side down.

   Swabs and pipettes should be placed in the appropriate disinfectant jars.

   Disposable plasticware should be placed in marked autoclave baskets.

   Used paper towels should be discarded.

*Julia Child, L. Bertholle, and S. Beck, 1961. *Mastering the Art of French Cooking,* vol. 1. New York: Alfred A. Knopf, Inc.

Safety Precautions

1. All bacteria are potential pathogens, so use aseptic technique. Perform all experiments as demonstrated so your culture remains in the desired container. Aseptic technique will prevent the participation of microbes from the environment in your experiment.
2. Keep your laboratory bench clean and free of extra books and papers, backpacks, coats, and purses.
3. At the beginning of each laboratory period, clean your workbench with disinfectant. Repeat this procedure before leaving the laboratory. The disinfectant used in this laboratory is ________.
4. Don't eat, drink, or smoke in the laboratory Don't put anything (for example, pencils or fingers) in your mouth or you may quickly acquire new microflora.
5. Accidents
   a. If you spill or break something, don't panic but don't ignore it. Spilled cultures should be disinfected before attempting to clean them up. Ask the instructor for assistance.
   b. If you cut or burn yourself, let the instructor know immediately; bacteria love open wounds.
6. Personal care
   a. Be careful of laboratory burners. Don't lean over them if you have long hair or a beard. Long hair should be tied back.
   b. Wear a lab coat or lab apron. If a culture is spilled, the coat is easier to sterilize than your clothes. Also, a lab coat will prevent permanent dyeing of your clothes with laboratory stains.
7. Keep your cultures current; discard old experiments.
8. Do not remove lab materials from the laboratory without permission.
9. Wash your hands with soap and water before leaving the laboratory.
10. If you got this far in the instructions, you'll probably do well in lab. Enjoy lab and make a new friend.

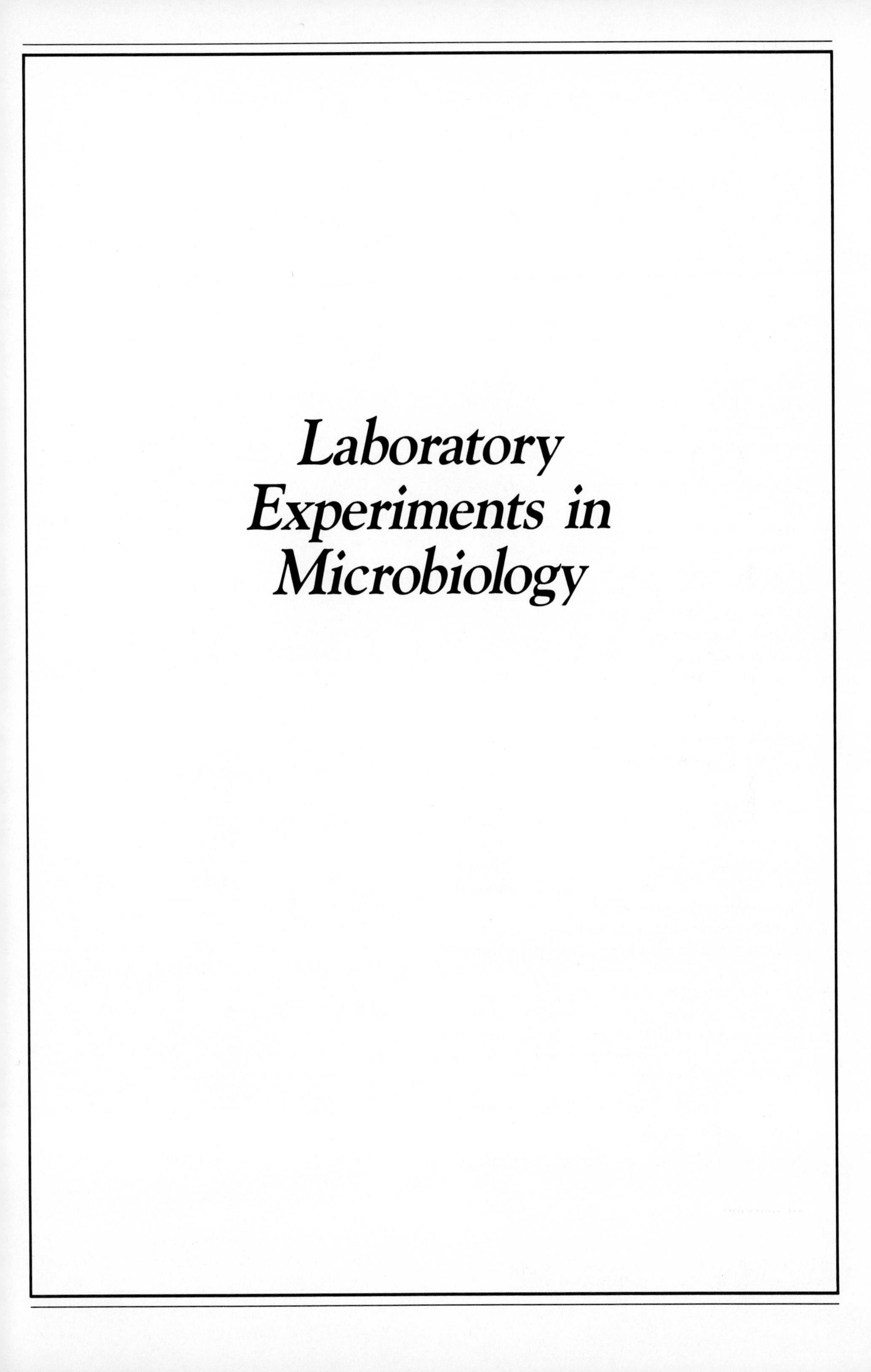

Laboratory Experiments in Microbiology

PART 1

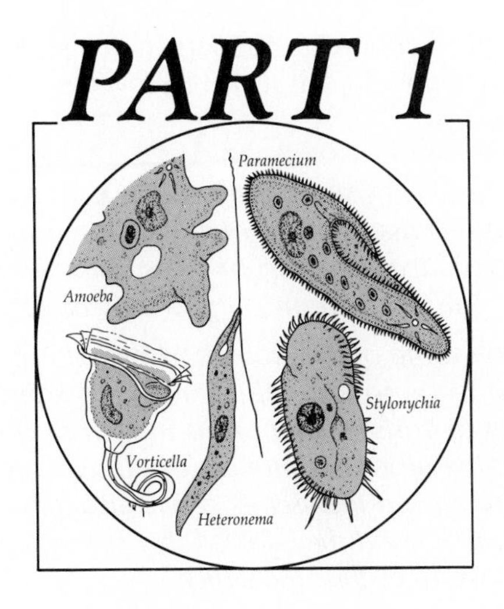

Microscopy

EXERCISES

Microscopes and microscopy (microscope technique) are introduced in Exercises 1 through 4. The microscope is a useful tool for a microbiologist. These exercises are designed to help you become familiar with and proficient in the use of the compound light microscope. This knowledge will be a valuable aid in later exercises.

Beginning students frequently become impatient with the microscope and forgo this opportunity to practice and develop their observation skills. Simple observation is a critical part of any science. Making discoveries by observation requires *curiosity* and *patience*. We cannot provide procedures for observation but we can offer this suggestion:

Make *careful* sketches to enhance effective observation. You need not be an artist to draw what you see. In your drawings pay special attention to the following:

1. Size relationships. For example, how big is a bacterium relative to a protozoan?
2. Spatial relationships. For example, where is one bacterium in relation to the others? Are they all together in chains?

3. Behavior. For example, are individual cells moving or are they all flowing in the liquid medium?
4. Sequence of events. For example, were cells active when you first observed them?

Looking at objects through a microscope is not easy at first, but with a little practice you, like Leeuwenhoek, will make discoveries in the microcosms of raindrops (see the figure).

> *Tho my teeth are kept usually very clean, nevertheless when I view them in a Magnifying Glass, I find growing between them a little white matter as thick as wetted flour: In this substance tho I do not preceive any motion, I judged there might probably be living creatures.*
>
> *I therefore took some of this flour and mixed it either with pure rain water wherein were no Animals; or else with some of my Spittle (having no air bubbles to cause a motion in it) and then to my great surprise perceived that the aforesaid matter contained very many small living Animals, which moved themselves very extravagantly. Their motion was strong and nimble, and they darted themselves thro the water or spittle, as a Jack or Pike does thro the water.*
>
> —Anton van Leeuwenhoek, 1684

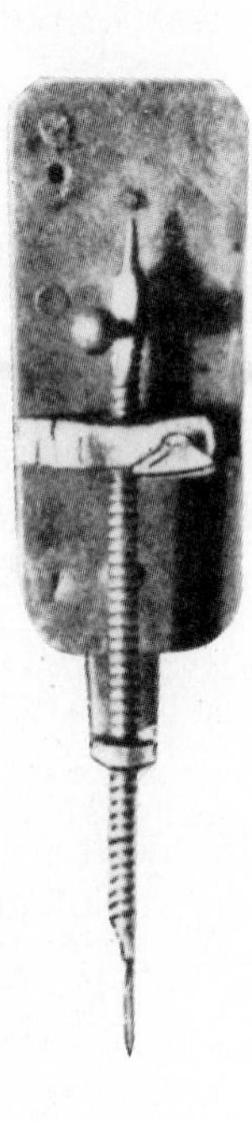

A simple microscope made by Anton van Leeuwenhoek to observe living organisms too small to be seen with the naked eye. The specimen was placed on the tip of the adjustable point and viewed from the other side through the tiny round lens. The highest magnification with his lenses was about 300×.

EXERCISE 1

Use and Care of the Microscope

The most important discoveries of the laws, methods and progress of nature have nearly always sprung from the examination of the smallest objects which she contains.

J. B. LAMARCK
PHILOSOPHIE ZOOLOGIQUE, 1809

Objectives

After completing this exercise you should be able to

1. Demonstrate correct use of the compound light microscope.
2. Diagram the path of light through a compound microscope.
3. Name the major parts of the compound microscope.
4. Identify the three basic morphologies of bacteria.

Background

Virtually all organisms studied in microbiology cannot be seen with the naked eye, but require the use of optical systems for magnification. The microscope was invented shortly before 1600 by either Zacharias Janssen of the Netherlands or Galileo Galilei of Italy. History has been unable to determine which scientist should be given credit. The microscope was not used to examine microorganisms until the 1680s, when a clerk in a dry goods store, Anton van Leeuwenhoek, examined scrapings of his teeth and any other substances he could find. The early microscopes, called **simple microscopes,** consisted of a biconvex lens and were essentially magnifying glasses. To see microbes, a **compound microscope,** which involves a two-lens system between the eye and the object, is required. This optical system magnifies the object, and an illumination system (sun and mirror or lamp) ensures that adequate light is available. A **brightfield microscope,** which has dark objects in a bright field, is used most often.

You will be using a brightfield compound microscope similar to that seen in Figure 1-1. The basic frame of the microscope consists of a **base,** a **stage** to hold the slide, an **arm** for carrying the microscope, and a **tube** for transmitting the magnified image. The stage may have two clips or a movable mechanical stage to hold the slide. The light source is in the base. Above the light source is a **substage condenser,** which consists of several lenses that concentrate light on the slide by focusing it into a cone, as shown in Figure 1-1. The condenser has an **iris diaphragm** that controls the angle and size of the cone of light. This ability to control the amount of light ensures that optimal light will reach the image. Above the stage, on one end of the body tube, is a revolving nosepiece holding three or four **objective lenses.** At the upper end of the tube is an **ocular** or **eyepiece lens** (10× – 12.5×). If a microscope has only one ocular lens, it is called a **monocular** microscope; a **binocular** microscope has two ocular lenses.

By moving the tube closer to the slide or the stage closer to the objective lens, using the coarse or fine adjustment knobs, one can focus the image. The larger knob, the **coarse adjustment,** is used for focusing with the low power objectives (4× and 10×), and the smaller knob, the **fine adjustment,** is used for high-power focusing and oil immersion. The coarse adjustment knob moves the lenses or the stage longer distances. The area seen through a microscope is called the **field of vision.**

The **magnification** of a microscope depends on the type of objective lens used with the ocular. Compound microscopes have three or four objective lenses mounted on a nosepiece: scanning (4×), low power (10×), high-dry (40–44×), and oil immersion (97–100×). The magnification of each lens is stamped on the barrel. The total magnification of the object is calculated by multiplying the magnification of the ocular (usually 10×) by the magnification of the objective

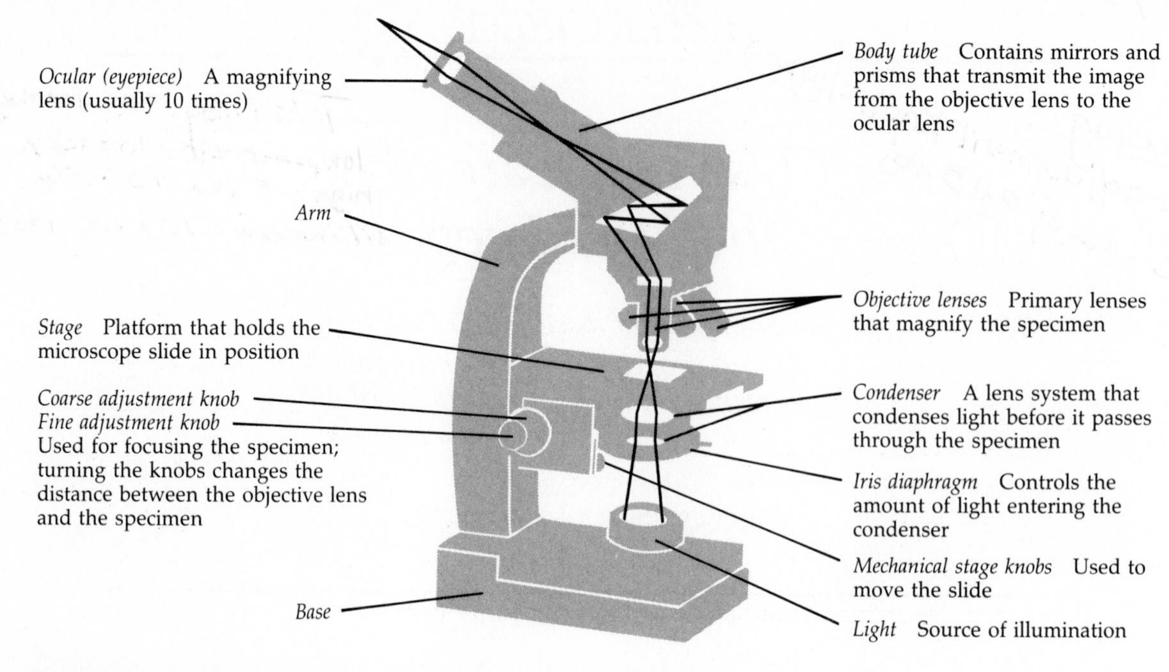

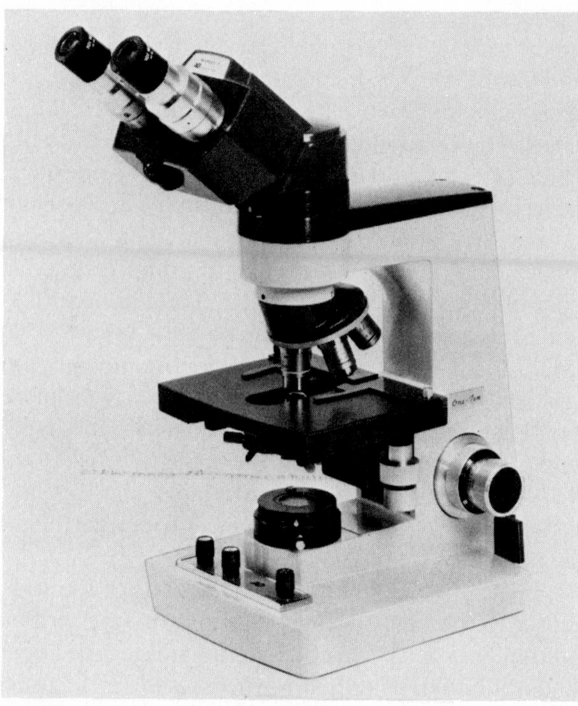

Figure 1-1.
The compound light microscope: its principal parts and their functions. Lines from the light source through the ocular lens illustrate the path of light.

lens. The most important lens in microbiology is the **oil immersion lens;** it has the highest magnification (97–100×) and must be used with immersion oil. Optical systems could be built to magnify much more than the 1000× magnification of your microscope, but the resolution would be poor.

The ability of lenses to reveal fine detail or two points distinctly separated is called **resolving power,** or **resolution.** An example of resolution is a car approaching you at night. At first only one light appears, but as it nears, you can distinguish two lights. The resolving power is a function of the wavelength of light used and a characteristic of the lens system called numerical aperture. Resolving power is the greatest when two objects are seen as distinct even though they are very close together. Resolving power is expressed in units of length; the smaller the distance, the better the resolving power.

$$\text{Resolving power} = \frac{\text{Wavelength of light used}}{2 \times \text{numerical aperture}}$$

Smaller wavelengths of light increase resolving power. The effect of decreasing the wavelength can be seen by electron microscopes (Exercise 3), which utilize electrons as a source of "light." The electrons have an extremely short wavelength and result in excellent resolving power. A light microscope has a resolving power of about 200 nanometers (nm), whereas an electron microscope has a resolving power of less than 0.2 nm. The numerical aperture is engraved on the side of each objective lens (usually abbreviated N.A.). If the numerical aperture increases, for example, from 0.65 to 1.25, the resolving power is improved. The **numerical aperture** is dependent on the maximum angle of the light entering the objective lens and on the **refractive index** (the amount the light bends) of the material (usually air) between the objective lens and the slide. This relationship is defined by the following:

N.A. = $N \sin \theta$
N = Refractive index of medium
θ = Angle between the most divergent light ray gathered by the lens and the center of the lens (Figure 1-2)

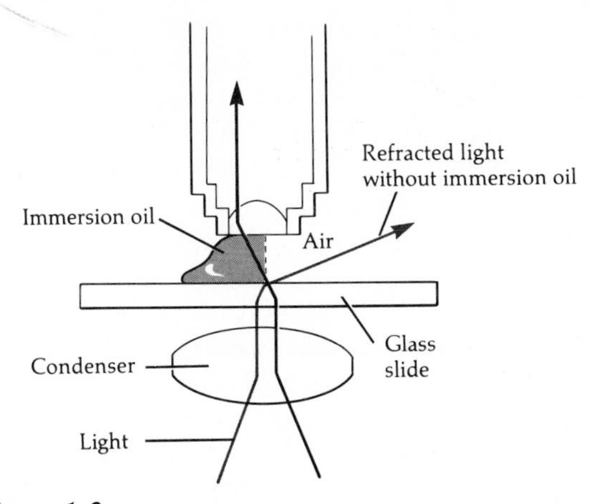

Figure 1-2
Refractive index. Since the refractive indices of the glass microscope slide and immersion oil are the same, the oil keeps the light rays from refracting.

As shown in Figure 1-2, light is refracted when it emerges from the slide because of the change in media as the light passes from glass to air. When immersion oil is placed between the slide and the oil immersion lens, the light ray continues without refraction because immersion oil has the same refractive index ($N = 1.52$) as glass ($N = 1.52$). This can be seen easily. When you look through a bottle of immersion oil, you cannot see the glass rod in it because of the identical N values of the glass and immersion oil. The result of using oil is that light loss is minimized and the lens focuses very close to the slide.

As light rays pass through a lens they are bent to converge at the **focal point,** where an image is formed (Figure 1-3*a*). When you bring the center of a microscope field into focus, the periphery may be fuzzy due to the curvature of the lens, resulting in multiple focal points. This is called **spherical aberration** (Figure 1-3*b*). Spherical aberrations can be minimized by the use of the iris diaphragm, which eliminates light rays to the periphery of the lens, or by a series of lenses resulting in essentially a flat optical system. Sometimes a multitude of colors, or **chromatic aberration,** is seen in the field. This is due to the prismlike effect of the lens as various wavelengths of white light pass through to a different focal point for each wavelength (Figure 1-3*c*). Chromatic aberrations can be minimized by the use of filters (usually blue) or lens systems corrected for red and blue light, called achromatic lenses, or lenses corrected for red, blue, and other wavelengths, called apochromatic lenses. Of course, the most logical, but most expensive, method of eliminating chromatic aberrations is to use a light source of one wavelength, or **monochromatic light.**

Compound microscopes require a light source. The light may be reflected to the condenser by a mirror under the stage. If your microscope has a mirror, the sun or a lamp may be used as the light source. Most newer compound microscopes have a built-in

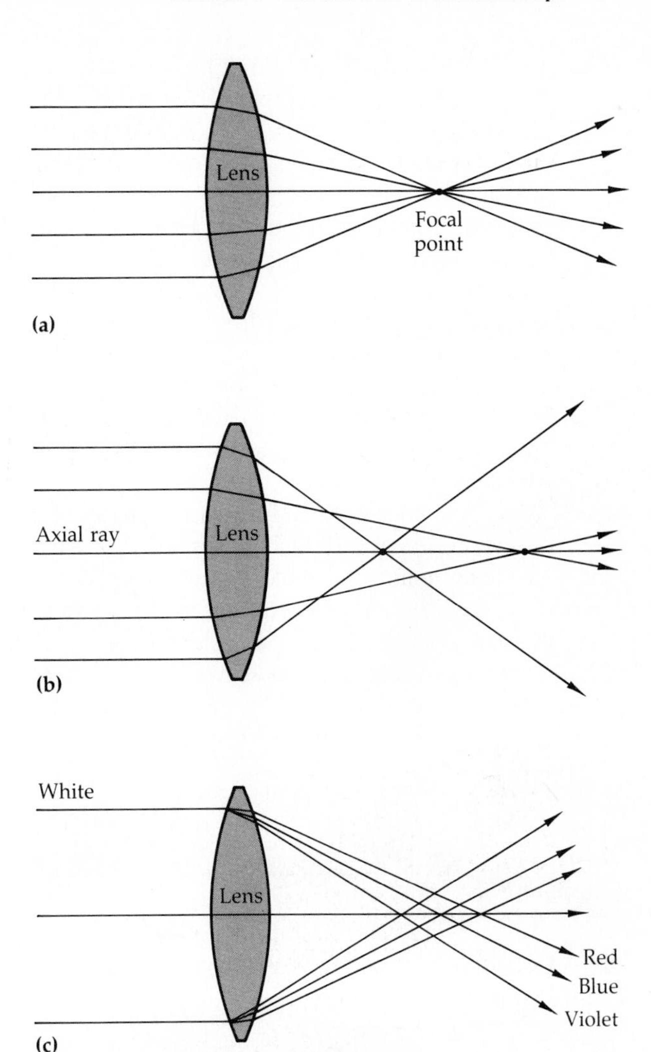

Figure 1-3
Focal point. **(a)** An image is formed when light converges at one point called the focal point. **(b)** Spherical aberration. Curved lenses result in light passing through one region of the lens having a different focal point than light passing through another part of the lens. **(c)** Chromatic aberration. Each wavelength of light may be given a different focal point by the lens.

illuminator in the base. The intensity of the light can often be adjusted with a transformer or rheostat.

The microscope is a very important tool in microbiology and it must be used carefully and correctly. Follow these guidelines *every* time you use your microscope.

General Guidelines

1. Carry the microscope with one hand beneath the base and one hand on the arm.
2. Do not tilt the microscope but instead adjust your stool so you can comfortably use your microscope.
3. Observe the slide with both eyes open. You will have less eye strain.
4. Always focus by moving the lens away from the slide.

5. Always focus slowly and carefully.
6. When using the low-power lens, the iris diaphragm should be barely open so that good contrast is achieved. More light is needed with higher magnification.
7. Before using the oil immersion lens, have your slide in focus under high power. *Always focus with low power first.*
8. Keep the stage clean and free of oil. Keep all lenses except the oil immersion lens free of oil.
9. Keep all lenses clean. Use *only* lens paper to clean them. Wipe oil off before putting your microscope away. Do not touch the lenses with your hands.
10. Clean the ocular lens carefully with lens paper. If dust is present, it will rotate as you turn the lens.
11. After use, remove the slide, wipe oil off, put the dust cover on, and return your microscope to the designated area.
12. When a problem does arise with your microscope, obtain help from the instructor. Do not use another microscope unless yours is declared "out of action."

Materials

Microscope

Immersion oil

Lens paper

Prepared slides of algae, fungi, protozoans, and bacteria.

Techniques Required

Part 1, Introduction

Procedure

1. Place your microscope on the bench squarely in front of you.
2. Obtain a slide of algae or fungi and place it on the stage.
3. Adjust the eyepieces on a binocular microscope to your own personal measurements.
   a. Look through the eyepieces and, using the thumb wheel, adjust the distance between the eyepieces until one circle of light appears.
   b. With the low-power (10×) objective in place, cover the left eyepiece with a small card and focus the microscope on the slide. When the right eyepiece has been focused, remove your hand from the focusing knobs and cover the right eyepiece. Looking through the microscope with your left eye, focus the left eyepiece by turning the eyepiece adjustment. Make a note of the number at which you focused the left eyepiece so you can adjust any binocular microscope for your own eyes.
4. Raise the condenser up to the stage. On some microscopes, the condenser can be focused by the following procedure:
   a. Focus with the 4× or 10× objective.
   b. Close the iris diaphragm so only a minimum of light enters the objective lens.
   c. Lower the condenser until the light is seen as a circle in the center of the field. On some microscopes the circle of light may be centered (Figure 1-4) using the centering screws found on the condenser.
   d. Raise the condenser up to the slide, lower it, and stop when the color on the periphery changes from pink to blue (1 or 2 mm below the stage).
   e. Open the iris diaphragm until the light just fills the field.
5. Diagram some of the cells on the slide under low power. Use a minimum of light by adjusting the ______________________.
6. When an image has been brought into focus with low power, the turret may be rotated to the next lens and the subject will remain almost in focus. All of the objectives (with the possible exception of the 4×) are **parfocal.** That is, when a subject is in focus with one lens, it will be in focus with all of the lenses. When you have completed your observations under low power, swing the high-dry objective into position and focus. Use the fine adjustment. Only a slight adjustment should be required. Why? ______________________
   More light is usually needed. Again, draw the general size and shape of some cells.
7. Move the high-dry lens out of position and place a drop of immersion oil on the area of the slide you are observing. Carefully click the oil immersion lens into position. It should now be immersed in the oil (Figure 1-5). Fine adjustment should bring the object into focus. Note the shape and size of the cells. Did the color of the cells change with the different lenses? ______________________ Did the size of the field change? ______________________

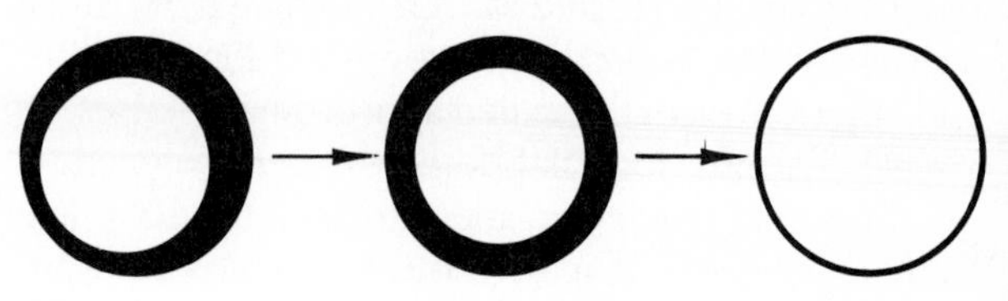

Figure 1-4
Using low power, lower the condenser until a distinct circle of light is visible. Center the circle of light using the centering screws. Open the iris diaphragm until the light just fills the field.

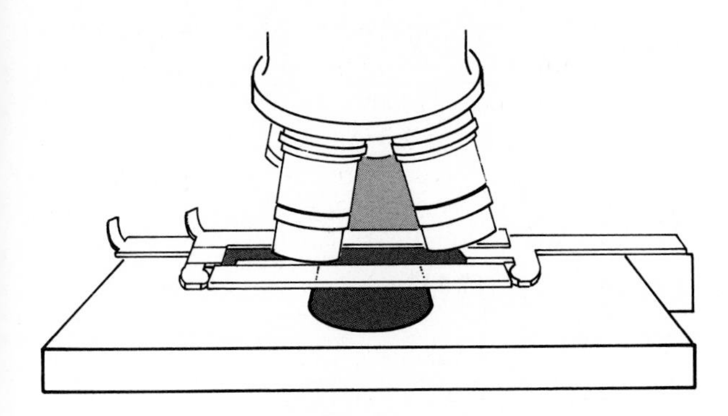

(a) Move the high-dry lens out of position.

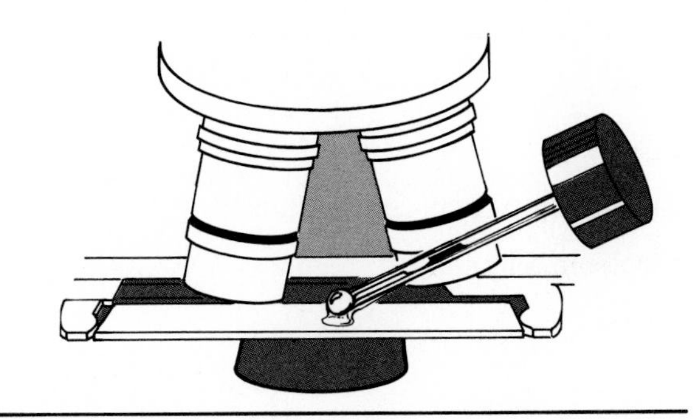

(b) Place a drop of immersion oil in the center of the slide.

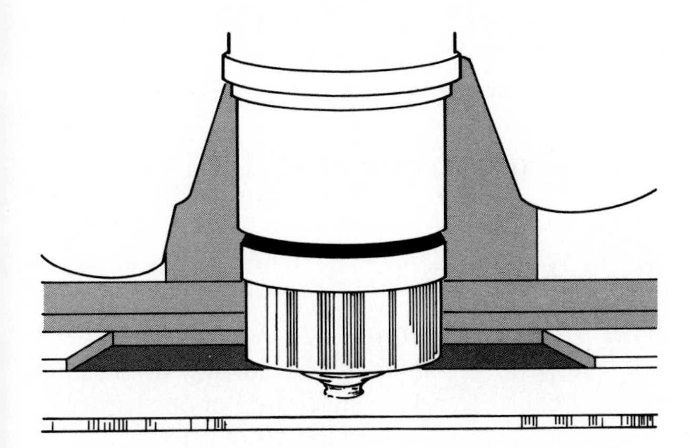

(c) Move the oil-immersion lens into position.

Figure 1-5

Using the oil immersion lens.

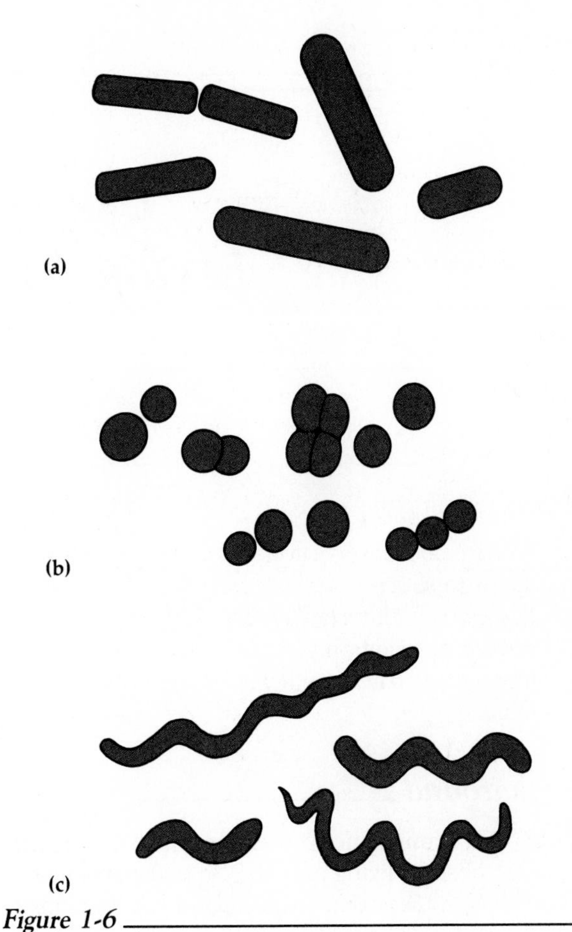

Figure 1-6

Basic shapes of bacteria. **(a)** Bacillus (pl. bacilli). **(b)** Coccus (pl. cocci). **(c)** Spirillum (pl. spirilla).

8. When your observations are completed, move the turret to bring a low-power objective into position. *Do not* rotate the high-dry (40×) objective through the immersion oil. Clean the oil off the objective lens with lens paper and clean off the slide with tissue paper or a paper towel. Remove the slide. Repeat this procedure with all the available slides. When observing the bacteria, note the three different morphologies or shapes shown in Figure 1-6.
9. Return your microscope to its cabinet.
10. Have a peaceful day while the term is still young.

Turn to the Laboratory Report for Exercise 1.

EXERCISE 2

Examination of Living Microorganisms

Objectives

After completing this exercise you should be able to

1. Prepare and observe wet mount slides and hanging-drop slides.
2. Distinguish different types of microbes in unstained preparations.
3. Distinguish true motility from Brownian movement.

Background

Anton van Leeuwenhoek was the first known individual to observe living microbes in a suspension. Unfortunately, he was very protective of his homemade microscopes and left no descriptions of how to make them. During his lifetime he kept "for himself alone" his microscopes and his method of observing "animalcules." Leeuwenhoek made a new microscope for each specimen. Directions for making a replica of Leeuwenhoek's microscope (see the figure on page 2) can be found in *American Biology Teacher.**

In this exercise you will examine, using wet mount techniques, different fluid environments to help you become aware of the numbers and varieties of microbes found in nature. The microbes that are seen will be exhibiting either Brownian movement or true motility. **Brownian movement** is not true motility but rather is movement caused by the molecules in the liquid striking an object and causing the object to shake or bounce. In Brownian movement the particles and microorganisms all vibrate at about the same rate and maintain their relative positions. Motile microorganisms move from one position to another. Their movement appears more directed than Brownian movement and occasionally the cells may roll or spin.

Many kinds of microbes such as protozoans, algae, and bacteria can be found in pond water and in infusions of organic matter. Direct examination by the hanging-drop method is very useful in determining size, shape, and movement.

*W. G. Walter and H. Via. 1968. "Making a Leeuwenhoek Microscope Replica." *American Biology Teacher 30* (6): 537–539.

Materials

Slides

Cover slips

Hanging-drop (depression) slide

Petroleum jelly

Pasteur pipettes

Cultures

Hay infusion, incubated 1 week in light

Hay infusion, incubated 1 week in dark

Peppercorn infusion

12–18-hour-old broth culture of *Bacillus*

Techniques Required

Exercise 1

Procedure

Wet Mount Technique

1. Suspend the infusions by stirring or shaking carefully. Transfer a small drop of one hay infusion, using a Pasteur pipette, to a slide.
2. Handle the cover slip by its edges like a phonograph record and place it on the drop.
3. Gently press on the cover slip with the end of a pencil.
4. Place the slide on the microscope stage and observe with low power. Adjust the iris diaphragm so a small amount of light is admitted. Concentrate your observations on the larger, more rapidly moving organisms. At this magnification, bacteria are barely discernible as tiny dots. Figure 2-1 may be helpful in identifying some of the microorganisms.
5. Examine with the high-dry lens, then increase the light and focus carefully. Bacteria should now be magnified sufficiently to see them.
6. After recording your observations, examine the slide under oil immersion. Some microorgan-

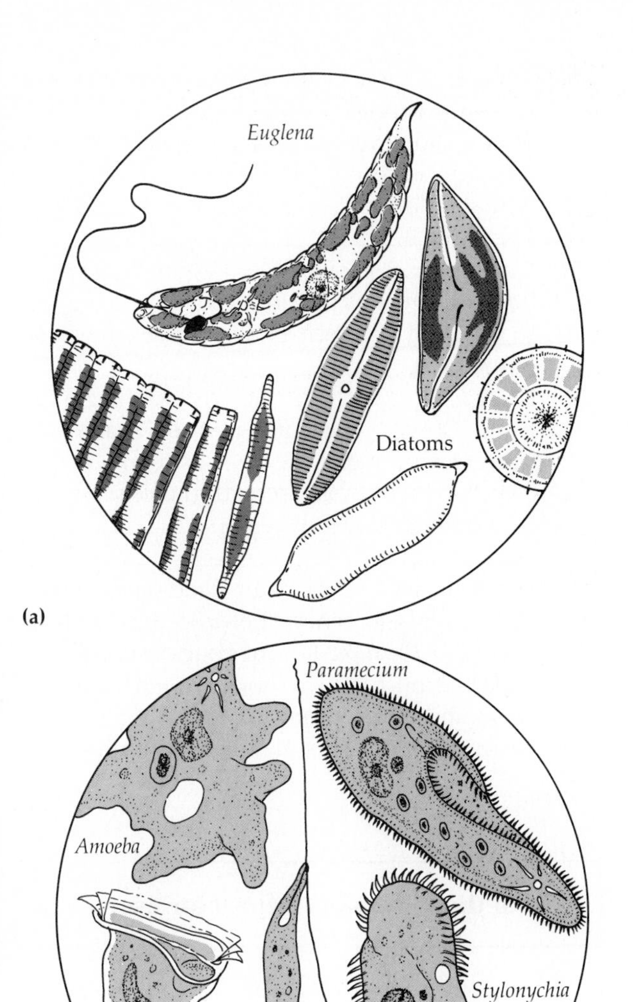

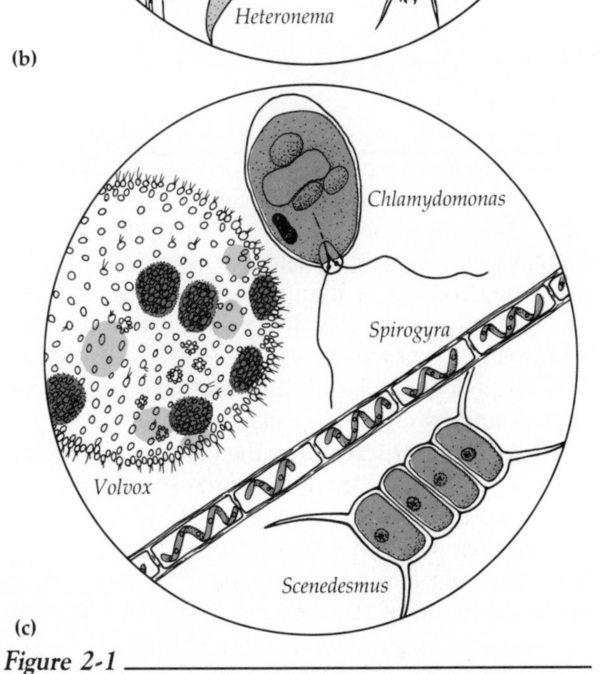

Figure 2-1

Some common protists and algae that can be found in infusions. **(a)** Algallike protists. **(b)** Animallike protists. **(c)** Algae.

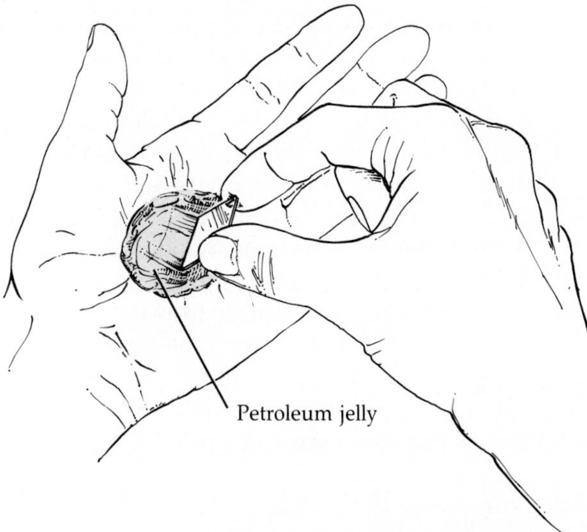

(a) Place a ring of petroleum jelly around the edge of a cover slip.

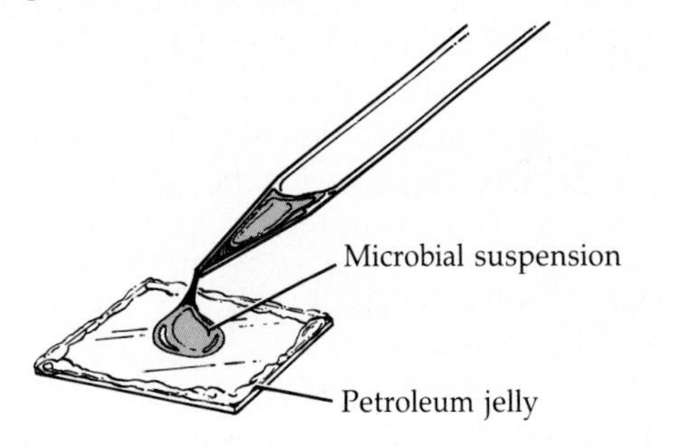

(b) Place a drop of an infusion in the center of the cover slip.

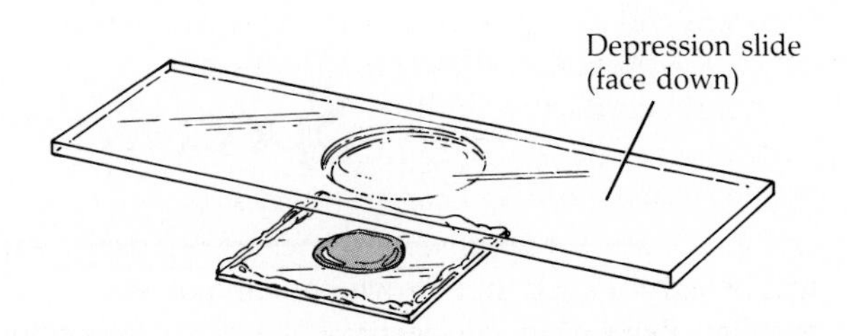

(c) Place the depression slide on the cover slip.

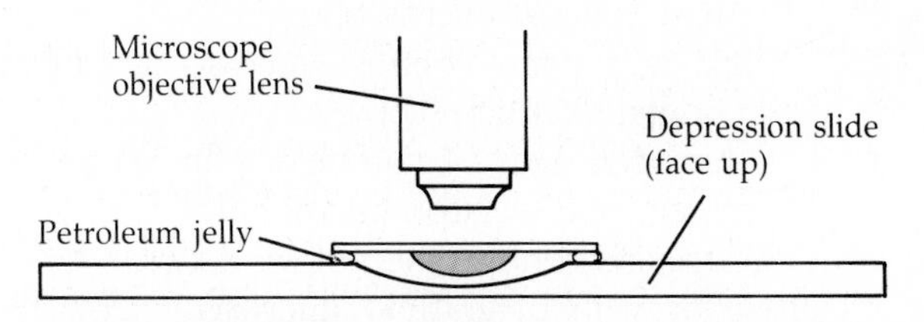

(d) Place the slide, cover slip up, on the microscope stage and observe under the low and high-dry objectives.

Figure 2-2

Preparing a hanging-drop slide.

isms are motile while others exhibit Brownian movement.

7. If you want to observe the motile organisms further, place a drop of alcohol or Gram's iodine at the edge of the cover slip and allow it to run under and mix with the infusion. What does the alcohol or iodine do to these organisms? _____ _____They can now be observed more carefully.
8. Record your observations, noting the relative size and shape of the organisms.
9. A wet mount should be made from the other hay infusion and observed, using the low and high-dry objectives. Record your observations.
10. Clean all the slides and return them to the slide box. Cover slips can be discarded.

Hanging-Drop Procedure

1. Obtain a hanging-drop slide.
2. Place a small amount of petroleum jelly on the palm of your hand and smear it into a circle about the size of a 50-cent coin.
3. Pick up a cover slip (by its edges) and carefully scrape the petroleum jelly with an edge of the cover slip. Repeat with the other three edges (Figure 2-2*a*), keeping the petroleum jelly on the same side of the cover slip.
4. Place the cover slip on a paper towel, with the petroleum jelly side up.
5. Transfer a drop of the peppercorn infusion to the cover slip.
6. Place a slide over the drop and quickly invert so the drop is suspended (Figure 2-2*b* and 2-2*c*). Why should the drop be hanging? _____________
7. Examine under low power (Figure 2-2*d*) by locating the edge of the drop and moving the slide so the edge of the cover slip crosses the center of the field.
8. Reduce the light with the iris diaphragm and focus. Observe the different sizes, shapes, and types of movement.
9. Switch to high-dry and record your observations. Do not focus down. Why not? _____________
10. When finished, clean your slide, and, using a new cover slip, repeat the procedure with the culture of *Bacillus*. Record your observations.
11. Wipe the oil from the objective lens with lens paper and return your microscope to its proper location. Clean your slides well and return them.

Turn to the Laboratory Report for Exercise 2.

EXERCISE 3

Specialized Microscopy

Objectives

After completing this exercise you should be able to

1. Use phase-contrast microscopes.
2. Explain how phase-contrast microscopy differs from brightfield microscopy.
3. List the advantages of darkfield, phase-contrast, fluorescent, and electron microscopy.
4. Compare and contrast the following types of microscopy with brightfield microscopy: phase-contrast, darkfield, fluorescent, and electron.

Background

Brightfield microscopy is of little value for observing unstained microorganisms. Since the optical properties of the organisms and their aqueous environment are very similar, very little contrast can be seen. Two other types of compound microscopes, however, are useful for observing living organisms: darkfield and phase-contrast microscopes. These microscopes optically increase contrasts between the organism and background by utilizing special condensers.

In **darkfield microscopy** the objects are light and the field is dark. In brightfield microscopy, light rays that strike the specimen are reflected away from the lens (Figure 3-1*a*). The darkfield condenser concentrates the light into a hollow cone of light at such an angle that none of the light rays reach the objective lens unless they pass through an object such as a cell to change their direction (Figure 3-1*b*). An opaque darkfield disk eliminates all central light rays. Thus, the objects appear brightly illuminated against a dark background. Darkfield microscopy allows the investi-

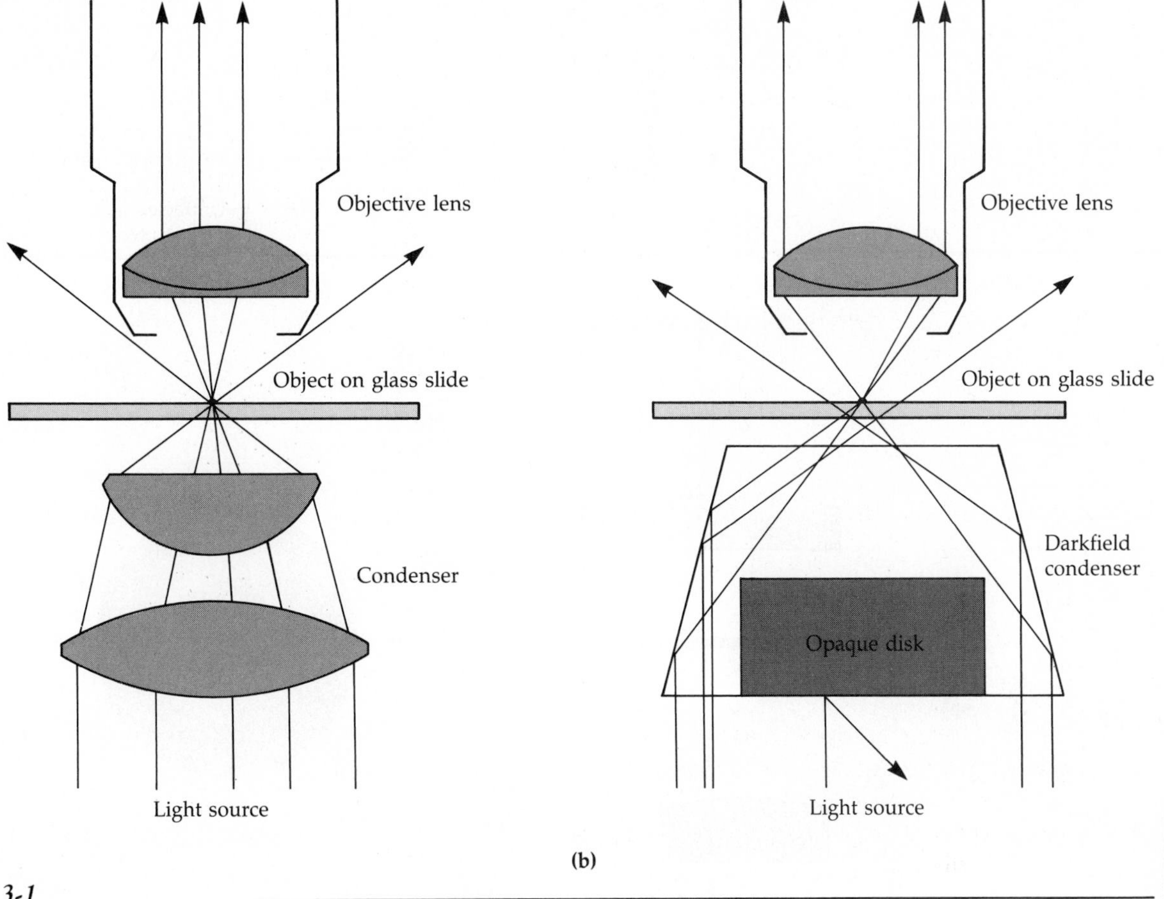

Figure 3-1

A comparison of brightfield and darkfield microscopy. **(a)** In brightfield microscopy, light is reflected away from the objective lens by the specimen. **(b)** In darkfield microscopy, only light rays that go through the object reach the objective lens.

gator to observe the shape and motility of unstained live organisms. Darkfield microscopy is valuable for observing the spirochete *(Treponema pallidum)* that causes syphilis. This bacterium is not stainable with conventional stains but can be observed in direct smears with darkfield microscopy.

In **phase-contrast microscopy,** small differences in the refractive properties (see Exercise 1) of the objects and the aqueous environment are transformed into corresponding variations of brightness. In phase-contrast microscopy a ring of light passes through the object, light rays are **diffracted** (retarded) and are out of phase with the light rays not hitting the object. The phase-contrast microscope enhances these phase differences so that the eye detects the difference as contrast (Figure 3-2) between the organisms and background and between structures within a cell. In phase-contrast microscopy, the organisms appear as degrees of brightness against a darker background. The advantage of phase-contrast microscopy is that structural detail within live cells can be studied.

A **fluorescent microscope** is a compound microscope with an ultraviolet or near-ultraviolet light source. Some chemicals naturally fluoresce; that is, when light of one wavelength strikes them and is absorbed, they give off visible light of another color. Fluorescent dyes called *fluorochromes* can be used to stain specimens that do not naturally fluoresce. In fluorescent microscopy, the object is seen as bright and luminescent against a dark background. Fluorescent dyes can be used either to label parts of microbes or to determine the identity of certain microbes that can then be visualized using a fluorescent microscope (Exercise 75).

In **electron microscopy,** resolution is greatly enhanced by utilizing a beam of electrons that has a shorter wavelength than visible light. Magnetic fields are used to focus the electrons. Whereas light is *absorbed* by objects in light microscopy, electrons are *scattered* by an object to provide contrast. The main advantage of an electron microscope is the high resolving power that can be attained because of the short wavelength of electrons.

A compound light microscope has a resolving power of 200 nm and magnification of 1000 times; an electron microscope can achieve resolution of 0.2 nm and magnification of 200,000 times. There are two types of electron microscopes: transmission electron microscopes (TEMs) and scanning electron microscopes (SEMs). Figure 3-3*a* shows the relatively straight path

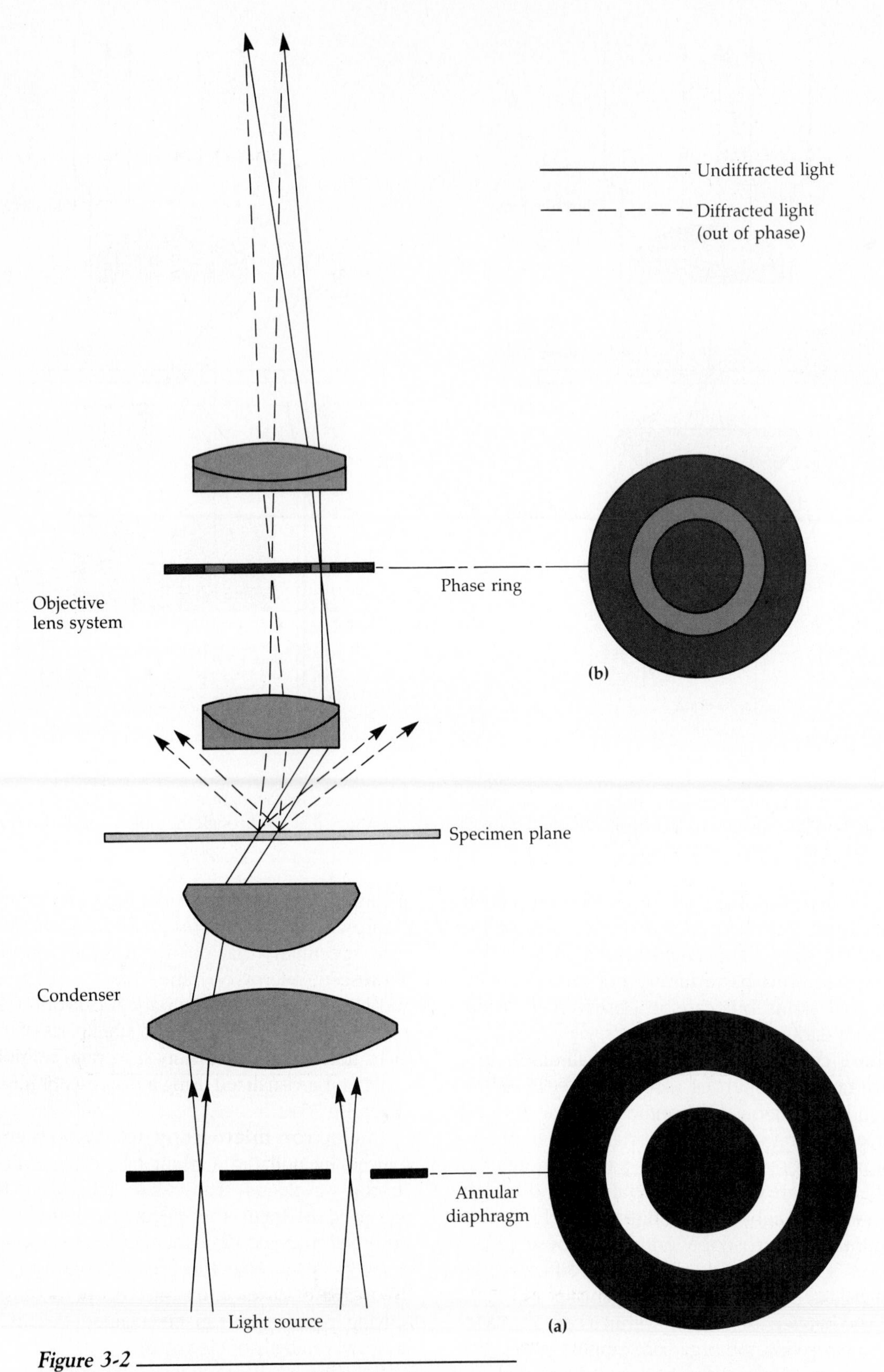

Figure 3-2

Phase-contrast microscope. A hollow cone of light is formed by the annular diaphragm **(a).** Diffracted rays are further retarded by the phase ring **(b)** in the objective lens.

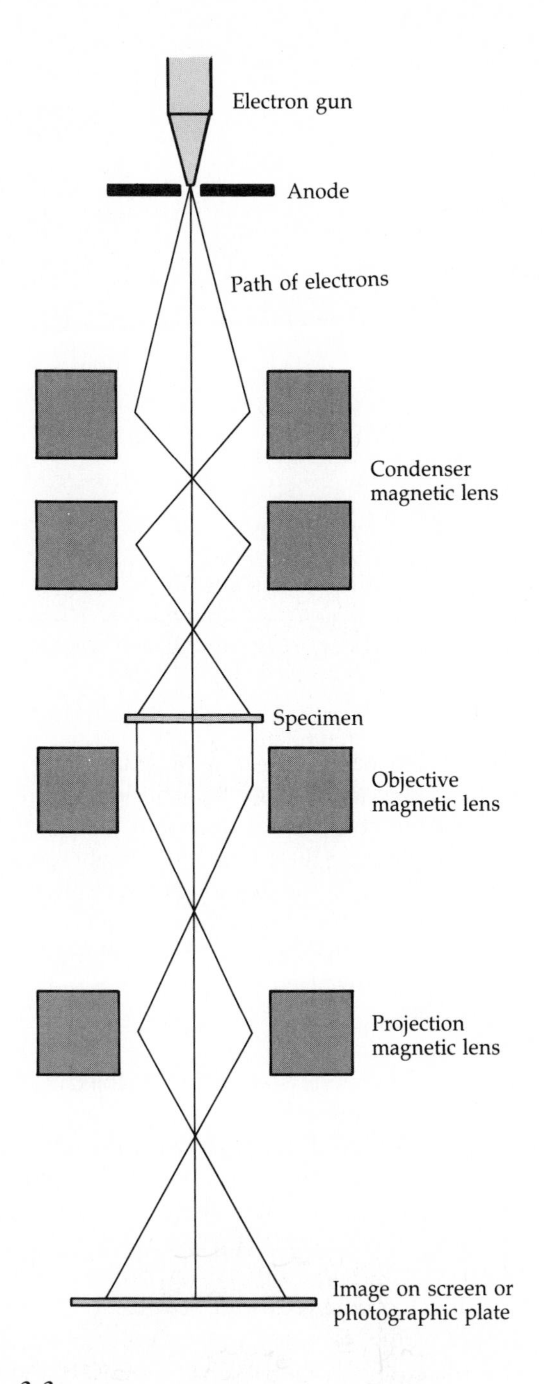

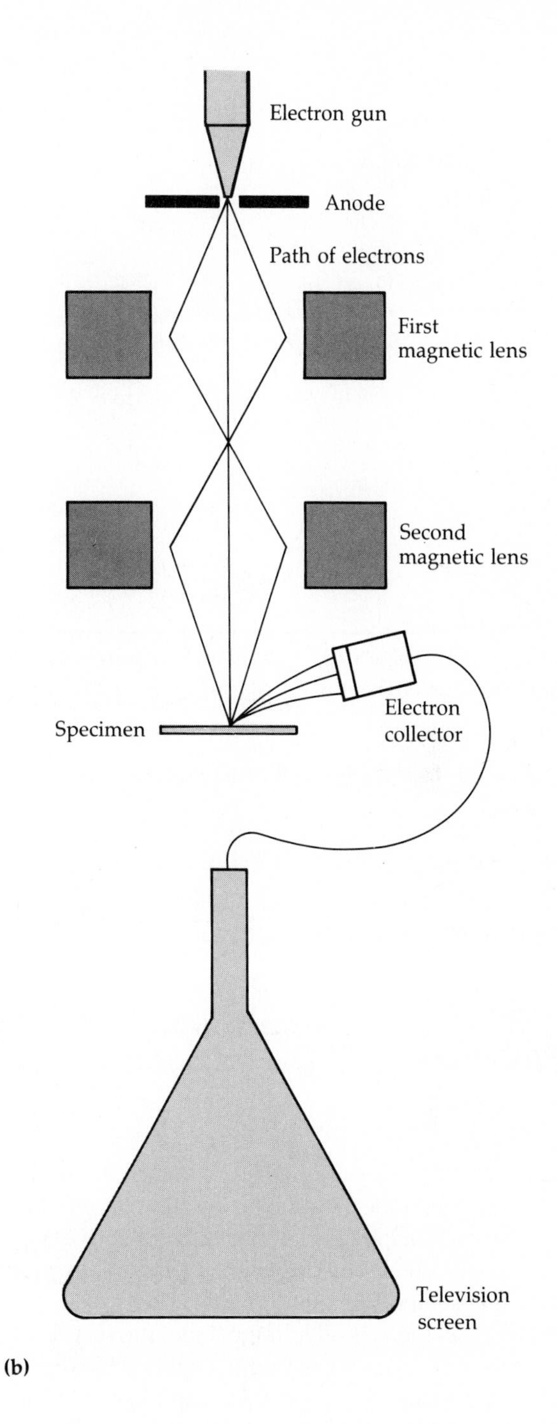

Figure 3-3

Electron microscopes. **(a)** The path of electrons in a transmission electron microscope. **(b)** The path of electrons in a scanning electron microscope.

of electrons in a **transmission electron microscope.** The electrons are passed through ultrathin sections (0.05 to 0.1 μm) of the specimen and the resulting image is formed on a photographic plate. In a **scanning electron microscope** (Figure 3-3*b*), the electrons are reflected from the specimen onto a televisionlike screen or a photographic plate to provide a three-dimensional image of the specimen's surface.

In this exercise, you will examine the image formed by a phase-contrast microscope.

Materials

Slides (very clean)

Cover slips

Phase-contrast microscope and centering telescope

Cultures

Pond water with algae

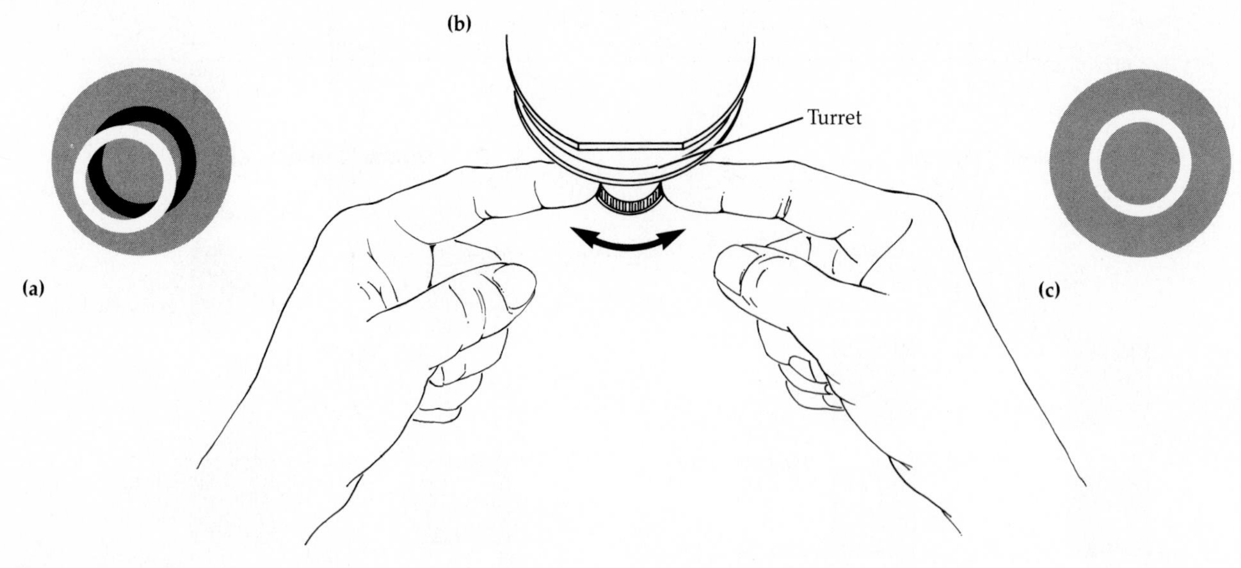

Figure 3-4

Adjustment of phase rings. **(a)** The two images seen before adjustment of the phase rings. By moving the wheel under the condenser turret **(b),** the images are made to coincide **(c).**

One of the following organisms: *Paramecium* or *Euglena*

Baker's yeast suspended in water

Techniques Required

Exercises 1 and 2

Procedure

1. Make a wet mount (Exercise 2) of any of the suspensions.
2. Place the slide on the stage and turn on the light.
3. Start with the 10× objective and move the condenser diaphragm to matching setting "10."
4. Focus on an obvious clump of material with the coarse and fine adjustments.
5. Close the iris diaphragm and move the condenser up and down until a light octagon comes into focus. Then open the diaphragm until light just fills the field.
6. Adjust the phase rings.
   - **a.** Replace the eyepiece (screw out) with the centering telescope and focus the telescope on the phase plate ring by revolving its head.
   - **b.** You will see two rings (Figure 3-4*a*): one, a bright image of the phase ring, and the other, the dark image of the phase plate in the objective. Adjust the knurled wheel under the condenser turret with your forefingers (Figure 3-4*b*) to make the bright image coincide with the dark ring (Figure 3-4*c*). Do not touch the diaphragm.
   - **c.** When the ring has been centered (Figure 3-4*c*), replace the telescope with the ocular lens.
7. Observe.
8. Focus the slide with the 40× objective and the 40× condenser turret.
9. Readjust the phase rings using the telescope, as done in step 6. Record your observations. Can you distinguish any of the organelles of the organisms? ____________
10. Focus the slide with the 100× objective and the 100× condenser turret, readjust the phase rings using the telescope. Diagram your observations.
11. Clean the slide. Make wet mounts of the other organisms. Observe. Are any of them motile?__ Compare your observations with those made in Exercise 2.

Turn to the Laboratory Report for Exercise 3.

EXERCISE 4

Measurement of Microbes

Objectives

After completing this exercise you should be able to

1. Estimate the size of organisms through a microscope.
2. Calibrate an ocular micrometer.
3. Measure cells using an ocular micrometer.

Background

You have learned that microorganisms are too small to be seen with the naked eye and must be observed through a microscope. Microbes vary greatly in size, and their size can be measured through a microscope.

Usually a microbe can be identified as to kingdom or phylum by observing morphologic characteristics, including size. Most bacteria fall within a range of 0.20 to 2.0 μm; a yeast cell, by comparison, is quite large (5.0 to 30 μm). It is useful to be able to estimate the size of organisms while observing them. We can do this by determining the size of the field of vision of the microscope.

To accurately measure microorganisms, the microscope (ocular lens) must be fitted with an ocular micrometer. The **ocular micrometer** is a glass disk with an arbitrary scale marked on it (Figure 4-1). The ocular micrometer must be calibrated for each set of ocular and objective lenses. This calibration is done with a stage micrometer. The **stage micrometer** is a 2 mm ruler etched into a glass slide (Figure 4-2). This 2 mm unit is divided into 0.1 mm units of length that are subdivided into 0.01 mm units at one end. One millimeter equals _____ μm.

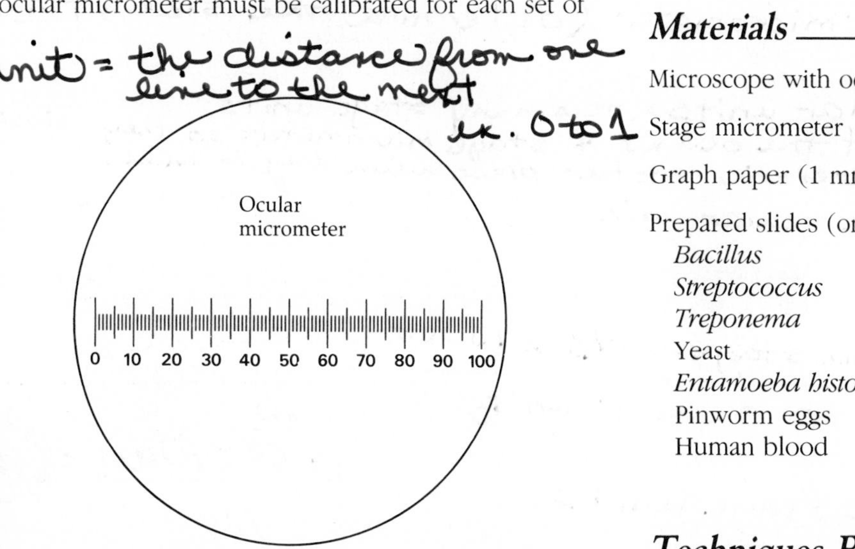

Figure 4-1
An arbitrary scale is marked on the ocular micrometer.

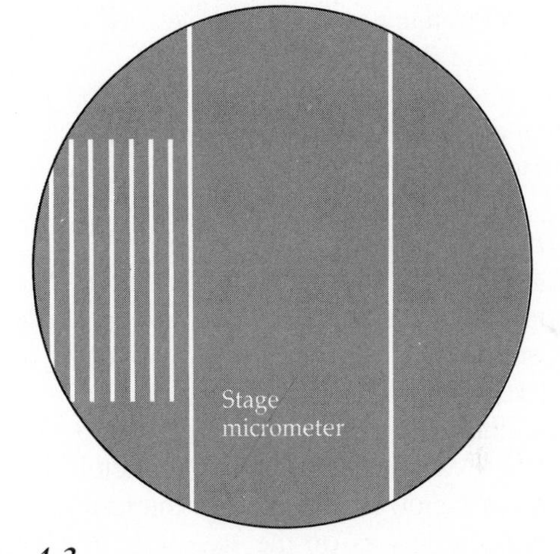

Figure 4-2
The stage micrometer is divided into 0.1 mm and 0.01 mm units.

Materials

Microscope with ocular micrometer

Stage micrometer

Graph paper (1 mm ruled) slide

Prepared slides (one of each):
- *Bacillus*
- *Streptococcus*
- *Treponema*
- Yeast
- *Entamoeba histolytica*
- Pinworm eggs
- Human blood

Techniques Required

Exercise 1

Procedure

Estimating Size

1. Place a graph paper slide on the microscope stage and observe it under low power. How many squares can you observe in the field? ______________ What is the diameter of the field in millimeters (mm)? ______________ In micrometers (μm)? ______________
2. Observe the graph paper slide under the high-dry and oil immersion objectives. Record your observations in the Laboratory Report.
3. Observe a slide of *Bacillus* under the low and high-dry objectives and estimate the size of the bacterium. Estimate the size of a yeast cell.

Accurate Measurements

1. Calibration of the ocular micrometer. Place the stage micrometer on the microscope stage and observe it under low power. Align the ocular micrometer so that a line is superimposed on a line on the stage micrometer (Figure 4-3). Count the spaces on the ocular micrometer until the lines coincide with the stage micrometer again. Count the spaces on the stage micrometer until the lines coincide.

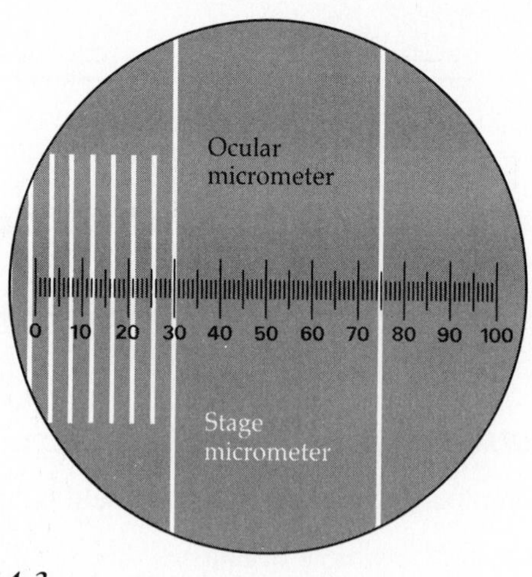

Figure 4-3

Superimpose the ocular micrometer on the stage micrometer to calibrate the ocular micrometer.

Calibrate the ocular micrometer for each objective lens on your microscope.

2. Measurement of microbes. To measure cells, remove the stage micrometer and place a slide of *Streptococcus* on the stage. Superimpose the ocular micrometer on the cells. The length of a chain of cells is determined by counting the number of ocular units filled by the chain. Make the measurements called for in the Laboratory Report.

If, for example, 45 ocular units are equal to 10 small, 0.01 mm stage units, the distance between the lines of the ocular micrometer at that magnification is equal to

$$\frac{10 \text{ stage units} \times 0.01 \text{ mm}}{45 \text{ ocular units}} = 0.0022 \text{ mm} = 2.2\ \mu\text{m}$$

Turn to the Laboratory Report for Exercise 4.

To calibrate the ocular micrometer (determine the size of 1 unit in μm)
① Determine how many ocular units = how many stage units
a) line up the "0" lines of the ocular + stage micrometers so they coincide
b) Look to the right to determine another place where they coincide.
? no. stage units = ? no. ocular units
2. To calculate
1 ocular unit in μm = no. stage units × 10 μm / no. ocular units
ex. 20 ocular units = 10 stage units
1 ocular unit = 10 × 10 μm / 20 = 5 μm

Low
50 × 10 μm / 100 = 5
High
10 × 10 / 85 = 1.2

PART 2

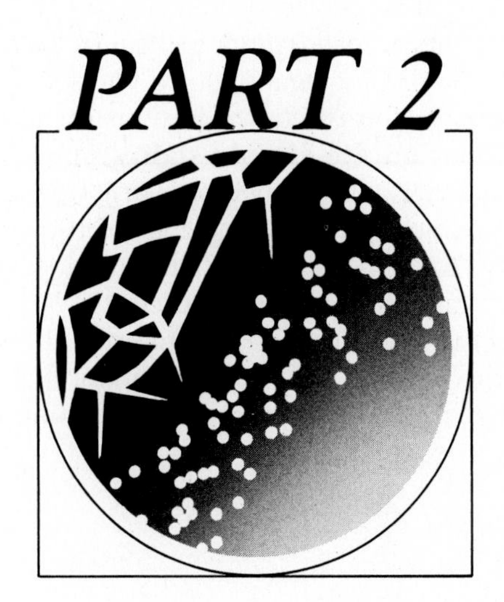

Staining Methods

EXERCISES

In 1877, Robert Koch wrote,

> *How many incomplete and false observations might have remained unpublished instead of swelling the bacterial literature into a turbid stream, if investigators had checked their preparations with each other?*

To solve this problem, he introduced into microbiology the procedures of air drying, chemical fixation, and staining with aniline dyes.

In addition to making a lasting preparation, staining bacteria enhances the contrast between bacteria and the surrounding material and permits observation of greater detail and resolution than wet mount procedures (Exercise 2). Microorganisms are prepared for staining by smearing them onto a microscope slide (Exercise 5).

In the exercises in Part 2, bacteria will be transferred from growth or culture media to microscope slides using an inoculating loop. An **inoculating loop** is a nichrome wire held with an insulated handle. Before it is used, the inoculating loop is sterilized by heating or **flaming.** That is, the loop is held in the flame of a burner (part *a* of the figure) or electric incinerator (part *b* of the figure) until it is red hot, and then it is allowed to cool so that bacteria picked up with the loop won't be killed. Allow the loop to cool without touching it, then insert the loop into a bacterial culture for a loopful of bacteria. The loopful can then be placed on a slide as shown in part *c* of the figure (see Exercise 5). After transferring a loopful of bacterial culture, the loop must be sterilized by flaming again before you set it down (parts *a* and *b* of the figure). Use of the loop is discussed further in Exercise 13.

Staining techniques may involve **simple stains,** in which only one reagent is used and all bacteria are usually stained similarly (Exercises 5 and 6) or **differential stains,** in which multiple reagents are used and bacteria react to the reagents differently (Exercises 7 and 8). Structural stains are used to identify specific parts of microorganisms (Exercise 9).

Most of the bacteria that are grown in laboratories are bacilli or cocci (see Figure 1-6). Bacilli and cocci will be used throughout this part. In Exercise 10, "Morphologic Unknown," you will be asked to determine whether an unknown culture is a bacillus or coccus.

Using an inoculating loop.

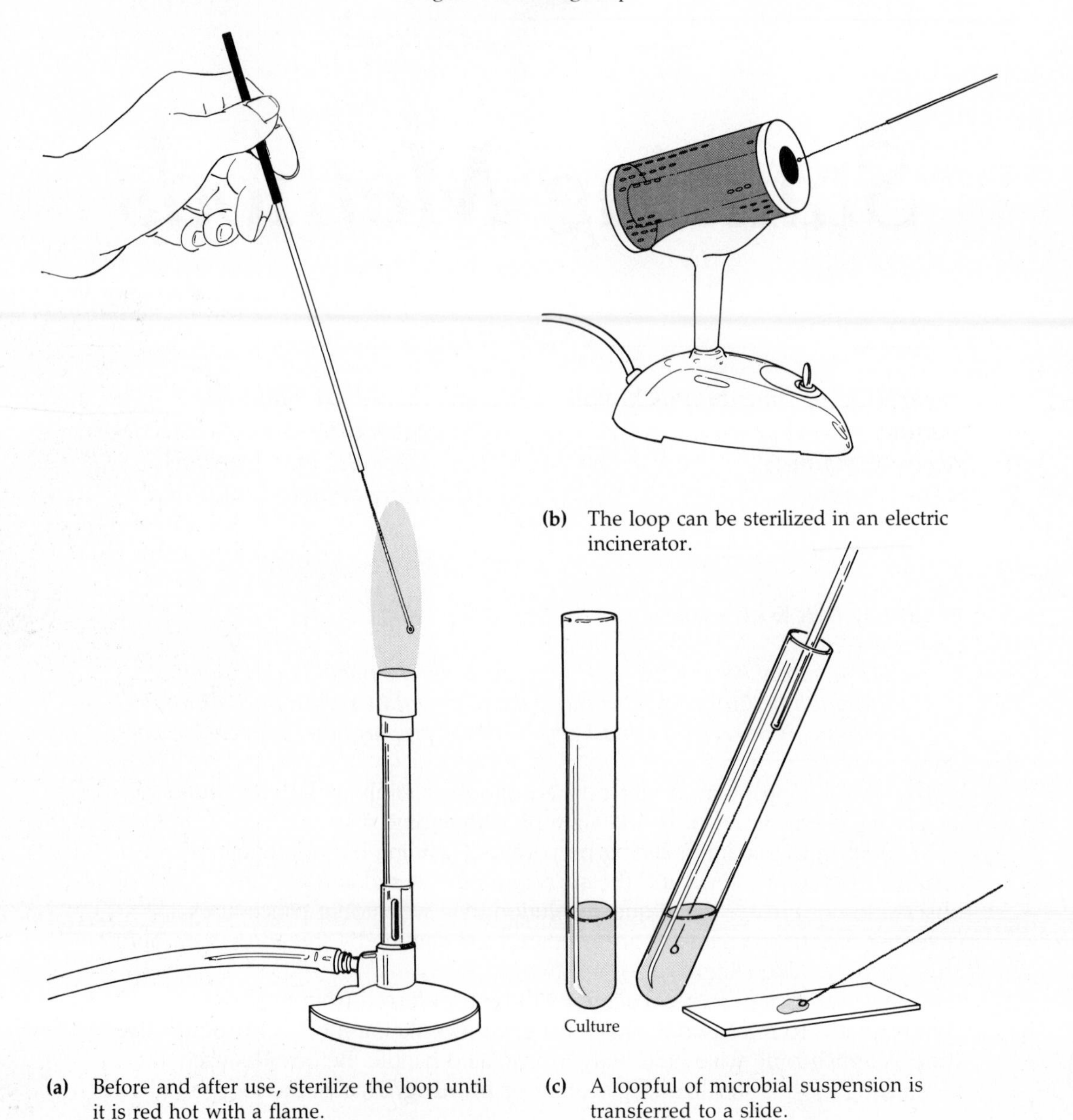

(a) Before and after use, sterilize the loop until it is red hot with a flame.

(b) The loop can be sterilized in an electric incinerator.

(c) A loopful of microbial suspension is transferred to a slide.

PLATE I ***Staining***

All photographs 300,000×.

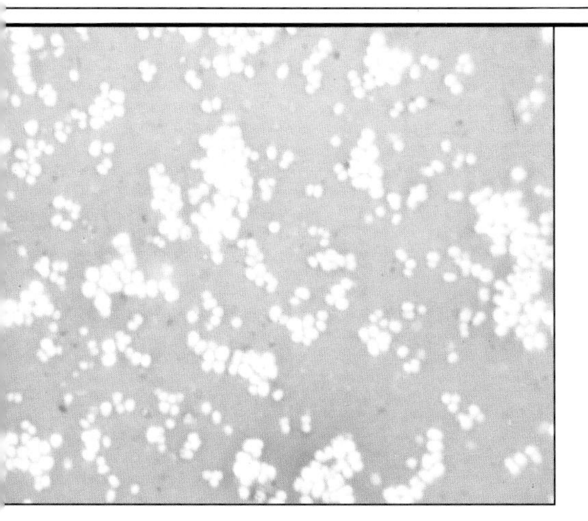

I-1. ***Negative stain.*** *Staphylococcus aureus* (*Exercise 4*).

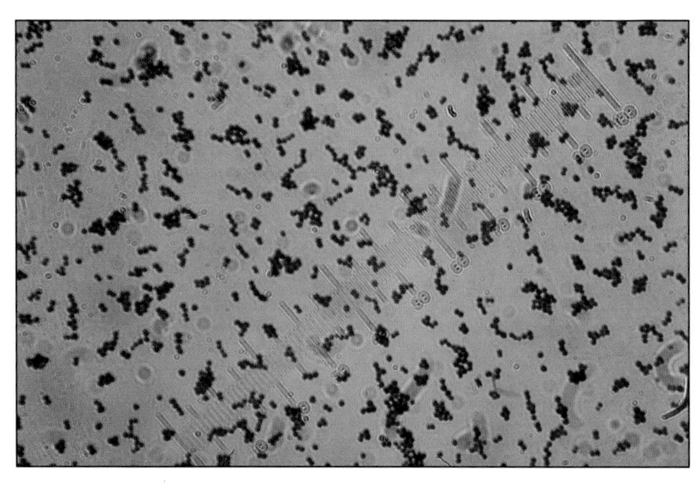

I-2. ***Gram-positive cocci.*** *Staphylococcus aureus* (*Exercise 5*). One ocular micrometer division equals 1 μm.

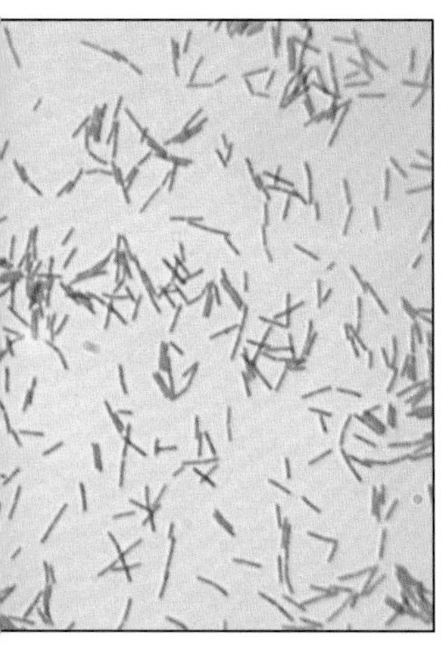

I-3. ***Gram-negative rods.*** *Escherichia coli* (*Exercise 5*).

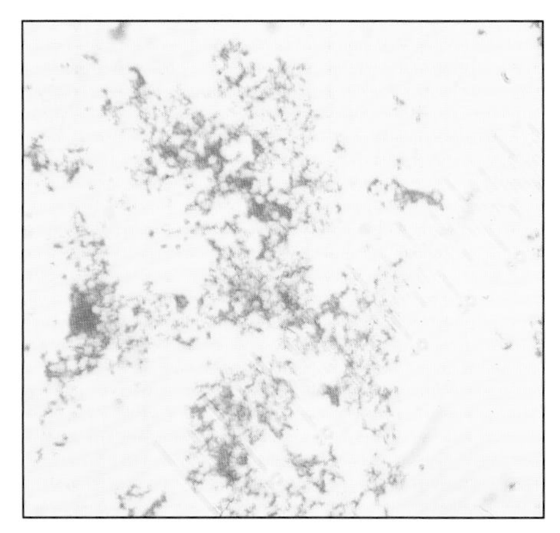

I-4. ***Ziehl-Neelsen acid-fast stain.*** *Mycobacterium phlei* (*Exercises 6 and 44*).

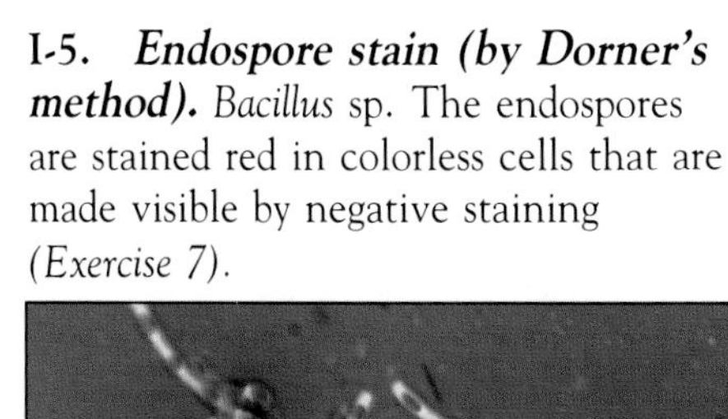

I-5. ***Endospore stain (by Dorner's method).*** *Bacillus* sp. The endospores are stained red in colorless cells that are made visible by negative staining (*Exercise 7*).

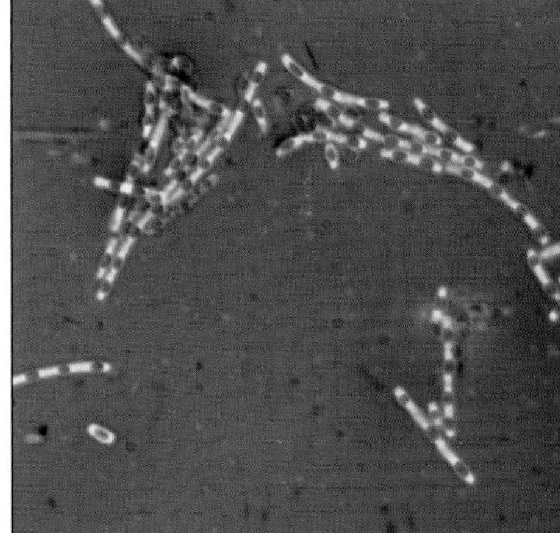

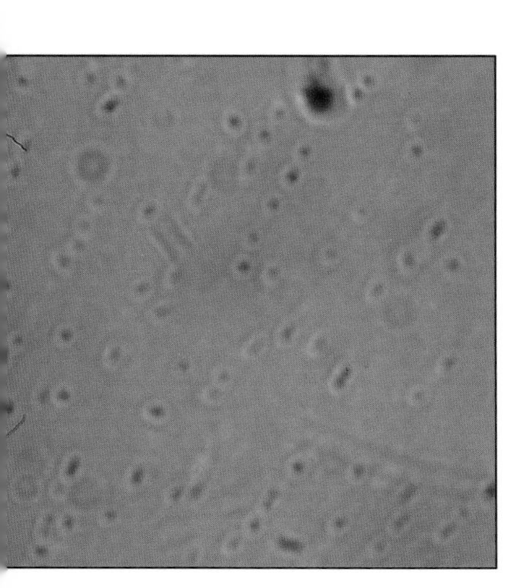

I-6. ***Capsule stain (by Gin's method).*** *Enterobacter aerogenes.* Unstained capsules surround red-stained cells (*Exercise 7*).

I-7. ***Flagella stain.*** Peritrichous flagella of *Proteus vulgaris* (*Exercise 7*).

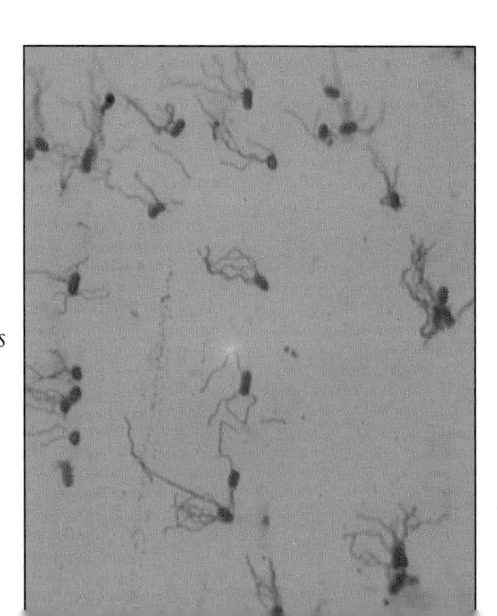

PLATE II

II-1. ***Reactions in O-F glucose medium.*** Bacterial growth is visible along the stab lines in each tube. There is no change in *1.* The blue color in *2* indicates metabolism of peptones. Glucose is fermented (anaerobically) in *3* and oxidized (aerobically) in *4* *(Exercise 13).*

II-2. ***Reactions in fermentation tubes.*** *1* is an uninoculated control. Growth and acid production from carbohydrate fermentation are seen in *2.* Acid and gas are produced from fermentation in *3* *(Exercise 14).*

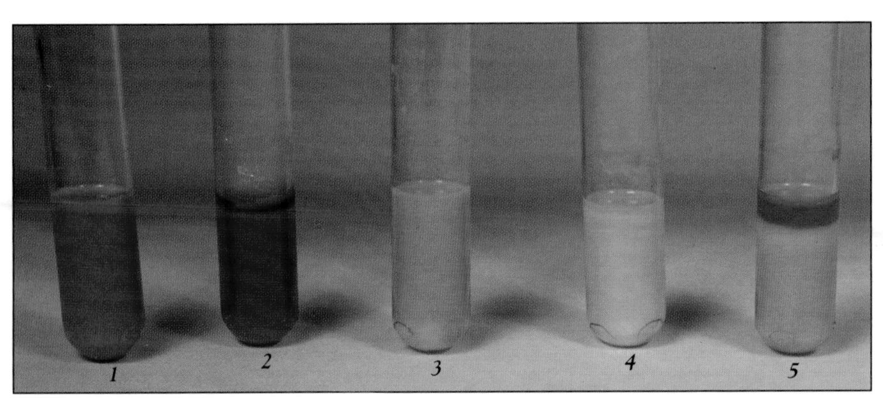

II-3. ***Reactions in litmus milk.*** *1* is the uninoculated control. *2* shows an alkaline reaction. Acid is produced in *3.* The litmus is reduced in *4.* The clear fluid layer in *5* is the result of peptonization *(Exercise 16).*

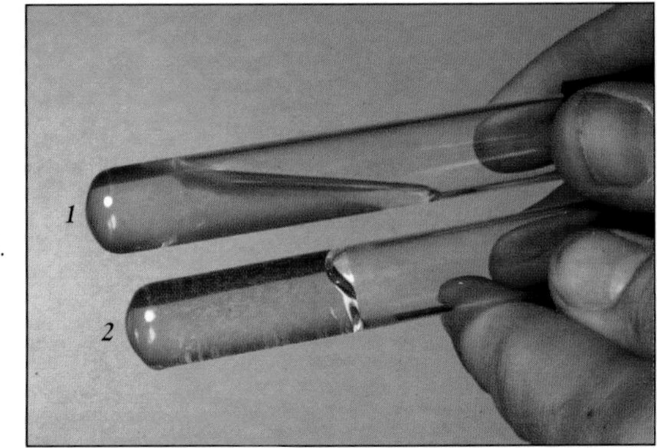

II-4. ***Gelatin hydrolysis.*** After hydrolysis (*1*), gelatin remains liquid. *2* is unhydrolyzed gelatin. *(Exercise 16).*

Fifteen biochemical tests are performed in Enterotube II. The bottom tube is uninoculated *(Exercise 19).*

Microbial Metabolism

II-6. ***Methyl red test.*** Red color (in *1*) after addition of methyl red indicates a positive test. *2* is methyl red-negative (*Exercise 14*).

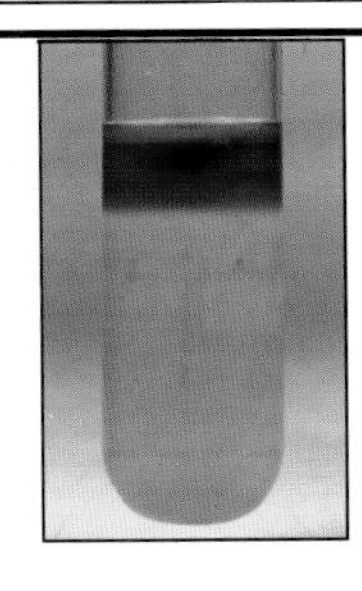

II-7. ***Voges-Proskauer test.*** A positive Voges-Proskauer test develops a wine-red color when exposed to oxygen (*Exercise 14*).

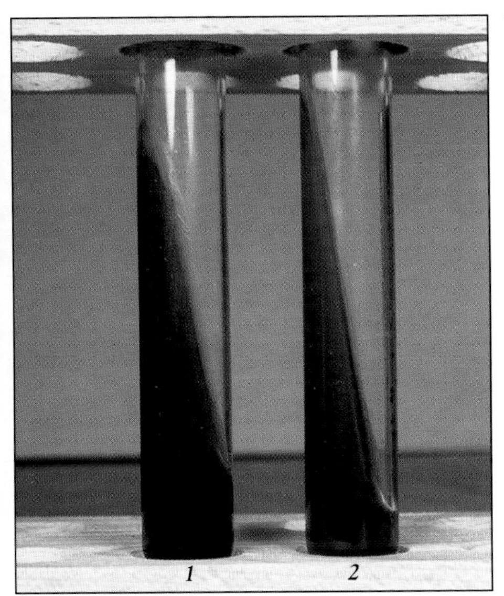

II-8. ***Citrate test.*** Utilization of citric acid as the sole carbon source in Simmons citrate agar causes the indicator to turn blue (*1*). *2* is citrate-negative (*Exercise 19*).

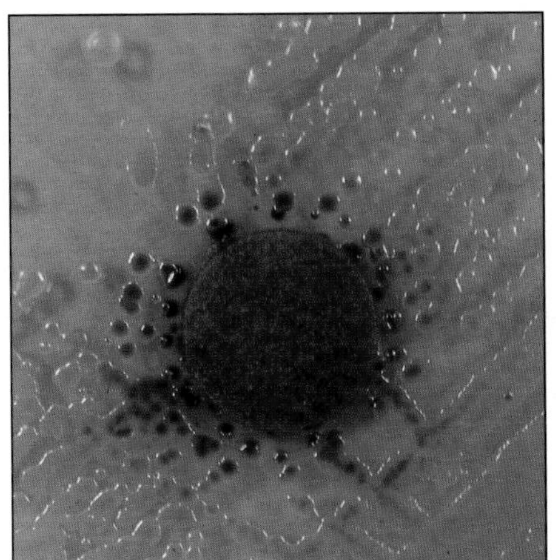

II-9. ***Oxidase test.*** Colonies of cytochrome oxidase-positive bacteria turn black as oxidase reagent diffuses from the oxidase test disk into the nutrient agar (*Exercise 18*).

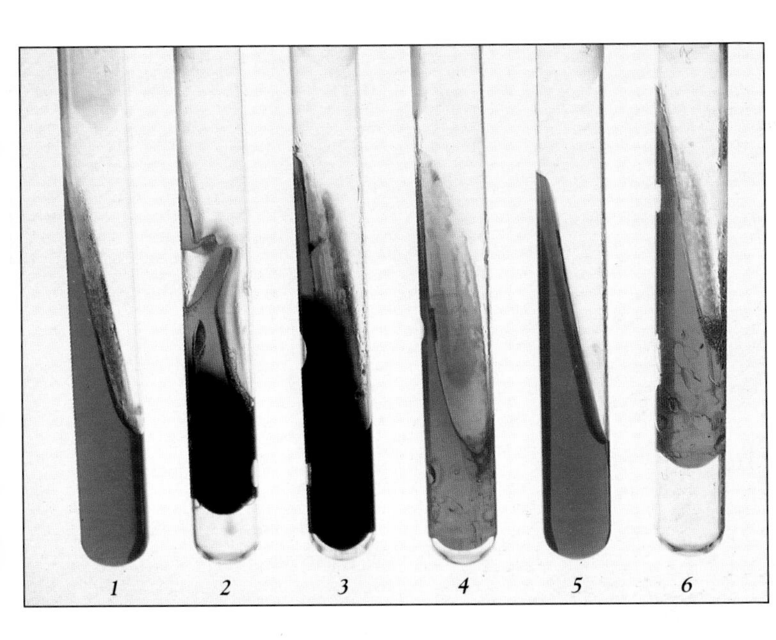

II-10. ***Reactions in triple sugar iron (TSI) agar.*** *1* shows growth with no fermentation or hydrogen sulfide (H_2S) production. *2* shows blackening due to H_2S and acid and gas from fermentation of glucose and sucrose and/or lactose. Sucrose and lactose were not fermented in ***3***, H_2S production masks the glucose fermentation reaction although gas is produced. Acid and gas are produced from glucose in ***4***. ***5*** is uninoculated. ***6*** shows acid and gas production from glucose and sucrose and/or lactose (*Exercise 46*).

PLATE III **Eucaryotes**

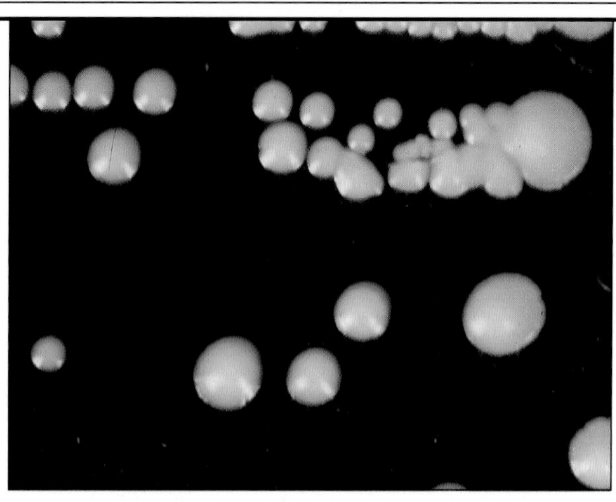

III-1. ***The yeast Saccharomyces cerevisiae produces circular, convex, glistening colonies on Sabouraud dextrose agar*** (7000×) *(Exercise 31).*

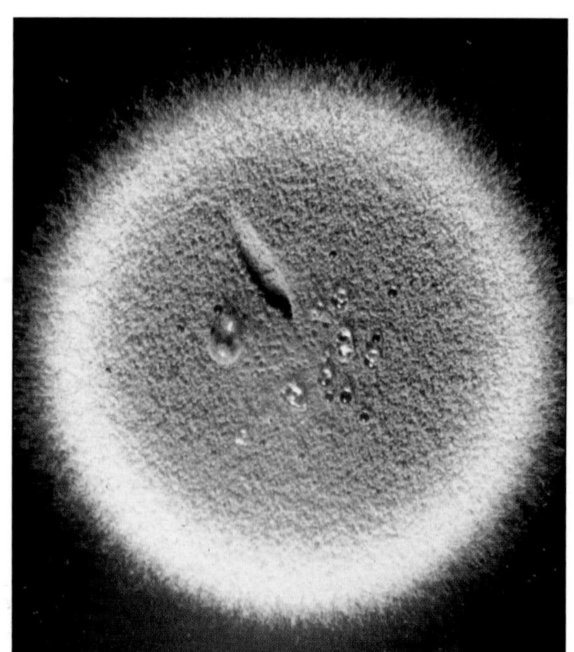

III-2. ***Rhizopus nigricans growing on styrofoam impregnated with a nutrient medium.*** Dark sporangia are visible at the tips c the sporangiophores *(Exercise 32).*

III-3. ***Penicillium has a white mycelium and green conidiospores.*** Note the crystallized penicillin on top of the mold colony *(Exercise 32).*

III-4. ***Red snow at Tioga Pass (Yosemite National Park) due to the green alga Chlamydomonas nivalis.*** The green color of chlorophyll is masked by red carotenoid pigments produced in the presence of intense light *(Exercise 33).*

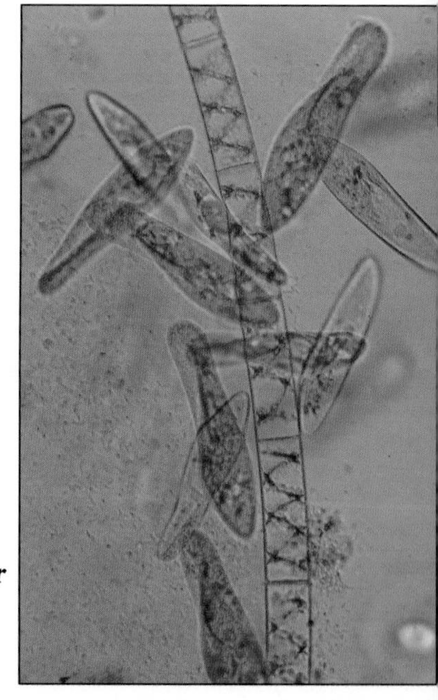

III-5. ***The green alga Spirogyra and protozoan Paramecium, in pond water*** (100,000×) *(Exercises 33 and 34).*

PLATE IV *Plate Cultures*

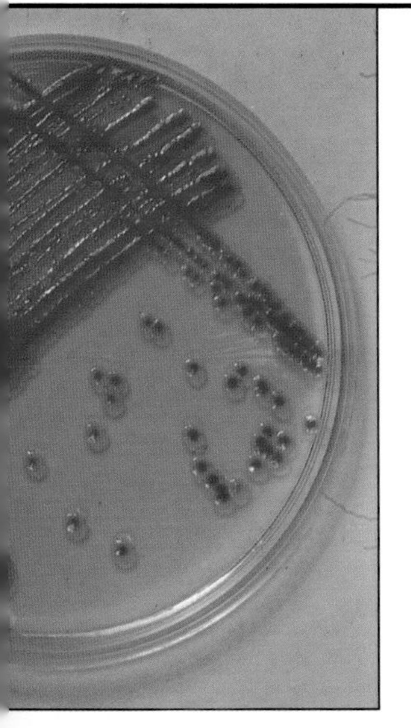

-1. ***Serratia marcescens*** **lonies on nutrient agar** **er incubation at 25°C** (*ercise 11*).

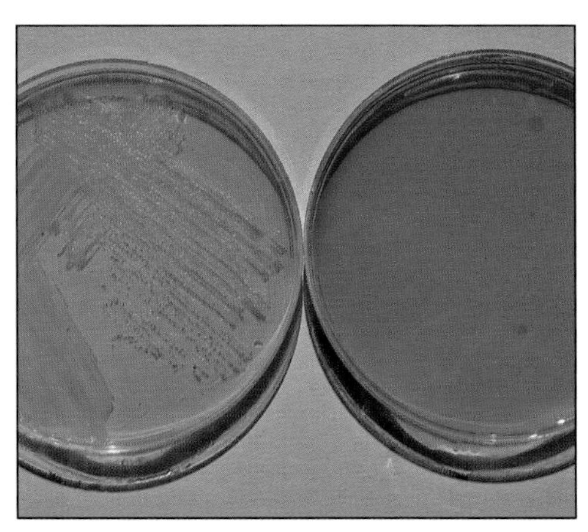

IV-2. ***Staphylococcus aureus*** **growing on mannitol salt agar.** The yellow color indicates that mannitol is fermented. The red plate is uninoculated (*Exercise 43*).

IV-3. ***Pseudomonas aeruginosa*** **produces a water-soluble blue pigment on Pseudomonas Agar P** (*Exercise 47*).

IV-4. **Colonies of *Escherichia coli* (shown) and *Citrobacter* develop a metallic green sheen on EMB agar** (*Exercises 46, 47, 49, and 50*).

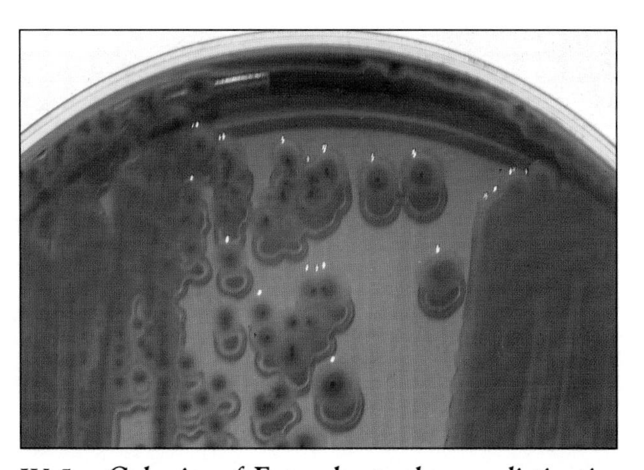

IV-5. **Colonies of *Enterobacter* have a distinctive blue "fish-eye" appearance on EMB agar** (*Exercises 46, 47, 49, and 50*).

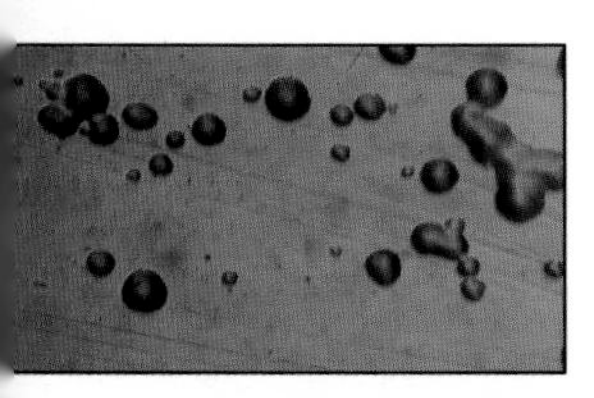

IV-6. **Nonlactose-fermenters such as *Proteus vulgaris* (shown) produce colorless colonies on EMB agar** (30×) (*Exercises 46, 47, 49, and 50*).

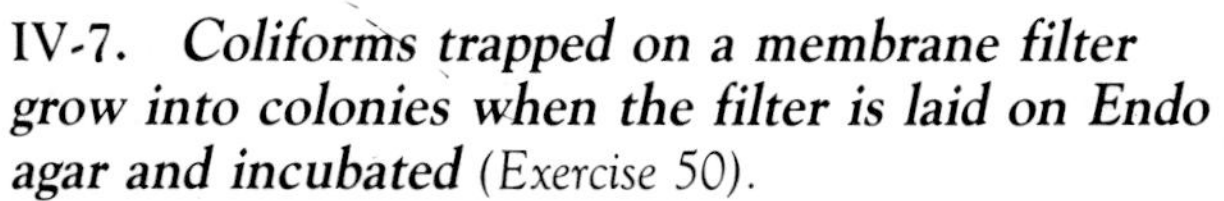

IV-7. **Coliforms trapped on a membrane filter grow into colonies when the filter is laid on Endo agar and incubated** (*Exercise 50*).

PLATE V Environmental Microbiology

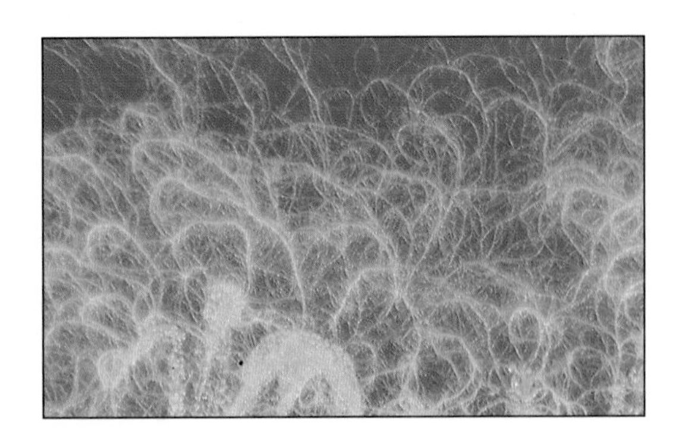

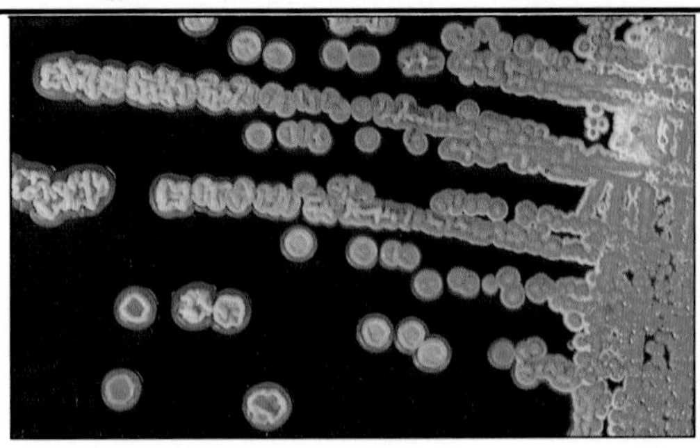

V-2. ***Actinomycete colonies.*** *Streptomyces griseus* colonies penetrate into the agar as well as extend above. Powdery appearance in the corner is due to conidiospores *(Exercise 9)*.

V-1. ***Bacillus cereus var. mycoides colonies.*** Easily recognized by their filamentous form. Bending of the filaments to the left or right is a strain characteristic *(Exercise 9)*.

◀ V-3. ***Root nodules.*** *Rhizobium leguminosarum* infected the roots of this vetch plant and caused the production of pink nodules *(Exercise 51)*.

V-4. ***Photosynthetic bacteria.*** Purple and green sulfur bacteria have colonized in this Winogradsky column. ▶

PLATE VI Pathology

VI-1. ***Numerous plaques (clearings) in this Escherichia coli culture are due to growth of a T-even bacteriophage*** (*Exercise 35*).

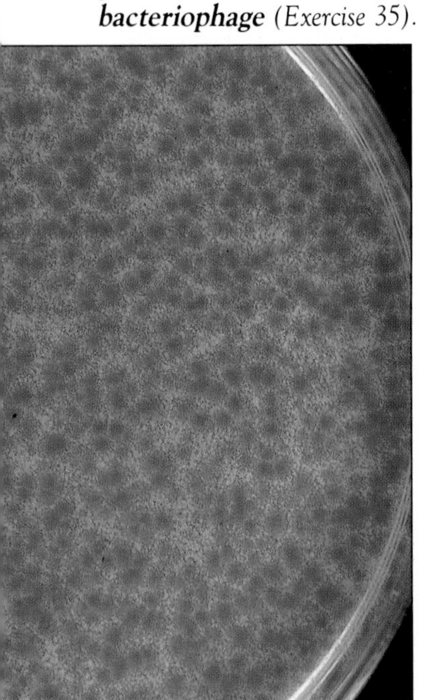

VI-2. ***Chlorosis (loss of green color) and plaques (spotting) in this tomato leaf are due to a tobacco mosaic virus infection.***

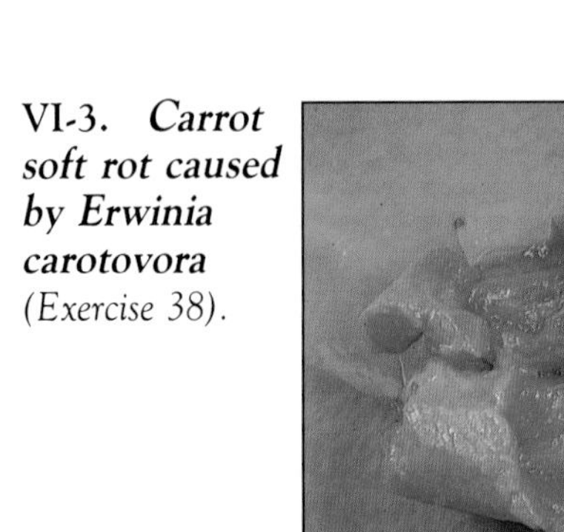

VI-3. ***Carrot soft rot caused by Erwinia carotovora*** (*Exercise 38*).

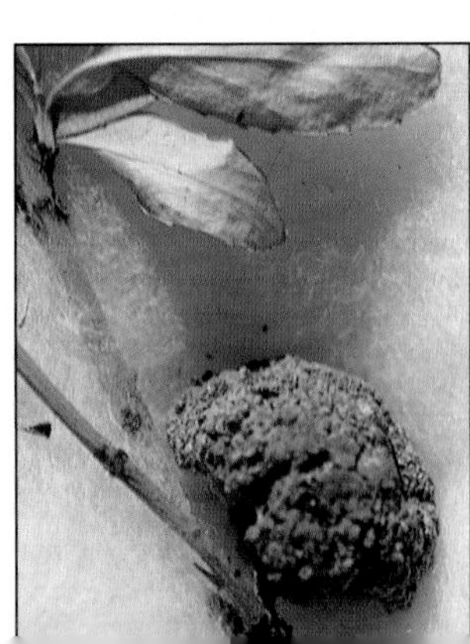

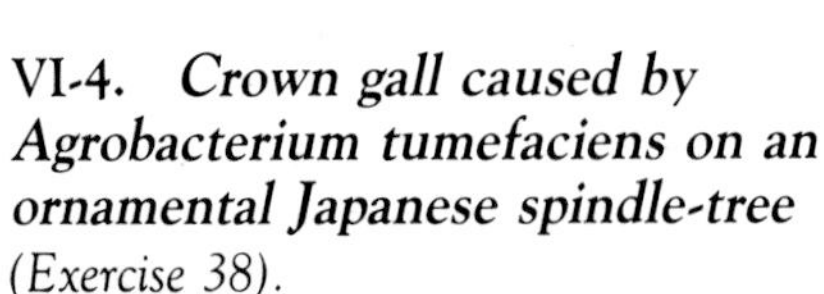

VI-4. ***Crown gall caused by Agrobacterium tumefaciens on an ornamental Japanese spindle-tree*** (*Exercise 38*).

PLATE VII Medical Microbiology

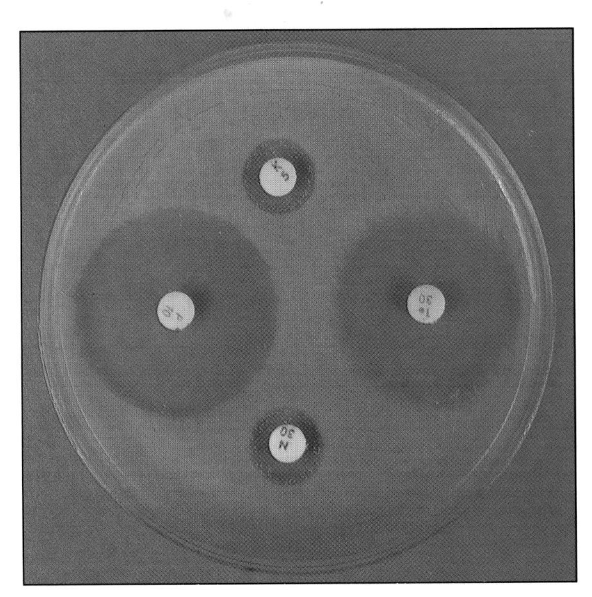

VII-1. ***Antibiotic sensitivity of this Staphylococcus aureus culture is demonstrated by the Kirby-Bauer agar diffusion test*** (*Exercise 26*).

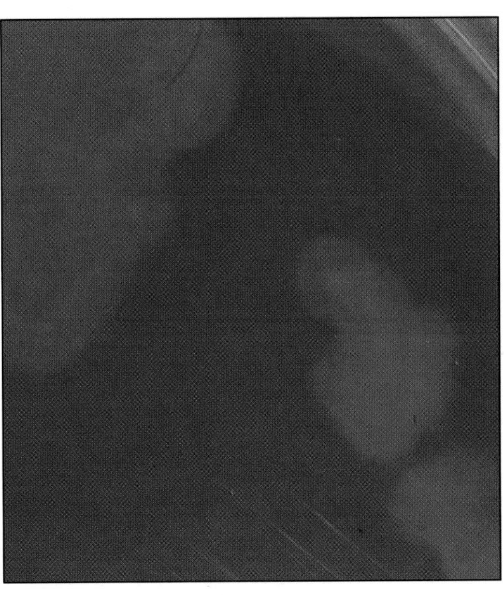

VII-2. ***Growth of Streptococcus pneumoniae on blood agar demonstrates alpha-hemolysis.*** Note the greenish color around the colonies (*Exercise 44*).

VII-3. ***Beta-hemolysis produces a clear area around the colonies of Streptococcus pyogenes on blood agar*** (*Exercise 44*).

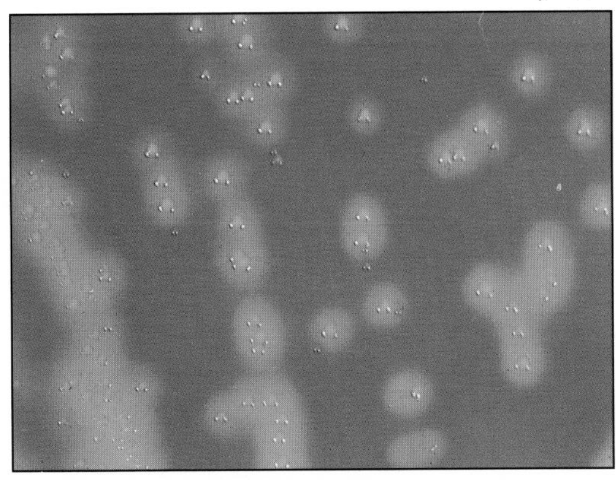

II-4. ***Trypanosoma cruzi in a blood mear*** (2000 ×). The third case of indigenous hagas' disease (American trypanosomiasis) in he United States occurred in California in 1982.

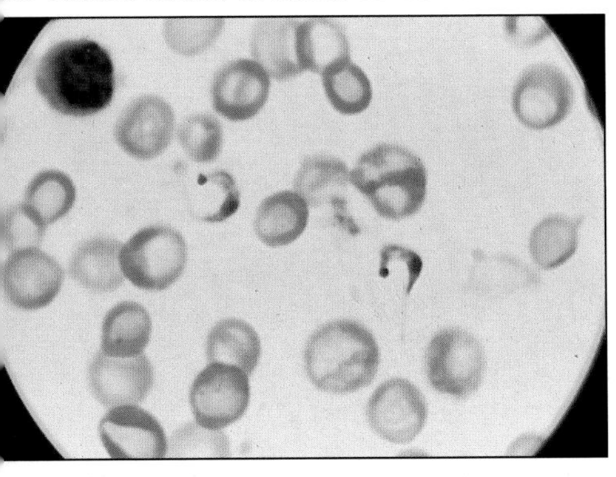

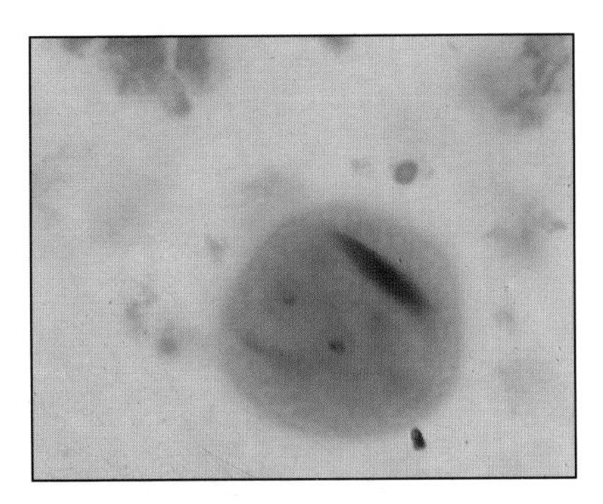

VII-5. ***Entamoeba coli cyst.*** The number of (round) nuclei and (rodlike) chromatoidal bars are used to identify species of *Entamoeba*.

PLATE VIII Immunology

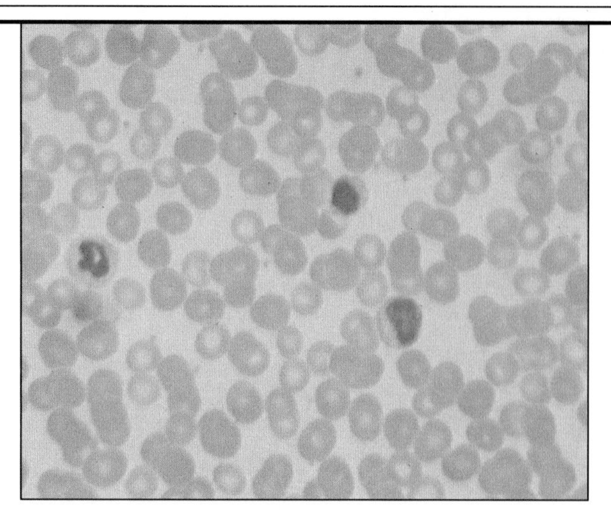

VIII-1. ***Wright's stain.*** Red blood cells stain light red; the light center of these cells is due to their biconcave shape. Nuclei of white blood cells stain dark blue and cytoplasm, light blue. These white blood cells (from left) are a neutrophil, lymphocyte, and monocyte.

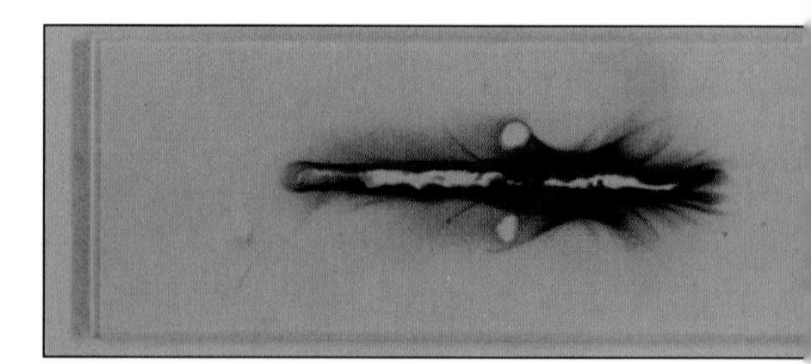

VIII-2. ***Immunoelectrophoresis.*** Curved bands show different serum proteins. Positive pole is on the left.

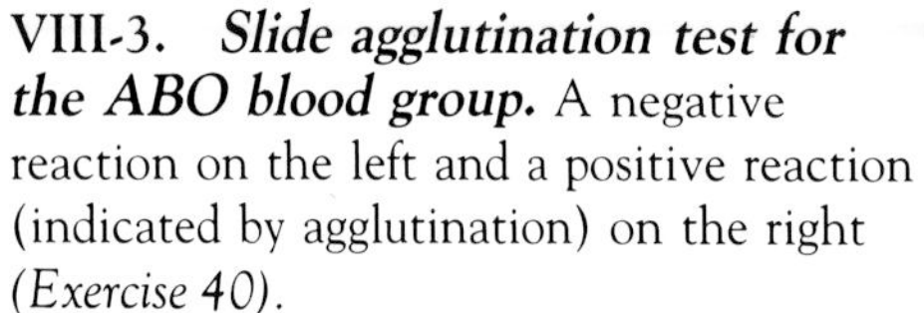

VIII-3. ***Slide agglutination test for the ABO blood group.*** A negative reaction on the left and a positive reaction (indicated by agglutination) on the right *(Exercise 40)*.

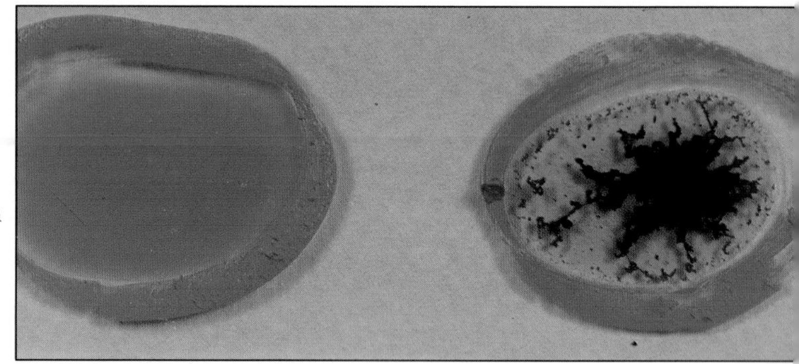

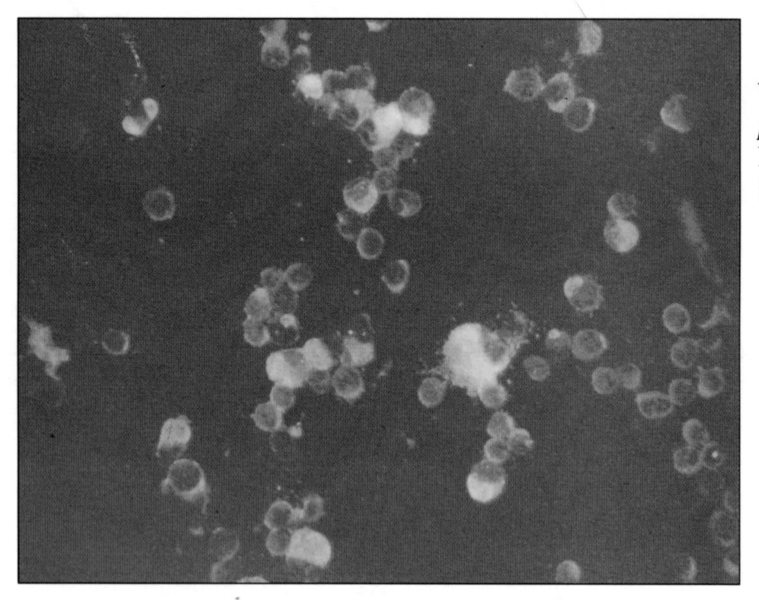

VIII-4. ***Fluorescent-antibody preparation of Chlamydia trachomatis.*** Large fluorescing cells contain colonies of intracellular *Chlamydia* *(Exercise 42)*.

Acknowledgments. Plates VII-4, VII-5, and VIII-4: Michael H. Nachtigall, San Mateo County Department of Public Health and Welfare. Plate VIII-2: Ted Johnson. All others: Christine L. Case.

EXERCISE 5

Preparation of Smears and Simple Staining

Objectives

After completing this exercise you should be able to

1. Make and heat fix a smear.
2. List the advantages of staining microorganisms.
3. Explain the basic mechanism of staining.
4. Perform a simple direct stain.

Background

Most stains used in microbiology are synthetic **aniline** (coal tar derivative) dyes derived from benzene. The dyes are usually salts, although a few are acids or bases, composed of charged colored ions. The ion that is colored contains a **chromophore** group. For example,

Methylene blue chloride ⇆ Methylene blue$^+$ + Cl$^-$
(Chromophore)

If the chromophore is on a positive ion like methylene blue, the stain is considered a **basic stain;** if on the negative ion, it is an **acidic stain.** Most bacteria are stained when a basic stain permeates the cell wall and adheres by weak ionic bonds to the negative charges of the bacterial cell.

Staining procedures that use only one stain are called **simple stains.** A simple stain that stains the bacteria is a **direct stain,** and a simple stain that stains the background, but leaves the bacteria unstained, is a **negative stain** (Exercise 6). Simple stains can be used to determine cell morphology, size, and arrangement.

Before bacteria can be stained, a smear must be made and heat fixed. A **smear** is made by spreading a bacterial suspension on a clean slide and allowing it to air dry. The dry smear is passed through a Bunsen burner flame several times to **heat fix** the bacteria. Heat fixing denatures bacterial enzymes, preventing autolysis. The heat also enhances the adherence of bacterial cells to the microscope slide.

Materials

Methylene blue

Wash bottle of distilled water

Slides (2)

Inoculating loop

Cultures (one of each)

Staphylococcus epidermidis slant

Bacillus megaterium broth

Techniques Required

Exercise 1, Part 2 Introduction

EXERCISE 5

Procedure

1. Smear preparation (Figure 5-1)
   a. Clean your slides well with abrasive soap or cleanser; rinse and dry. Handle clean slides by the end or edge. Use a marking pencil to make a dime-sized circle on each slide. Circles should be on the bottom of the slide so they will not wash off. Label each slide according to the bacterial culture used.
   b. For the bacterial culture on solid media: place 1 or 2 loopfuls of distilled water in the center of the circle on one slide, using the inoculating loop. Which bacterium is on a solid medium? __________
   c. Heat your loop to redness and allow it to cool (see the figure on page 18). Using the cooled loop, scrape a *small* amount of the culture off the slant. If you hear the sizzle of boiling water when you touch the agar with the loop, reflame your loop and begin again. Why? __________

      Try not to gouge the agar. Emulsify the organism in the drop of water and spread the suspension to fill a majority of the circle. The smear should look like diluted skim milk. Flame your loop again.
   d. Make a smear of bacteria from the broth culture on the other slide. *Do not use water* as

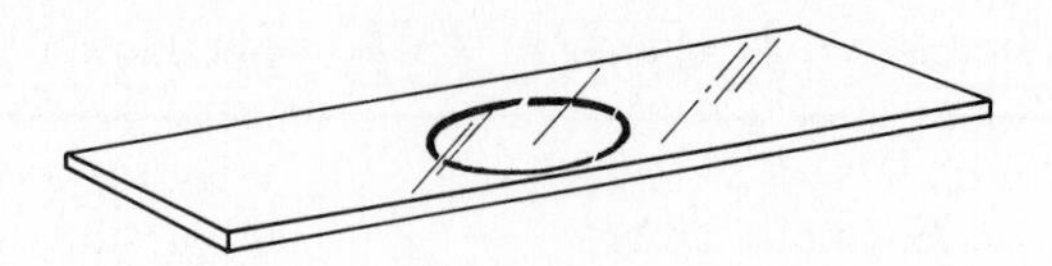

(a) Mark the smear area with a marking pencil on the underside of a clean slide.

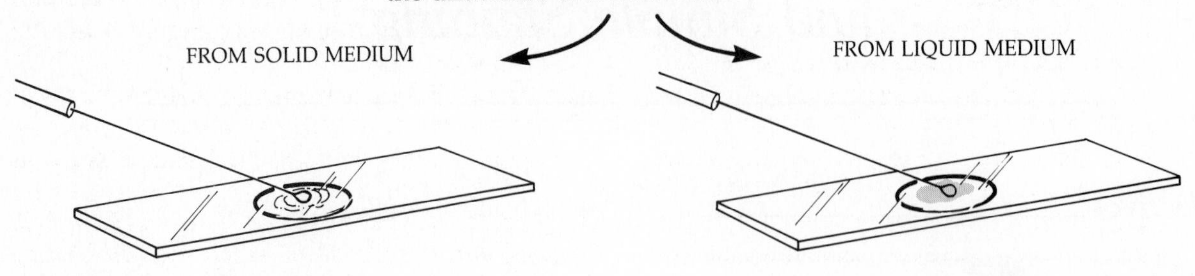

(b) Place 1 or 2 loopfuls of water on the slide.

(d) Place 2 or 3 loopfuls of the liquid culture on the slide with a sterile loop.

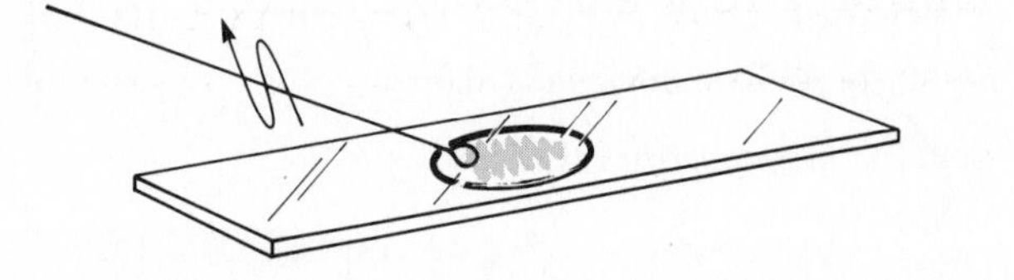

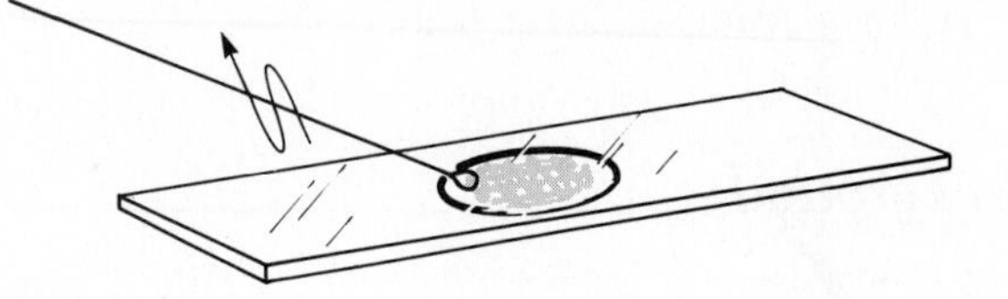

(c) Transfer a very small amount of the culture with a sterile loop. Mix with the water on the slide.

(e) Spread the organisms within the ring.

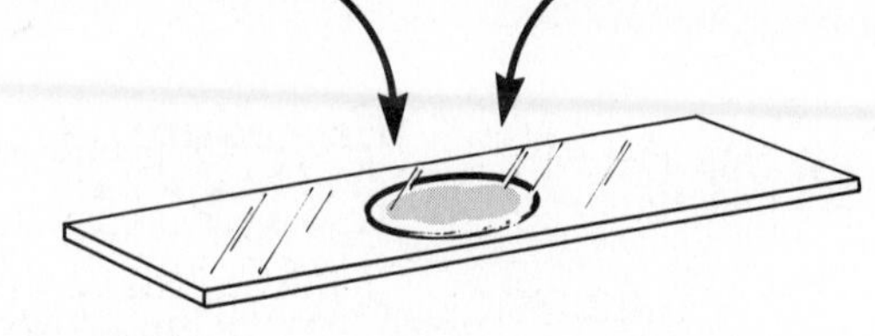

(f) Allow the smear to air dry at room temperature.

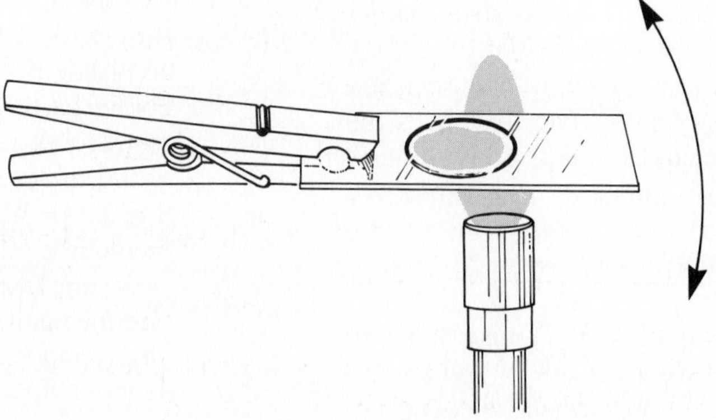

(g) Pass the slide over the flame of a burner 2 or 3 times.

Figure 5-1
Preparing bacterial smears.

the bacteria are already suspended in water. Flick the tube of broth culture lightly with your finger to resuspend sedimented bacteria and place 2 or 3 loopfuls of the culture in the circle. Flame your loop between each loopful. Spread the culture within the circle.

e. Flame your loop.
f. Let the smears dry. *Do not* blow on the slides as this will move the bacterial suspension; *do not* flame the slide as flaming will distort the cells' shapes.
g. Heat fix the smears by passing the slides through a flame 2 or 3 times. Do not heat fix until the smear is completely dry. Why? ____

____

2. Staining
   a. Do one slide at a time. Use a clothespin to hold the slide or place it on a staining rack.
   b. Cover the smear with methylene blue and leave for 30 to 60 seconds.
   c. Carefully wash the excess stain off with distilled water from a wash bottle. Let the water run down the tilted slide.
   d. Gently blot the smear with a paper towel or absorbent paper and let it dry.
3. Examine your stained smears microscopically using the low, high-dry, and oil immersion objectives. Put the oil *directly* on the smear; cover slips are not needed. Record your observations with labeled drawings.
4. Blot the oil from the objective lens with lens paper and return your microscope to its proper location. Clean your slides well or save them as described in step 5.
5. Stained bacterial slides can be stored in a slide box. Remove the oil from the slide by blotting with a paper towel. Any residual oil won't matter.

Turn to the Laboratory Report for Exercise 5.

EXERCISE 6

Negative Staining

Objectives

After completing this exercise you should be able to

1. Explain the application and mechanism of the negative stain technique.
2. Prepare a negative stain.

Background

The **negative stain** technique does not stain the bacteria but stains the background. The bacteria will appear clear against a stained background. No heat fixing or strong chemicals are used, so the bacteria are less distorted than in other staining procedures. The negative stain technique is very useful in situations where other staining techniques don't clearly indicate cell morphology or size, or the presence of a capsule.

The negative stain does not stain the bacteria due to the ionic repulsion of the bacteria (negative charge) and the acidic stain (negative charge).

Materials

Nigrosine

Clean slides (4)

Distilled water

Cultures (one of each)

Escherichia coli

Staphylococcus aureus

Techniques Required

Exercises 1 and 5

Procedure (Figure 6-1)

1. Slides must be clean and grease-free. See Exercise 5, Procedure step 1*a*.

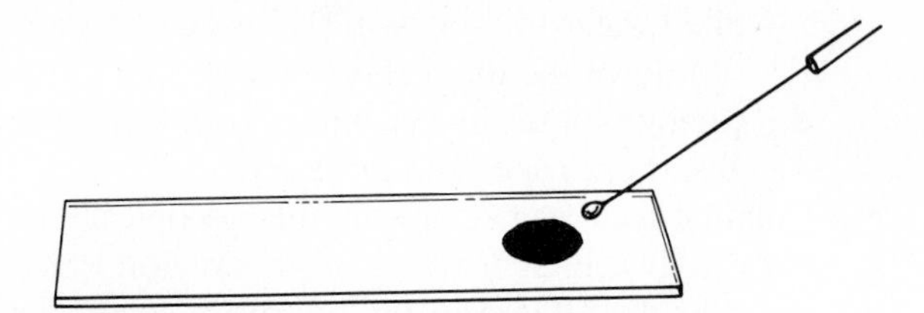

(a) Place a small drop of nigrosine near one end of a slide. Mix a loopful of broth culture in the drop. When the organisms are taken from a solid medium, mix a loopful of water in the nigrosine.

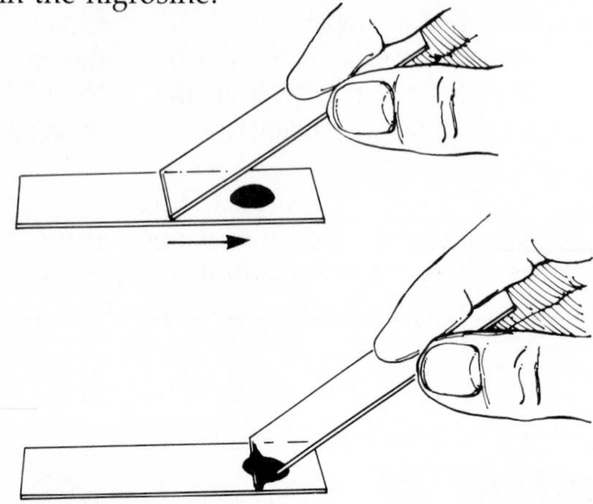

(b) Draw a second slide across the surface of the first until it contacts the drop. The drop will spread across the edge of the top slide.

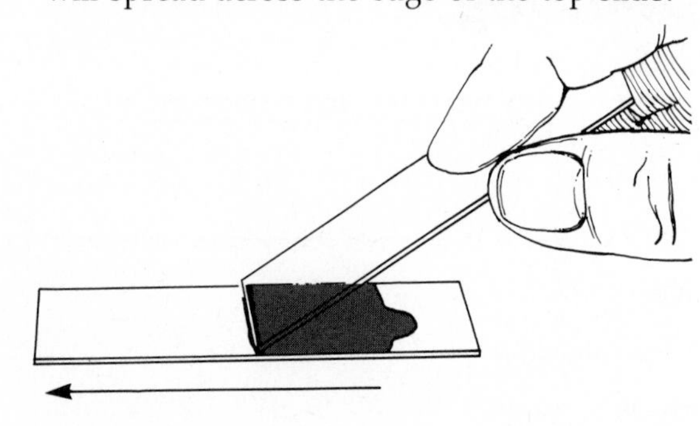

(c) Push the top slide to the left along the entire surface of the bottom slide.

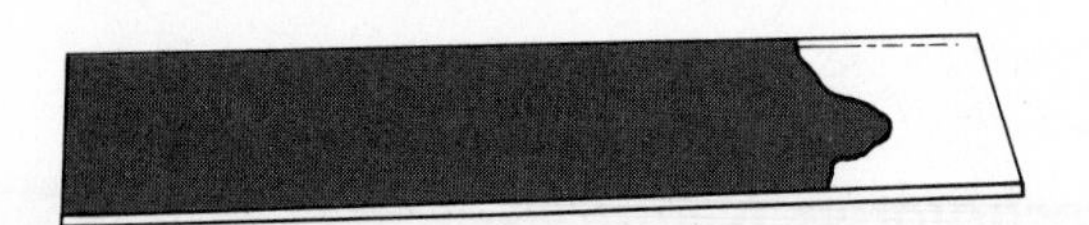

(d) Let the smear air dry.

Figure 6-1
The negative stain.

2. **a.** Place a *small* drop of nigrosine at the end of the slide. For cultures on solid media, add a loopful of distilled water and emulsify a small amount of the culture in the nigrosine–water drop. For broth cultures, mix a loopful of the culture into the drop of nigrosine (Figure 6-1*a*). Do not spread the drop or let dry.
   b. Using the end edge of another slide, spread the drop out, as shown in Figure 6-1*b* and *c*, to give a smear varying from opaque black to gray. The angle of the spreading slide will determine the thickness of the smear.
   c. Let the smear dry (Figure 6-1*d*). *Do not heat fix.*
   d. Prepare a negative stain of the other culture.
3. Examine the stained slides microscopically using the low, high-dry, and oil immersion objectives (Figure 6-2). As a general rule, if a few short rods are seen with small cocci, the morphology is rod-shaped. The apparent cocci are short rods viewed from the end or products of the cell division of small rods.
4. Wash your slides. Wipe the oil off your microscope and return it.

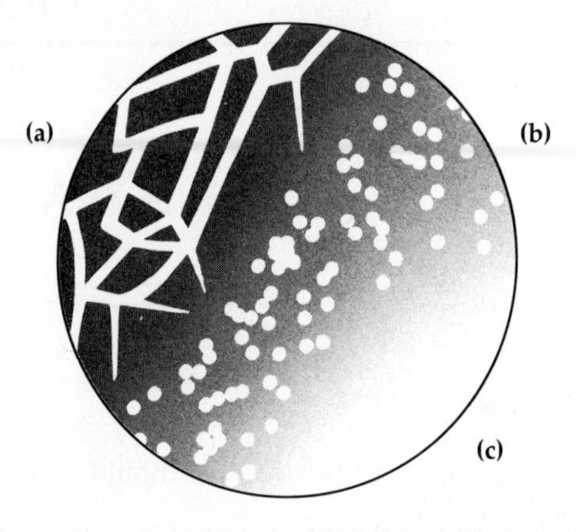

Figure 6-2
A negative stain as viewed under a microscope. **(a)** This part of the smear is too heavy. **(b)** Colorless cells are visible here. **(c)** Too little stain is in this area of the smear.

Turn to the Laboratory Report for Exercise 6.

b. Wash the slide carefully with distilled water from a wash bottle. Do not squirt water directly onto the smear.
c. Without drying, cover the smear with Gram's iodine for 30 seconds.
d. Without washing, decolorize with 95% ethyl alcohol. Let the alcohol run through the smear until no large amounts of purple wash out (usually a few seconds). The degree of alcohol decolorizing depends on the thickness of the smear. This is a critical step. *You should not over decolorize.* However, experience is the only way you will be able to determine how long to decolorize.
e. Immediately wash with distilled water. Why?

____

f. Add safranin for 30 seconds.
g. Wash with distilled water and blot the slide with a paper towel or absorbent paper. Let dry.

3. Examine the stained slide microscopically using the low, high-dry, and oil immersion objectives. Put the oil directly on the smear. Record your results. Do they agree with those given in your textbook? __________ If not, try to determine why. Some common sources of Gram staining errors are
   a. The loop was too hot.
   b. Excessive heat was applied during heat fixing.
   c. Decolorizing agent (alcohol) was left on the smear too long.

   Now, stain the remaining two slides.
4. If time allows, clean another slide.
   a. Scrape the base of your teeth and gums with a sterile toothpick and make a smear in 2 or 3 loopsful of distilled water.
   b. Heat fix.
   c. Gram stain, observe, and describe your results.
5. Design your own differential stain using different reagents but the same general scheme. Does it work? ______________________________

Turn to the Laboratory Report for Exercise 7.

EXERCISE 8

EXERCISE 8

Acid-Fast Staining

Objectives

After completing this exercise you should be able to

1. Provide the application of the acid-fast procedure.
2. Explain what is occurring during the acid-fast staining procedure.
3. Perform and interpret an acid-fast stain.

Background

The **acid-fast stain** is a differential stain. In 1882 Paul Ehrlich discovered that *Mycobacterium tuberculosis* (the causative agent of tuberculosis) retained the primary stain even after washing with an acid-alcohol mixture. (We hope you can appreciate the phenomenal strides that were made in microbiology in the 1880s. Most of the staining and culturing techniques used today originated during that time.) Most bacteria are decolorized by acid-alcohol, with only the families Mycobacteriaceae and Nocardiaceae of the order Actinomycetales *(Bergey's Manual,* Part 17*) being acid-fast. The acid-fast technique has great value as a diagnostic procedure because both *Mycobacterium* and *Nocardia* contain **pathogenic** (disease-causing) species.

The cell walls of acid-fast organisms contain a wax-like lipid called *mycolic acid,* which renders the cell wall impermeable to most stains. The cell wall is so impermeable, in fact, that a clinical specimen is usually treated with strong sodium hydroxide to remove debris and contaminating bacteria prior to culturing mycobacteria.

Ziehl introduced the use of carbolfuchsin (containing 5% phenol) to replace aniline dyes in staining. Today, the **Ziehl–Neelsen procedure** is the most

*R. D. Buchanan and N. E. Gibbons, eds. 1974. *Bergey's Manual of Determinative Bacteriology.* 8th ed. Baltimore: Williams and Wilkins. You will find references to this very important work throughout this manual.

widely used acid-fast staining technique. In the Ziehl–Neelsen procedure, the smear is flooded with carbolfuchsin, which has a high affinity for a chemical component of the bacterial cell. The smear is heated to facilitate penetration of the stain into the bacteria. The stained smears are washed with an acid-alcohol mixture that easily decolorizes most bacteria except the acid-fast microbes. Methylene blue is then used as a counterstain to enable the investigator to observe the non–acid-fast organisms.

The mechanism of the acid-fast stain is thought to involve the relative solubility of carbolfuchsin and impermeability of the cell wall. Fuchsin is more soluble in carbolic acid (phenol) than in water, and carbolic acid solubilizes more easily in lipids than in acid-alcohol. When the carbolfuchsin is added, the lipids are stained red and the tenacious cell wall prevents the stained globules from leaving.

Materials

Acid-fast staining reagents:
- Ziehl's carbolfuchsin
- Acid-alcohol
- Methylene blue

Wash bottle of distilled water

Slides (2)

Boiling water bath

Cultures (one of each)

Mycobacterium phlei

Escherichia coli

Demonstration slides

Acid-fast sputum slides

Techniques Required

Exercises 1 and 5

Procedure

1. Prepare and heat fix a smear (Exercise 5) of one culture.
2. Cover the smear with a small piece of absorbent paper.
3. Add a copious amount of Ziehl's carbolfuchsin.
4. Heat by setting the slide on a boiling water bath (Figure 9-4*c*). Steam for 5 minutes, adding more stain as it is needed.
5. Discard the paper and wash the slide well with distilled water; then decolorize for 15 to 20 seconds with acid-alcohol. Wash with distilled water.
6. Counterstain for about 30 seconds with methylene blue.
7. Wash with distilled water and blot dry. Prepare an acid-fast stain of the other culture.
8. Observe the acid-fast-stained slides microscopically and record your results. Observe the demonstration slides.

Turn to the Laboratory Report for Exercise 8.

EXERCISE 9

Structural Stains (Endospore, Capsule, and Flagella)

Objectives

After completing this exercise you should be able to

1. Prepare and interpret endospore, capsule, and flagella stains.
2. Recognize the different types of flagellar arrangements.

Background

Structural stains can be used to identify and study the structure of bacteria. Currently, most of the fine structural details are examined using an electron microscope (Exercise 3), but historically, staining techniques have given much insight into bacterial fine structure. We will examine a few structural stains that

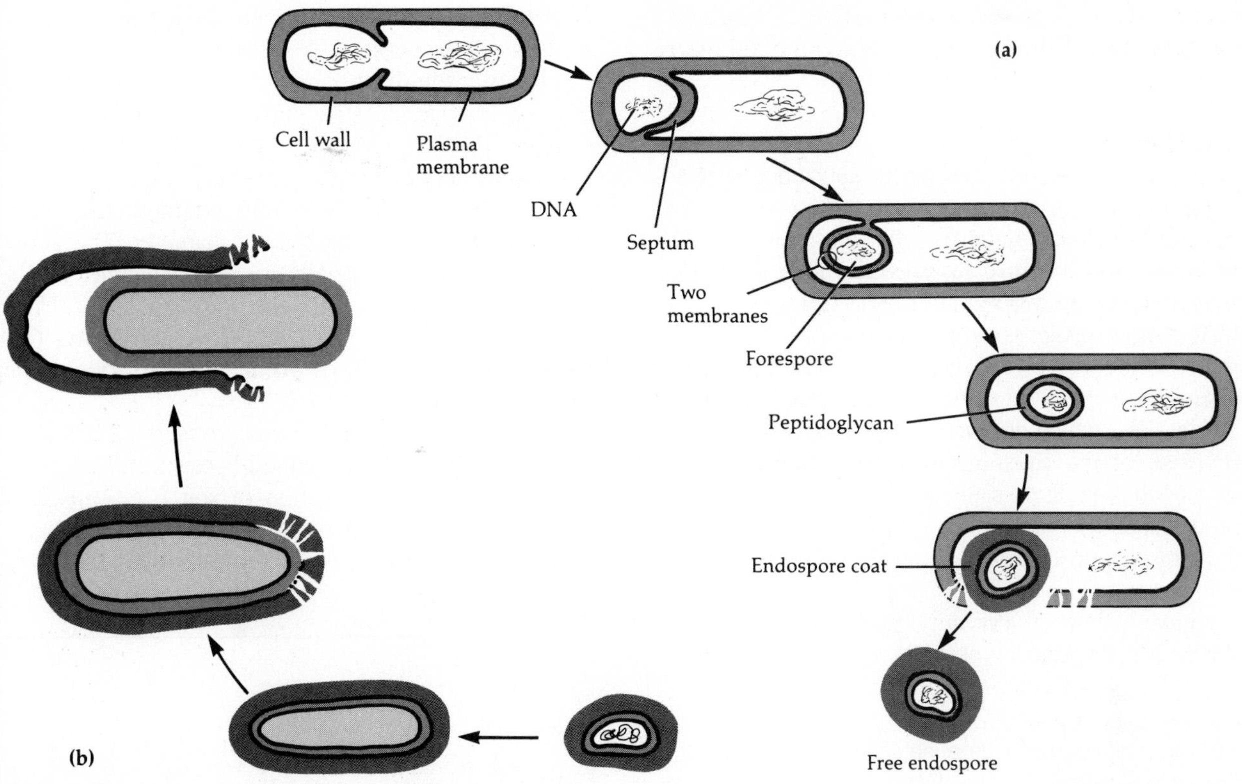

Figure 9-1

(a) Sporogenesis. The process of endospore formation.
(b) Germination of an endospore to a vegetative cell.

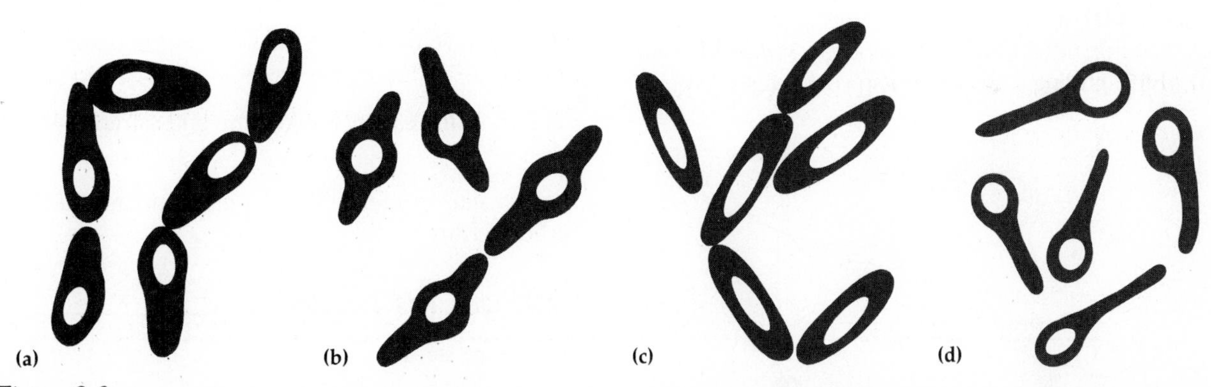

Figure 9-2

Some examples of bacterial endospores. **(a)** Subterminal endospores *(Bacillus subtilis).* **(b)** Central, swollen endospores *(Clostridium perfringens).* **(c)** Central endospores *(Bacillus polymyxa).* **(d)** Terminal, swollen endospores *(Clostridium tetani).*

are still useful today. These stains are used to observe endospores, capsules, and flagella.

Endospores

Endospores are formed by members of five genera included in *Bergey's Manual,* Part 15. *Bacillus* and *Clostridium* are the most familiar genera. Endospores are metabolically inert and are resistant to heating, various chemicals, and many harsh environmental conditions. Endospores are not for reproduction. Once an endospore forms in a cell, the cell will disintegrate (Figure 9-1). Endospores can remain dormant for long periods of time. However, an endospore may return to its vegetative or growing state.

Endospores do not form in response to a harsh environment, but they allow a bacterium to survive in a harsh environment. Perhaps the reasons for sporulation are expressed best by Cook: *"Bacteria form spores because they form spores."*

Taxonomically, it is very helpful to know if a bacterium is an endospore-former and to know the position of the endospores (Figure 9-2). Endospores are impermeable to most stains, so heat is usually applied to drive the stain into the endospore. Once stained, the endospores do not readily decolorize. We will use the **Schaeffer–Fulton** (malachite green) endospore stain.

Capsules

Many bacteria secrete a slippery substance that adheres to their surfaces and forms a viscous coat. This structure is called a **capsule** when it is round or oval in shape, and a **slime layer** when it is irregularly shaped and loosely bound to the bacterium. The ability to form capsules is genetically determined, but the size is influenced by the medium on which the bacterium is growing. Most capsules are composed of polysaccharides, which are water-soluble and uncharged. Because of the capsule's nonionic nature, simple stains will not adhere to it. Most capsule staining techniques stain the bacteria and the background, leaving the capsules unstained—essentially, a "negative" capsule stain.

Capsules have a very important role in the **virulence** (disease-causing ability) of some bacteria. For example, when bacteria such as *Streptococcus pneumoniae* have a capsule, the body's white blood cells cannot phagocytize (Exercise 70) the bacteria efficiently and disease occurs. When *S. pneumoniae* lack a capsule, they are easily engulfed and are not virulent.

Flagella

Many bacteria are **motile,** which means they have the ability to move from one position to another in a directed manner. Myxobacteria (*Bergey's Manual,* Part 2) exhibit gliding motion, and spirochaetes (*Bergey's Manual,* Part 5) undulate using axial filaments, but most motile bacteria possess flagella.

Flagella, the most common means of motility, are thin proteinaceous structures that originate in the cytoplasm and project out from the cell wall. They are very fragile and are not visible with a light microscope. They can be stained by carefully coating them using a mordant, which increases their diameter. The presence and location of flagella are helpful characteristics in the identification and classification of bacteria. Flagella are of two main types: **peritrichous** (all around the bacterium) or **polar** (at one or both ends of the cell) (Figure 9-3).

Motility may be determined by observing hanging-drop preparations of unstained bacteria (Exercise 2), flagella stains, or inoculation of soft (or semisolid) agar (Exercise 13). If time does not permit doing flagella stains, observe the demonstration slides.

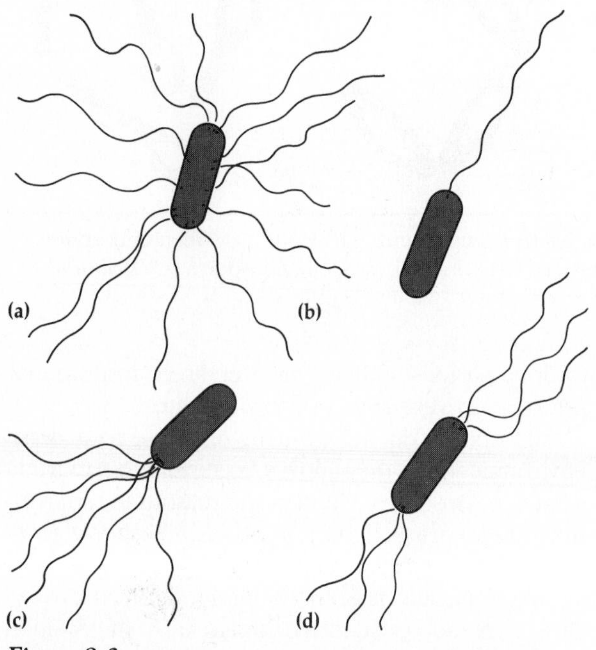

Figure 9-3
Flagellar arrangements. **(a)** Peritrichous flagella. Types of polar flagella: **(b)** Monotrichous flagella, **(c)** Lophotrichous flagella, and **(d)** Amphitrichous flagella.

Materials

Slides

Cover slip

Paper towels

Wash bottle of distilled water

Forceps

Scalpel

Endospore stain reagents: malachite green and safranin

Capsule stain reagents: nigrosine and safranin; India ink.

Flagella stain reagents: flagella mordant and Ziehl's carbolfuchsin

Cultures (as needed)

Endospore stain

Bacillus megaterium (24-hour)

Bacillus subtilis (24-hour)

Bacillus subtilis (72-hour)

Capsule stain

Streptococcus mutans

Enterobacter aerogenes

Flagella stain

Proteus vulgaris (18-hour)

Demonstration Slides

Endospore stain

Capsule stain

Flagella stain

Techniques Required

Exercises 1, 5, and 6

Procedures

Endospore Stain (Figure 9-4)

Be careful. The malachite green has a messy habit of ending up everywhere. But most likely you will end

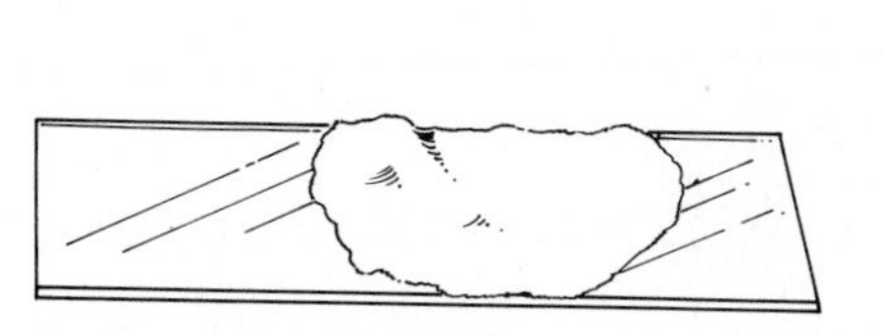

(a) Place a piece of absorbent paper over the smear.

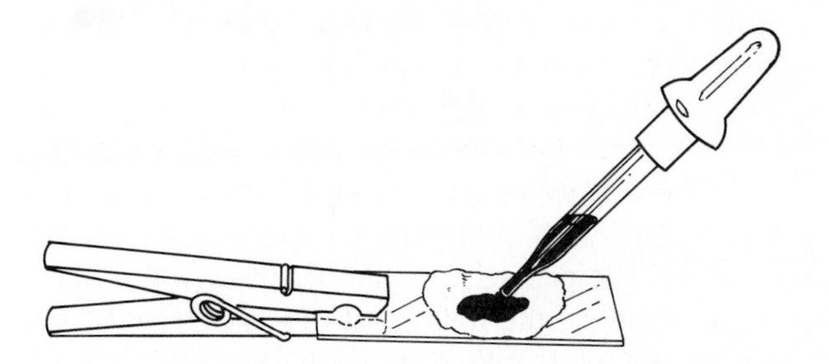

(b) Cover the paper with malachite green.

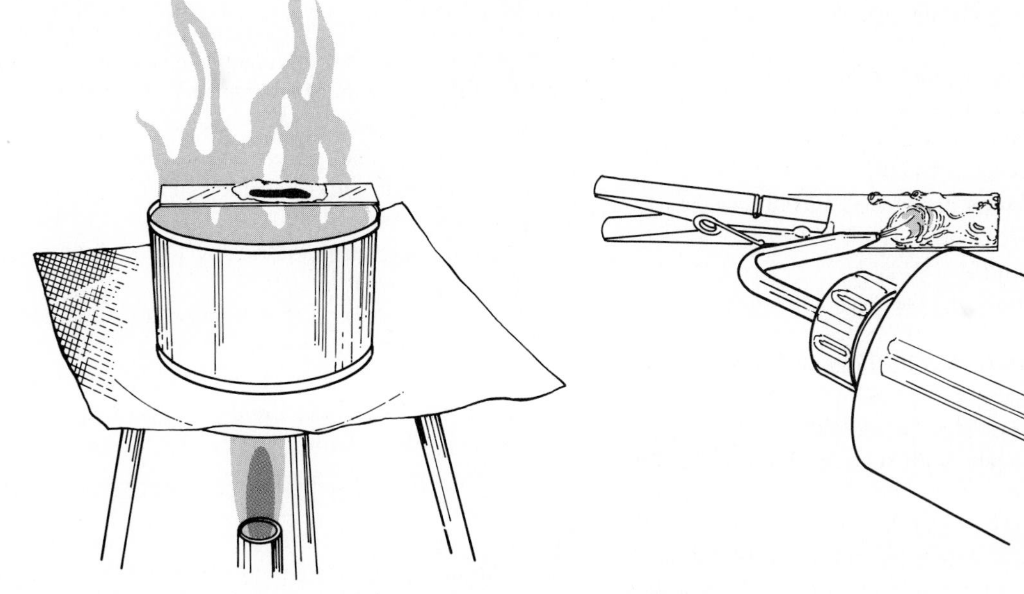

(c) After steaming for 5 minutes, wash the smear with water.

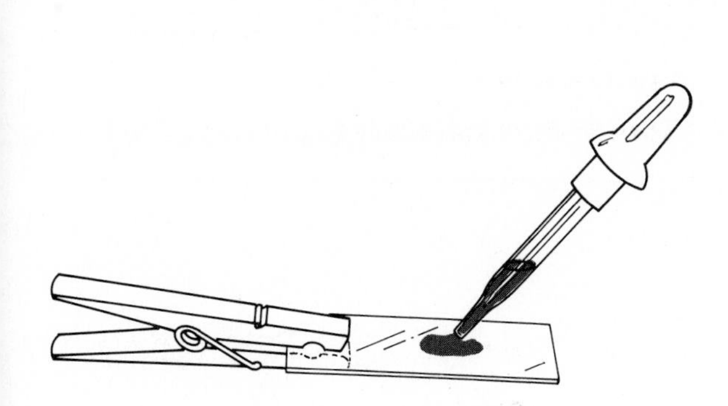

(d) Cover the smear with safranin for 30 seconds.

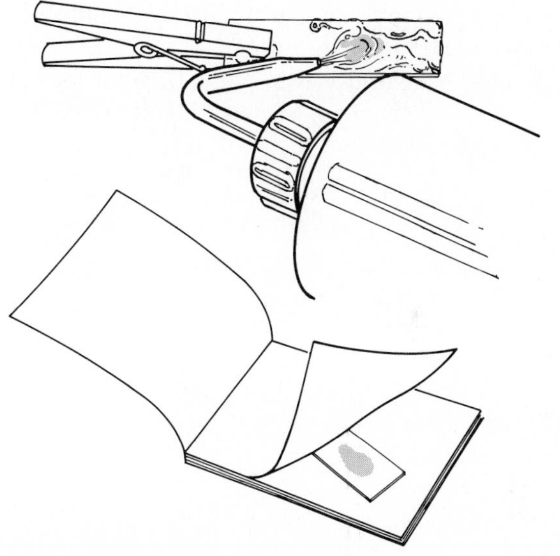

(e) Wash with water and blot dry.

Figure 9-4

The endospore stain.

up with green fingers no matter how careful you are. Here's your chance to develop a green thumb!

1. Clean three slides.
   a. Make smears of the three *Bacillus* cultures, air dry, and heat fix (Exercise 5).
   b. Tear out small pieces of paper towel and place on the slides. The paper should be smaller than the slide.
   c. Flood the smear and paper with malachite green, steam for 5 minutes. Add more stain as it is needed. *Keep it wet.* What is the purpose of the paper? ______
   d. Remove the towel and discard carefully. *Do not put it in the sink.* Wash the stained smears well with distilled water.
   e. Counterstain with safranin for 30 seconds.
   f. Wash with distilled water and blot dry.
2. Observe microscopically and record your observations.
3. Observe the demonstration slides of bacterial endospores.

Capsule Stain

Use one of the cultures designated and do one of the following procedures.

1. Wet mount method
   a. Add 1 loopful of a bacterial culture to a clean slide.
   b. Add a small amount of India ink until the mixture is dark gray.
   c. Place a cover slip on the slide and examine microscopically with reduced light using the oil immersion lens. Look for areas of clearing around the bacteria.
   d. Capsules will not be seen unless the ink and culture are in the right proportion. Repeat the procedure using different proportions if capsules are not seen. Describe your results. Are any of the bacteria with capsules motile? ______
2. Gin's method
   a. Make a negative stain of the bacteria with nigrosine (Exercise 6). Remember to use a very clean slide.
   b. Let dry. Heat fix briefly.
   c. Cover with safranin for 10 to 20 seconds, then carefully let the safranin run off.
   d. Prop the slide up at a 45° angle on a paper towel and let it dry.
   e. Observe using the oil immersion objective; the background will be black, the bacteria red, and the capsule clear. Record your results.
3. Observe the demonstration slides of bacterial capsules.

Flagella Stain

1. Flagella stains require special precautions to avoid damaging the flagella. Scrupulously clean slides are essential, and the culture must be handled carefully to prevent flagella from coming off the cells.
2. Without touching the bacterial culture, use a scalpel and forceps to cut out a piece of agar on which *Proteus* is growing. *Do not touch the culture with your hands.* Gently place the agar, culture side down, on a clean glass slide. Then carefully remove the agar with the forceps. Place the piece of agar in the Petri plate.
3. Allow the organisms adhering to the slide to air dry. *Do not heat fix.*
4. Cover the slide with flagella mordant and allow it to stand for 10 minutes.
5. Gently rinse off the stain with distilled water.
6. Cover the slide with Ziehl's carbolfuchsin for 5 minutes. Rinse gently with distilled water. What color should the flagella be? ______
   The cells? ______
7. Allow the rinsed smear to air dry *(do not blot)* and examine microscopically for flagella.
8. Observe the demonstration slides illustrating various flagellar arrangements.

Turn to the Laboratory Report for Exercise 9.

EXERCISE 10

Morphologic Unknown

Dishonesty is knowing but ignoring the fact that the data are contradictory. Stupidity is not recognizing the contradictions.

AUTHOR UNKNOWN

Objective

After completing this exercise you should be able to

1. Identify the morphology and staining characteristics of an unknown organism.

Background

To identify a bacterium, it is very important to determine morphology, arrangement, Gram reaction, and structural details. You will be given an unknown culture of bacteria. Determine its morphologic and structural characteristics. The culture contains one species and is less than 24 hours old.

Materials

Staining reagents

Culture

24-hour unknown slant culture of bacteria # ______

Techniques Required

Exercises 1 and 2, and 5 through 9

Procedure

1. Record the number of your unknown.
2. Determine the morphology, Gram reaction and arrangement of your unknown, perform a Gram stain and, if needed, an endospore stain, acid-fast stain, flagella stain, hanging-drop technique, or capsule stain. When are the latter needed? ______
______
3. Tabulate your results in the Laboratory Report.

Turn to the Laboratory Report for Exercise 10.

PART 3

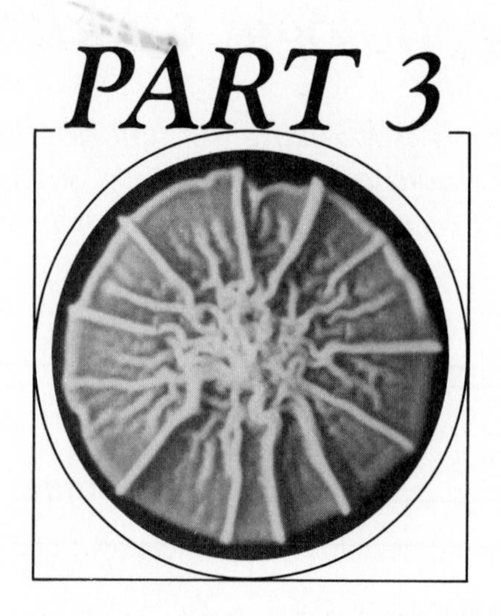

Cultivation of Bacteria

EXERCISES

In the previous exercises you saw that bacteria are grown on culture media in a laboratory. Bacteria must be cultured in order to characterize and identify them. This statement from the first edition of *Bergey's Manual* summarizes the need for cultivating bacteria:

> *The earlier writers classified the bacteria solely on their morphologic characters. A more detailed classification was not possible because the biologic characters of so few of the bacteria had been determined. With the accumulation of knowledge of the biologic characters of many bacteria it was realized that it is just as incorrect to group all rods under a single genus as to group all quadruped animals under one genus.*

Before attempting to culture bacteria, consideration must be given to their nutritional requirements. Bacteria require sources of energy, carbon, nitrogen, minerals, and growth factors. Most bacteria are **chemoheterotrophs,** which require organic compounds for carbon and energy sources. Bacteria exhibit a wide range of nutritional requirements, and, in many instances, enriching nutrients such as milk, serum, blood, or tomato juice must be added for fastidious organisms that require many growth factors. **Growth factors** are organic compounds such as vitamins or amino acids that are incorporated into a cell without alteration.

The first bacterial cultures were grown in *broth* (liquid) media such as infusions and blood. Koch attempted to culture bacteria on potato slices when he realized the need for *solid media.* When some bacteria wouldn't grow on potato, he added gelatin to the liquid but the gelatin liquefied under standard incubation conditions. The wife of Walter Hess, a colleague of Koch, suggested the use of agar as a solidifying agent. Today, agar is the most commonly used solidifying agent in culture media (see the figure).

The exercises in Part 3 emphasize the use of aseptic technique. **Aseptic technique** is prevention of unwanted microorganisms in laboratory and medical procedures.

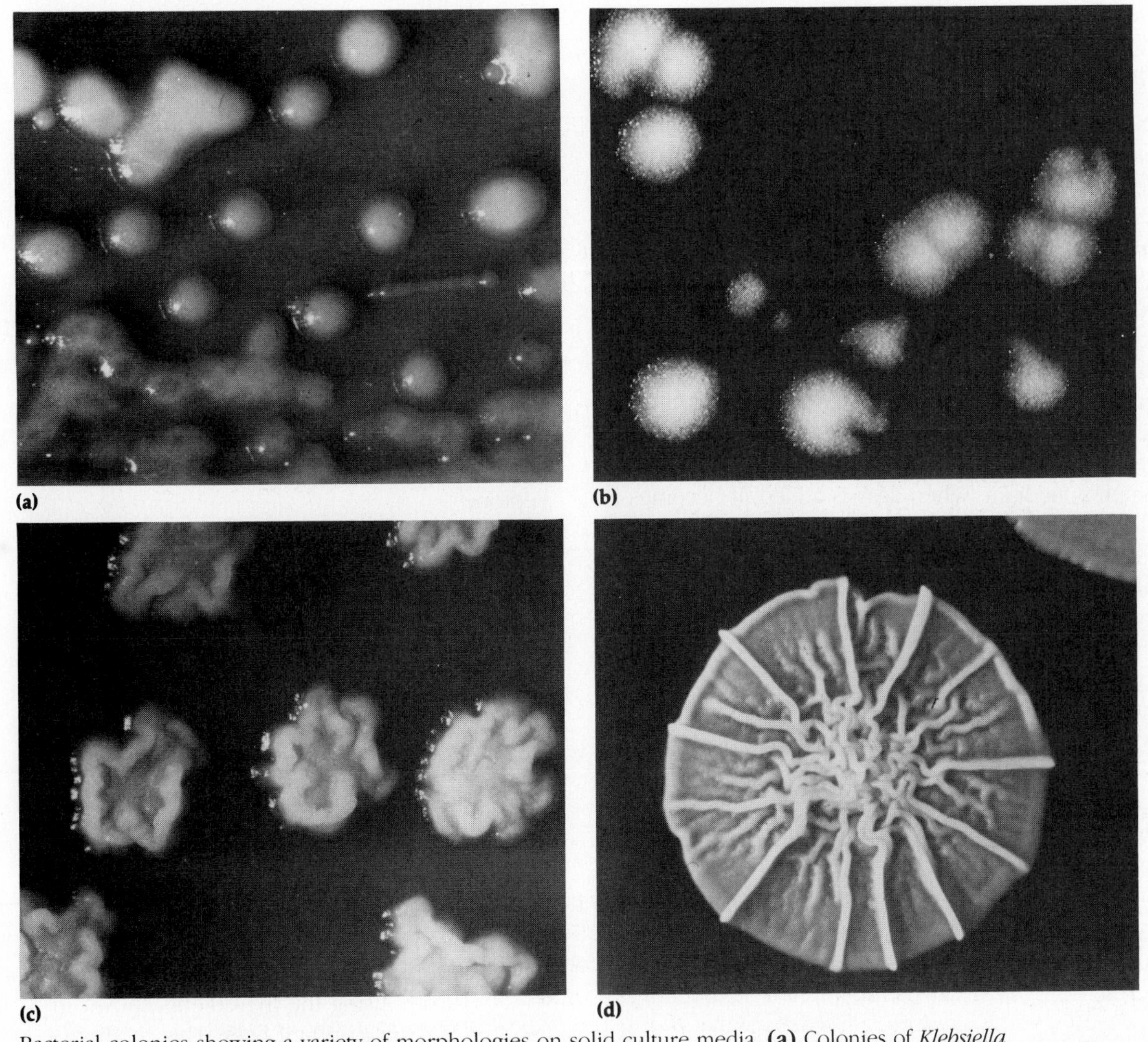

Bacterial colonies showing a variety of morphologies on solid culture media. **(a)** Colonies of *Klebsiella pneumoniae* on MacConkey agar. **(b)** Colonies of *Bacillus anthracis* on artificial medium. **(c)** Rough colonies of *Mycobacterium smegmatis* on Penassay agar. **(d)** Colony of *Micrococcus luteus* on blood agar.

EXERCISE 11

Culture Media Preparation

Objectives

After completing this exercise you should be able to

1. Differentiate between complex and chemically defined media.
2. Provide a rationale for sterilization of culture media.
3. Explain why agar is used in culture media.
4. Prepare a complex and a chemically defined medium.

Background

When a medium is selected for culturing bacteria, macronutrients, an energy source, and any necessary growth factors must be provided. A medium whose exact chemical composition is known is called a **chemically defined medium.**

Most chemoheterotrophic bacteria are routinely grown on **complex media,** that is, media for which the exact chemical composition varies slightly from batch to batch. Organic carbon, energy, and nitrogen sources are usually supplied by protein in the form of meat extracts and partially digested proteins called *peptones.* **Nutrient broth** is a commonly used liquid complex medium. When agar is added, it becomes a solid medium called **nutrient agar.**

Agar, an extract from marine red algae, has some unique properties that make it useful in culture media. Few microbes can degrade agar, so it remains solid during microbial growth. It liquefies at 100°C and remains in a liquid state until cooled to 40°C. Once the agar has solidified, it can be incubated at temperatures up to 100°C and remain solid.

Media must be sterilized after preparation. The most common method of sterilizng culture media that are heat stable is **steam sterilization,** or **autoclaving,** using steam under pressure. During this process, material to be sterilized is placed in the autoclave and heated to 121°C at 15 pounds of pressure (15 psi) for 15 minutes.

Materials

250-ml Erlenmeyer flasks with caps or cotton plugs (2)

100-ml graduated cylinder (1)

Distilled water

1-ml pipette (1)

10-ml pipette (1)

Weighing paper

Triple beam balance

Stirring rod

Tongs or glove

Hot plate or burner and ring stand

Test tube with cap (1)

Sterile Petri plates (4)

Reagents

Glucose

Sodium chloride (NaCl)

Ammonium dihydrogen phosphate ($NH_4H_2PO_4$)

Dipotassium phosphate (K_2HPO_4)

Magnesium sulfate ($MgSO_4$), 2% stock solution

Peptone

Beef extract

Agar

Demonstration

Use of the autoclave

Techniques Required

Appendices A and C

Procedure

A. Preparation of culture media

1. Prepare 100 ml of glucose-minimal salts broth, using the ingredients shown in Table 11-1, in a 250-ml flask. This medium is prepared by adding the first four reagents listed to 50 ml distilled water. Measure the water with a graduated cylinder. To this, add 1 ml of

Table 11-1
Glucose-minimal Salts Broth

Ingredient	Amount/100 ml
Glucose	0.5 g
Sodium chloride (NaCl)	0.5 g
Ammonium dihydrogen phosphate ($NH_4H_2PO_4$)	0.1 g
Dipotassium phosphate (K_2HPO_4)	0.1 g
Magnesium sulfate ($MgSO_4$)	0.02 g
Distilled water	100 ml

2% $MgSO_4$ stock solution, using a 1-ml pipette (Appendix A). Add 49 ml distilled water to bring up to volume. Stopper the flask, label it with your name and "glucose broth," and place it in the autoclave (or designated autoclave basket).

2. Prepare 100 ml of nutrient broth (see Table 11-2). Put 100 ml distilled water in a 250-ml flask. To this, add peptone. Weigh out the necessary amount of beef extract. Toss weighing paper and extract into the water. Why is this permissible and, in fact, necessary? ________

____

Remember to retrieve the paper after the beef extract has washed off, and discard it properly. Add NaCl. Stir to dissolve.

3. Pipette 10 ml nutrient broth into a test tube and cap it. Label with your name and "nutrient broth."
4. Add the appropriate amount of agar to the remaining *90 ml* nutrient broth. How much agar will you add? ________________
Use this space for your calculation.

Bring to a boil and continue boiling carefully until all the agar is dissolved. *Be careful:* Do not let the solution boil over. Stir often to prevent burning and boiling over.

Table 11-2
Nutrient Agar

Ingredient	Amount/100 ml
Peptone	0.5 g
Beef extract	0.3 g
Sodium chloride (NaCl)	0.8 g
Agar	1.5 g
Distilled water	100 ml

5. Stopper the flask, label "nutrient agar," and place the flask and tube in the autoclave (or designated autoclave baskets).
6. The instructor will demonstrate use of the autoclave.
7. After autoclaving, allow the flasks and tube to cool to room temperature or proceed to part B. What effect does the agar have on the culture medium? ________________

B. Pouring plates

Allow the flask of nutrient agar to cool to about 45°C (warm to the touch). If the agar has solidified, it will have to be reheated to liquefy. To what temperature will it have to be heated? ________

____

The sterile nutrient agar must be poured into Petri plates *aseptically,* that is, without letting microbes into the nutrient medium. *Read the following procedure before beginning* so that you can work quickly and efficiently.

1. Set four sterile, unopened Petri plates in front of you with the cover (larger half) on top. Have a lighted laboratory burner within reach on your workbench.
2. Hold the flask at an angle, remove the stopper with the fourth and fifth fingers of your other hand. Heat the mouth of the flask by passing it briefly through the flame (Figure 11-1*a*). Why is it necessary to keep the flask at an angle through this procedure? ________________

____

3. Remove the cover from the first plate with the hand holding the plug. Quickly and neatly pour melted nutrient agar into the plate until the bottom is just covered to a depth of approximately 5 mm (Figure 11-1*b*). Keep the flask at an angle and replace the plate cover; move on to the next plate until all the agar is poured.
4. When all the agar is poured, gently swirl the agar in each plate to cover any empty spaces; do not allow the agar to touch the sides or covers of the plates.

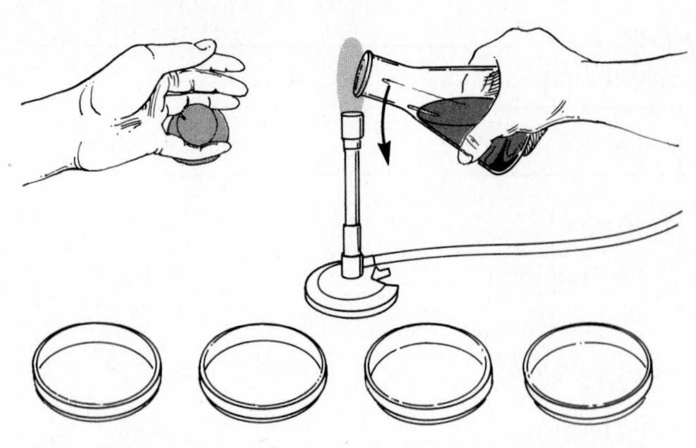

(a) Remove the stopper and flame the mouth of the flask.

(b) Remove the cover from one plate and pour nutrient agar into the plate bottom.

(c) Flame the surface of the nutrient agar.

Figure 11-1

Petri plate pouring.

5. To break bubbles and provide a smooth surface, lift the cover off one plate at a time, and pass the flame over the plate quickly, then replace the cover (Figure 11-1*c*). Repeat this procedure with each plate.
6. To decrease condensation, leave the Petri plate covers slightly ajar for about 15 minutes until the agar solidifies.
7. Place the empty flask in the discard area.
8. Your instructor will tell you where to store the media for use in future exercises.

Turn to the Laboratory Report for Exercise 11.

EXERCISE 12

Microbes in the Environment

Whatever is worth doing at all is worth doing well.

CHESTERFIELD

Objectives

After completing this exercise you should be able to

1. Explain the importance of aseptic technique.
2. Describe colony morphology using accepted descriptive terms.

Background

Microbes are everywhere: they are found in the water we drink, the air we breathe, and the earth we walk on. They live in and on our bodies. Microbes occupy ecological niches on all forms of life and in most environments. There are microbes in the air, in the water, and on the workbenches. In most situations, these ubiquitous microorganisms are harmless. However, in microbiology, work must be done carefully to avoid contaminating sterile media and materials with these microbes.

Aseptic technique is used in microbiology to prevent contamination by unwanted microorganisms. Culture media are sterilized after preparation (Exercise 11) to kill microbes that are present in the air and water, and on the glassware. In Exercise 11, sterile nutrient agar was dispensed into sterile Petri plates in a manner that would prevent the entry of microbes.

In this exercise, microbes will be **inoculated,** that is, microbes will be intentionally introduced, onto the nutrient agar and into the nutrient broth prepared in Exercise 11. The bacteria that are inoculated into culture media grow (increase in number) during an **incubation period.** After suitable incubation, liquid media are **turbid,** or cloudy, due to bacterial growth. On solid media, colonies will be visible to the naked eye (see the figure on page 33). A **colony** is a population of cells that arises from a single bacterial cell. Although many species of bacteria give rise to similar appearing colonies, each different appearing colony is usually a different species.

Materials

Media prepared in Exercise 11 can be used.

Petri plates containing nutrient agar (4)

Tube containing nutrient broth (1)

Tube containing sterile cotton swabs (1)

Tube containing sterile water (1)

Techniques Required

None

Procedure

First Period

1. *Design your own experiment.* The purpose is to sample your environment and your body. Use your imagination. Here are some suggestions.
   a. You may use the lab, a washroom, or any place on campus for the environment.
   b. One nutrient agar plate might be left open to the air for 30 to 60 minutes.
   c. A plate could be inoculated from an environmental surface such as the floor or workbench by wetting a cotton swab in sterile water, swabbing the environmental surface, and then swabbing the surface of the agar. Why is the swab first moistened in sterile water? ______________________ After inoculation, the swab should be discarded in the container of disinfectant.
2. Inoculate two plates from the environment. Inoculate one nutrient broth tube using a swab as described in step 1*c*. After swabbing the agar surface, place the swab in the broth and leave it there during incubation.

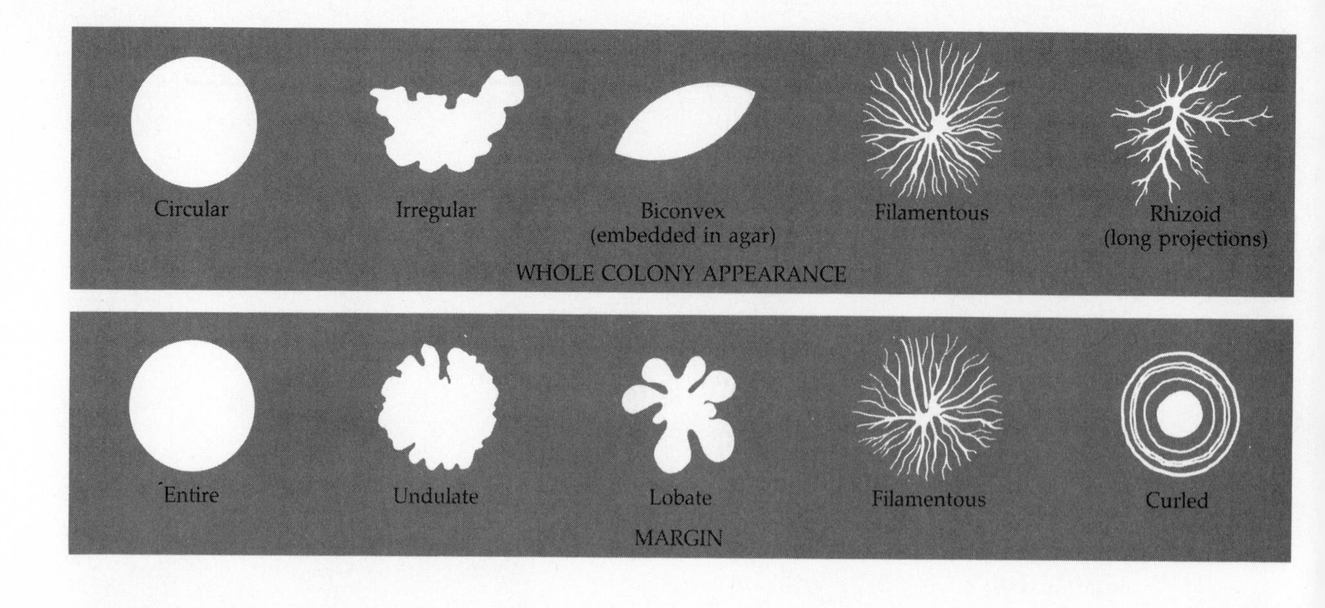

Figure 12-1
Colony descriptions.

3. The plates and tube should be incubated at the approximate temperature of the environment sampled.
4. Inoculate two plates from your body. You could
   a. Place a hair on the agar.
   b. Obtain an inoculum by swabbing (see step 1*c*) part of your body.
   c. Touch the plate.
5. Incubate bacteria from your body at or close to your body temperature. What is human body temperature? ______________ °C
6. Incubate all plates, inverted so water will condense in the lid instead of the surface of the agar. Why is condensation on the agar undesirable? ______________

   ____
7. Incubate all inoculated media until the next laboratory period.

Second Period

1. Observe and describe the resulting growth on the plates. Note each different appearing colony and describe the colony morphology using the characteristics given in Figure 12-1. Determine the approximate number of each type of colony. When many identical colonies are present, *TMTC* (too many to count) can be recorded as the number of colonies.
2. Describe the appearance of the nutrient broth. Is it uniformly cloudy or **turbid?** Look for clumps of microbial cells; this is called **flocculant.** Is there a membrane or **pellicle** across the surface of the broth? See whether microbial cells have settled on the bottom of the tube forming a **sediment.**
3. Discard the plates properly and save the tube of broth for Exercise 14.

Turn to the Laboratory Report for Exercise 12.

EXERCISE 13

Transfer of Bacteria: Aseptic Technique

Study without thinking is worthless;
thinking without study is dangerous.

CONFUCIUS

Objectives

After completing this exercise you should be able to

1. Provide the rationale for aseptic technique.
2. Differentiate among the following: broth, slant, and deep.
3. Aseptically transfer bacteria from one form of culture medium to another.

Background

In the laboratory, bacteria must be cultured to facilitate identification and to examine their growth and metabolism.

All culture media are **sterilized,** that is, rendered free of all life, prior to use. Sterilization is usually accomplished using an autoclave (Exercise 11). Containers of culture media such as test tubes or Petri plates should not be opened until you are ready to work with them, and even then, they should not be left open.

Petri plates containing nutrient media were used in Exercises 11 and 12; culture media can be prepared in other forms, depending on the desired use. **Broth cultures** provide large numbers of bacteria in a small space and are easily transported. **Agar slants** are test tubes containing solid culture media that were left at an angle while the agar solidified. Agar slants, like Petri plates, provide a solid growth surface, but slants are easier to store and transport than Petri plates. Agar is allowed to solidify in the bottom of a test tube to make an **agar deep.** Deeps are often used to grow bacteria that prefer less oxygen than is present on the surface of the medium. Semisolid agar deeps containing 0.5% to 0.7% agar instead of the usual 1.5% agar can be used to determine whether a bacterium is motile (Exercise 9). Motile bacteria will move away from the point of inoculation, giving an inverted "Christmas tree" appearance.

EXERCISE 13

A bacterial culture is transferred from one culture medium to another in order to keep it alive and to study its growth. Transferring must be done without introducing unwanted microbes called **contaminants** into the media. Transfer techniques that minimize contaminants are called **aseptic techniques.** Microorganisms are aseptically **inoculated,** or introduced, into various forms of culture media.

Transfer and inoculation are usually performed with a sterile, heat-resistent, noncorroding nichrome wire attached to an insulated handle. When the end of the wire is bent into a loop, it is called an **inoculating loop;** when straight, it is an **inoculating needle** (Figure 13-1). For special purposes, cultures may also be transferred with sterile cotton swabs, pipettes, glass rods, or syringes. These techniques will be introduced in later exercises.

Whether an inoculating loop or needle is used depends on the form of the medium; after completing

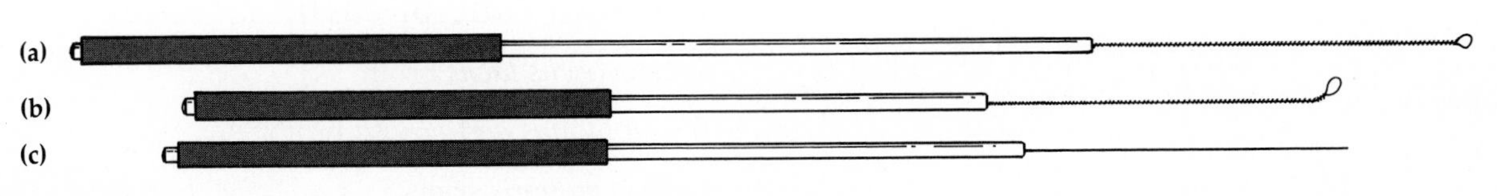

Figure 13-1

(a) An inoculating loop. **(b)** A variation of the inoculating loop in which the loop is bent at a 45° angle. **(c)** An inoculating needle.

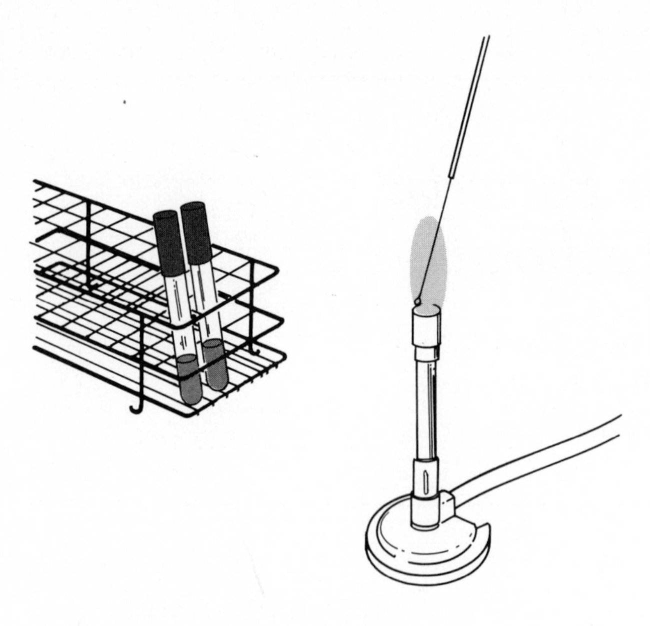

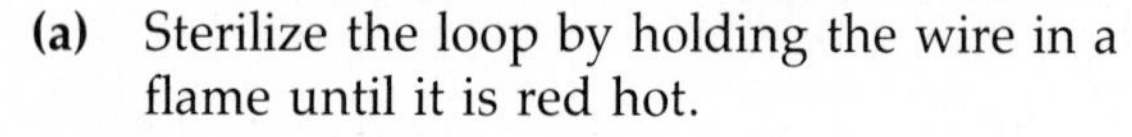

(a) Sterilize the loop by holding the wire in a flame until it is red hot.

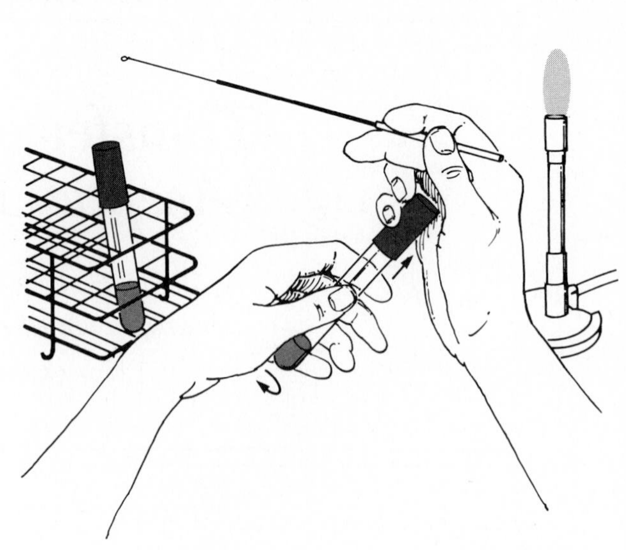

(b) While holding the sterile loop and the bacterial culture, remove the cap as shown.

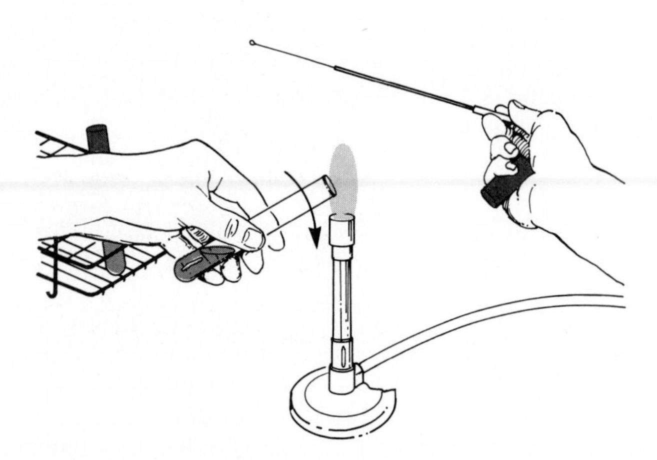

(c) Briefly heat the mouth of the tube in a burner flame before inserting the loop for an inoculum.

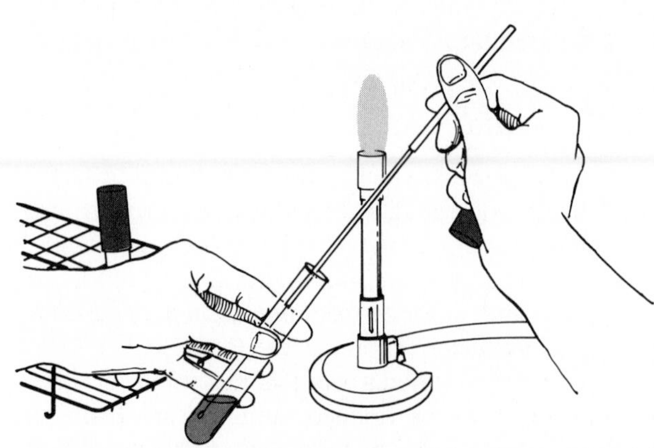

(d) Get a loopful of culture, heat the mouth of the tube, and replace the cap.

Figure 13-2
Inoculating procedures.

this exercise, you will be able to decide when an instrument is to be used.

Materials

Tubes containing nutrient broth (3)

Tubes containing nutrient agar slants (3)

Tubes containing nutrient semisolid agar deeps (3)

Inoculating loop

Inoculating needle

Test tube rack

Gram staining reagents

Cultures (one of each)

Streptococcus lactis broth

Pseudomonas aeruginosa broth

Proteus vulgaris slant

Techniques Required

Exercise 7

Procedure

1. Work with only one of the bacterial cultures at a time, to prevent any mixups or cross-contamination. Begin with one of the broth cultures, and gently tap the bottom of the broth culture to re-suspend the sediment.
2. To inoculate nutrient broth, hold the inoculating loop in your dominant hand and one of the broth cultures of bacteria in the other hand.
   - **a.** Sterilize the loop by holding the wire in a Bunsen burner flame (Figure 13-2*a*). Heat to redness. Why? ______
   - **b.** Holding the loop like a pencil, curl the little finger of the same hand around the cap of the broth culture. Gently pull the cap off the tube while turning the culture tube (Figure 13-2*b*). If cotton stoppers are used, simply grasp the stopper with your finger. Do not set the cap down. Why not? ______
   - **c.** Holding the tube at an angle, pass the mouth of the tube through the flame (Figure 13-2*c*). What is the purpose of "flaming" the mouth of the tube? ______
     Always hold culture tubes and uninoculated tubes at an angle to minimize the amount of dust that could fall into them.
   - **d.** Immerse the sterilized, cooled loop into the broth culture to obtain a loopful of culture (Figure 13-2*d*). Why must the loop be cooled first? ______
     Remove the loop, and while holding the loop, flame the mouth of the tube and recap by turning the tube into the cap. Place the tube in your rack.
   - **e.** Remove the cap from a tube of sterile nutrient broth as previously described and flame the mouth of the tube. Immerse the inoculating loop into the sterile broth. Flame the mouth of the tube and replace the cap. Return the tube to the test tube rack.
   - **f.** Reflame the loop until it is red and let it cool. Some individuals prefer to hold several tubes in their hands at once (Figure 13-3). *Do not* attempt holding and transferring between multiple tubes until you have mastered aseptic transfer techniques.
3. Obtain a nutrient agar slant. Repeat steps 2*a* through 2*d*, and inoculate the slant by moving the loop gently across the agar surface from the bottom of the slant to the top, being careful not to gouge the agar (Figure 13-4). Flame the mouth of the tube and replace the cap. Flame your loop and let it cool.
4. Obtain a nutrient agar semisolid deep, and using your inoculating needle, repeat steps 2*a* through 2*d*. Inoculate the semisolid agar deep by plunging the needle straight down the middle of the deep, then pull out through the same stab as shown in Figure 13-5. Flame the mouth of the tube and replace the cap. Flame your needle and let it cool.
5. Using the other broth culture, inoculate a broth culture, agar slant, and semisolid agar deep, as described in steps 2, 3, and 4, using your inoculating *needle*.
6. To transfer *Proteus vulgaris,* flame your loop and allow it to cool. Flame the mouth of the tube and carefully scrape a small amount of the culture from the agar. Flame the mouth of the tube and replace the cap. Inoculate a broth and slant as described in steps 2 and 3. Inoculate a semisolid agar deep with an inoculating needle as described in step 4.

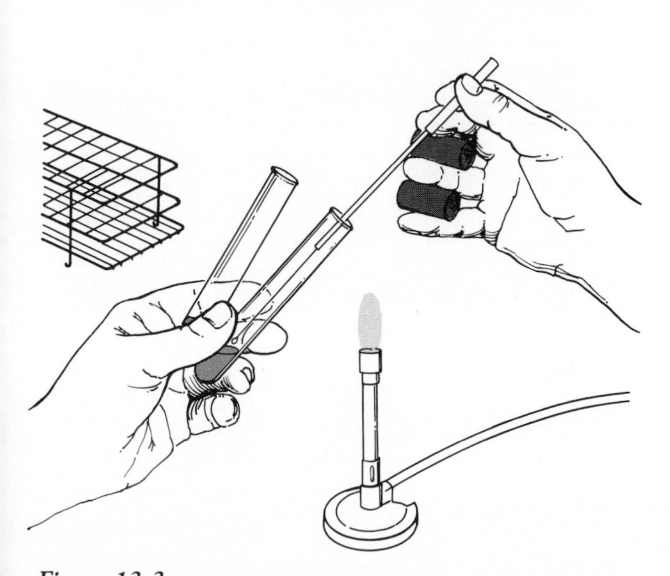

Figure 13-3

Experienced laboratory technicians can transfer cultures aseptically holding multiple test tubes.

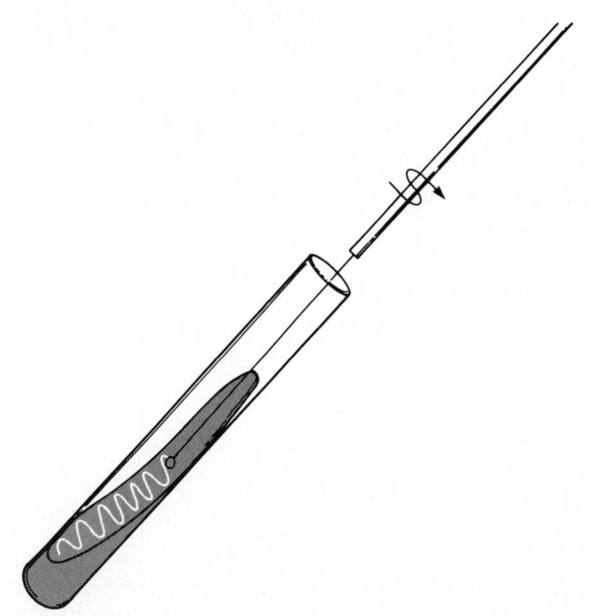

Figure 13-4

Inoculate a slant by streaking back and forth across the surface of the agar.

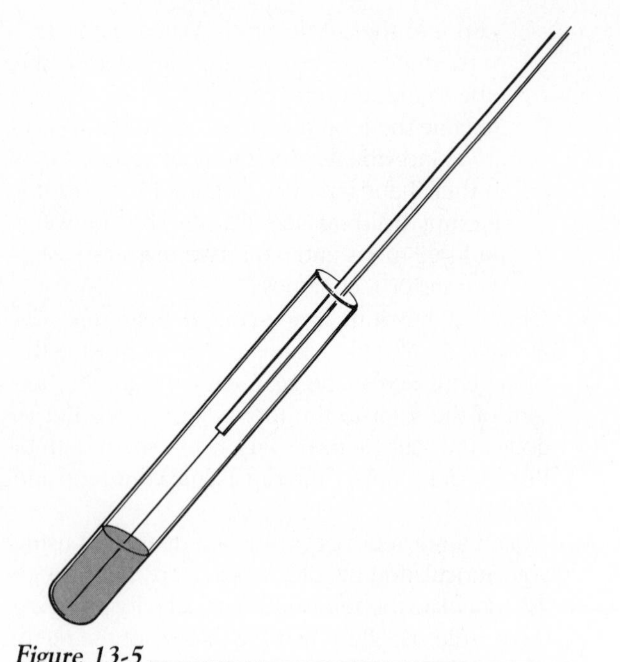

Figure 13-5

An agar deep is inoculated by stabbing into the agar with a needle.

7. Incubate all tubes at 35°C until the next period.
8. Make a smear of the *Streptococcus* broth culture and heat fix (Exercise 5). Keep it in your drawer.
9. Record the appearance of each culture, using Figure 13-6. Make a smear of *Streptococcus* from the slant. Gram stain both smears and compare them.

Turn to the Laboratory Report for Exercise 13.

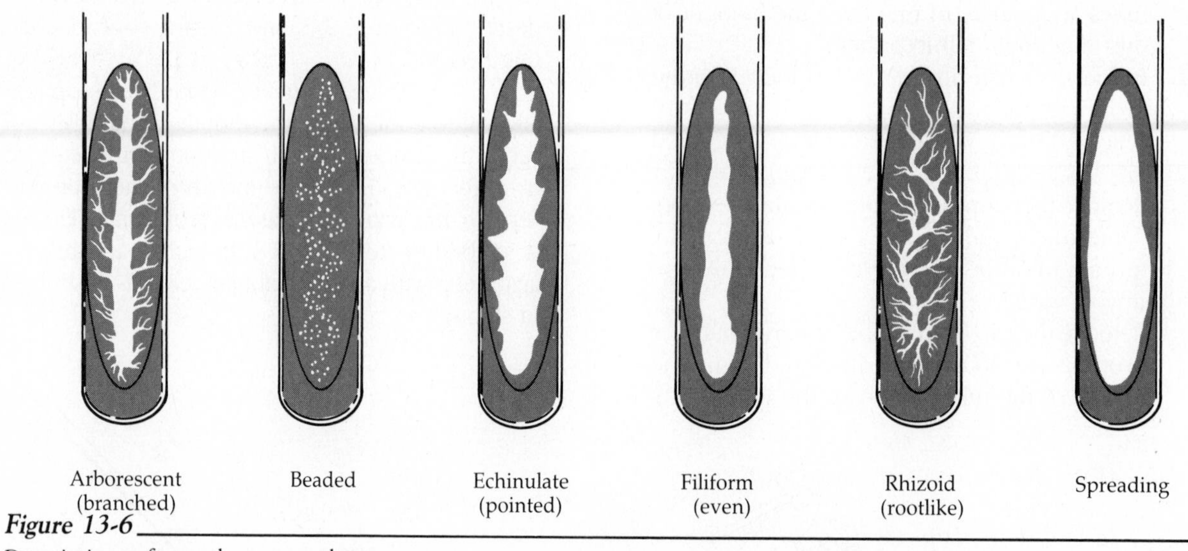

Figure 13-6

Descriptions of growth on agar slants.

EXERCISE 14

Isolation of Bacteria by Dilution Techniques

A little experience often upsets a lot of theory.

CADMAN

Objectives

After completing this exercise you should be able to

1. Isolate bacteria by the streak plate and pour plate techniques.
2. Prepare and maintain a pure culture.

Background

In nature most microbes are not found growing in isolation but in an environment that contains many different organisms. Unfortunately, mixed cultures are of little use in studying microorganisms because of the difficulty they present in determining which organism is responsible for the observed activity. A **pure culture,** that is, one containing a single kind of microbe, is required in order to study concepts such as growth characteristics, pathogenicity, metabolism, and antibiotic susceptibility. Since bacteria are too small to separate directly without sophisticated micromanipulation equipment, indirect methods of separation must be used.

In the 1870s, Lister attempted to obtain pure cultures by performing serial dilutions until each of his containers theoretically contained one bacterium. However, success was very limited and **contamination,** the presence of unwanted microorganisms, was common. In 1880 Koch prepared solid media, and after that bacteria could be separated by dilution and trapped in the solid media. An isolated bacterium grows into a visible colony that consists of one kind of bacterium.

Currently, there are three dilution methods commonly used for the isolation of bacteria: the streak plate, the spread plate, and the pour plate. In the **streak plate technique,** a loop is used to streak the mixed sample many times over the surface of a solid culture medium in a Petri plate. Theoretically, by streaking the loop repeatedly over the agar surface, the bacteria fall off the loop one by one and each cell develops into a colony. The streak plate is the most common isolation technique in use today.

In the **spread plate technique,** a small amount of a previously diluted specimen is spread over the surface of a solid medium using a bent glass rod (shaped like a hockey stick). The spread plate technique will be used in Exercises 26 and 29. In the **pour plate technique,** multiple dilutions of a sample are made, and a small amount of each dilution is mixed with melted agar and poured into Petri plates. Some of the plates will have separated colonies due to dilutions.

In this exercise, a loop is used to perform the dilutions so the technique will be qualitative. Pour plates will be used quantitatively in Exercises 51 and 54.

Materials

Petri plates containing nutrient agar (2)

Tubes containing melted nutrient agar (3)

Sterile Petri plates (3)

Nutrient agar slant (1)

250-ml beaker (1)

Cultures (one of each)

Mixed broth culture of bacteria

Turbid nutrient broth from Exercise 12

Techniques Required

Exercises 11 through 13

Procedure

Streak Plate

1. Label the bottoms of two nutrient agar plates to correspond to the two broth cultures: mixed culture and turbid broth from Exercise 12.
2. Flame the inoculating loop to redness, allow it to cool, and aseptically obtain a loopful of one broth culture.
3. The streaking procedure may be done with the

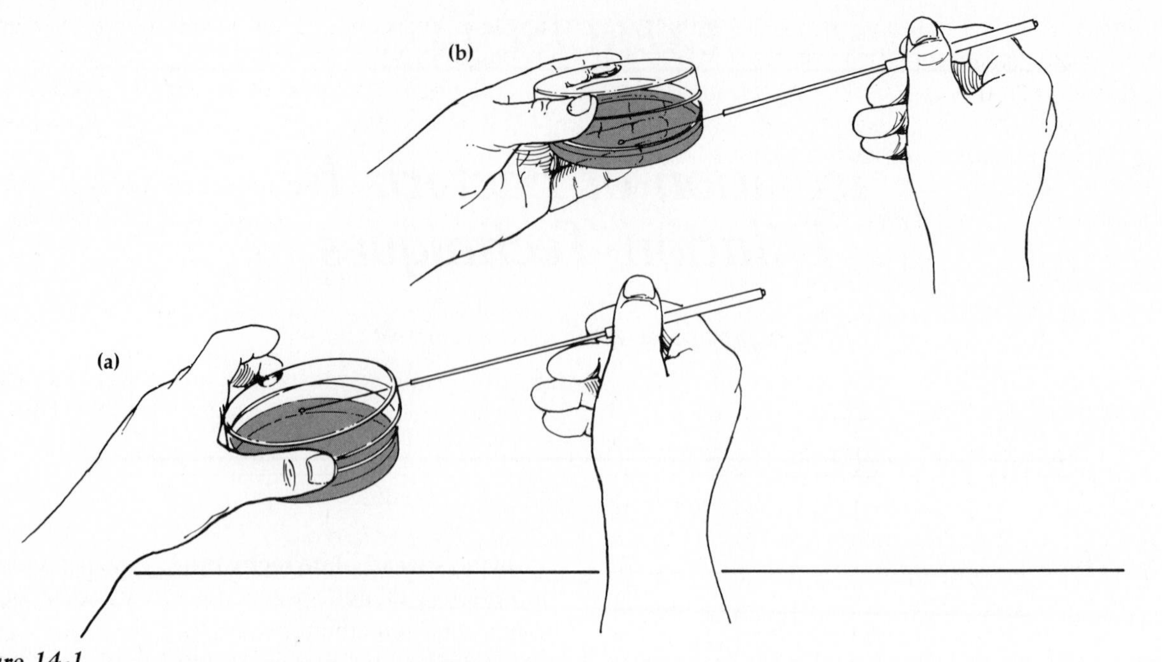

Figure 14-1

Inoculation of a solid medium in a Petri plate. Lift one edge of the cover while **(a)** the plate rests on the table, or **(b)** is held.

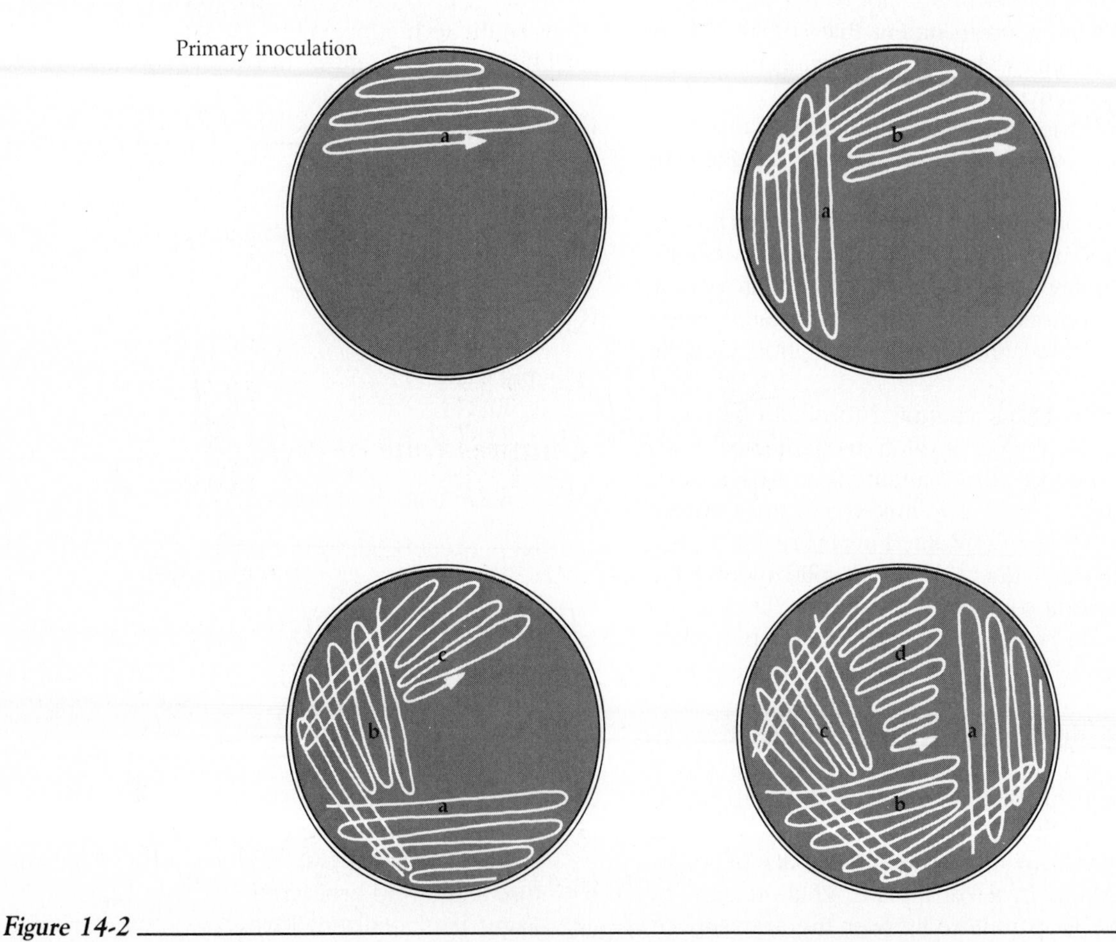

Figure 14-2

Streak plate procedure for pure culture isolation of bacteria. The direction of streaking is indicated by the arrows. Between each section, the loop is sterilized and reinoculated with a fraction of the bacteria by going back across the previous section.

Petri plate on the table (Figure 14-1*a*) or held in your hand (Figure 14-1*b*). To streak a plate (Figure 14-2), lift one edge of the Petri plate cover and streak the first sector **(a)** by making as many streaks as possible without overlapping previous streaks. **(b)** Flame your loop and let it cool. Turn the plate so the next sector is on top and streak it in the same manner. **(c)** Flame your loop, turn the plate again, and streak the third sector as before. **(d)** Flame your loop, streak through one area of the third sector, and streak the remaining area of the agar surface. Flame your loop before setting it down. Why? ____________________

____

4. You should streak two plates: one of the mixed culture provided and one of the turbid broth from Exercise 12. Each plate should be labeled on the bottom with your name, lab. section, date, and source of the inoculum.
5. Incubate the plates in an inverted position in the 35°C incubator until discrete, isolated colonies develop (usually 24 to 48 hours). Why inverted?

____

6. After incubation, record your results. (Refer to Figure 12-1).
7. Prepare a subculture of one colony. Sterilize your needle by flaming. Let it cool. Why use a needle instead of a loop? ____________________
To subculture, touch the center of a small isolated colony and then aseptically streak a sterile nutrient agar slant (Figure 13-4). How can you tell whether or not you touched only one colony and whether or not you have a pure culture? __________

____
8. Incubate the slant at 35°C until good growth is observed. Describe the growth pattern (Figure 13-6).

Pour Plate (Figure 14-3)

1. Label the bottoms of three sterile Petri plates with your name, lab. section, and date. Label one plate "1," another "2," and the remaining plate, "3."
2. Fill a beaker with hot (45–50°C) water (about 3 to 6 cm) and place three tubes of melted nutrient agar in the beaker.

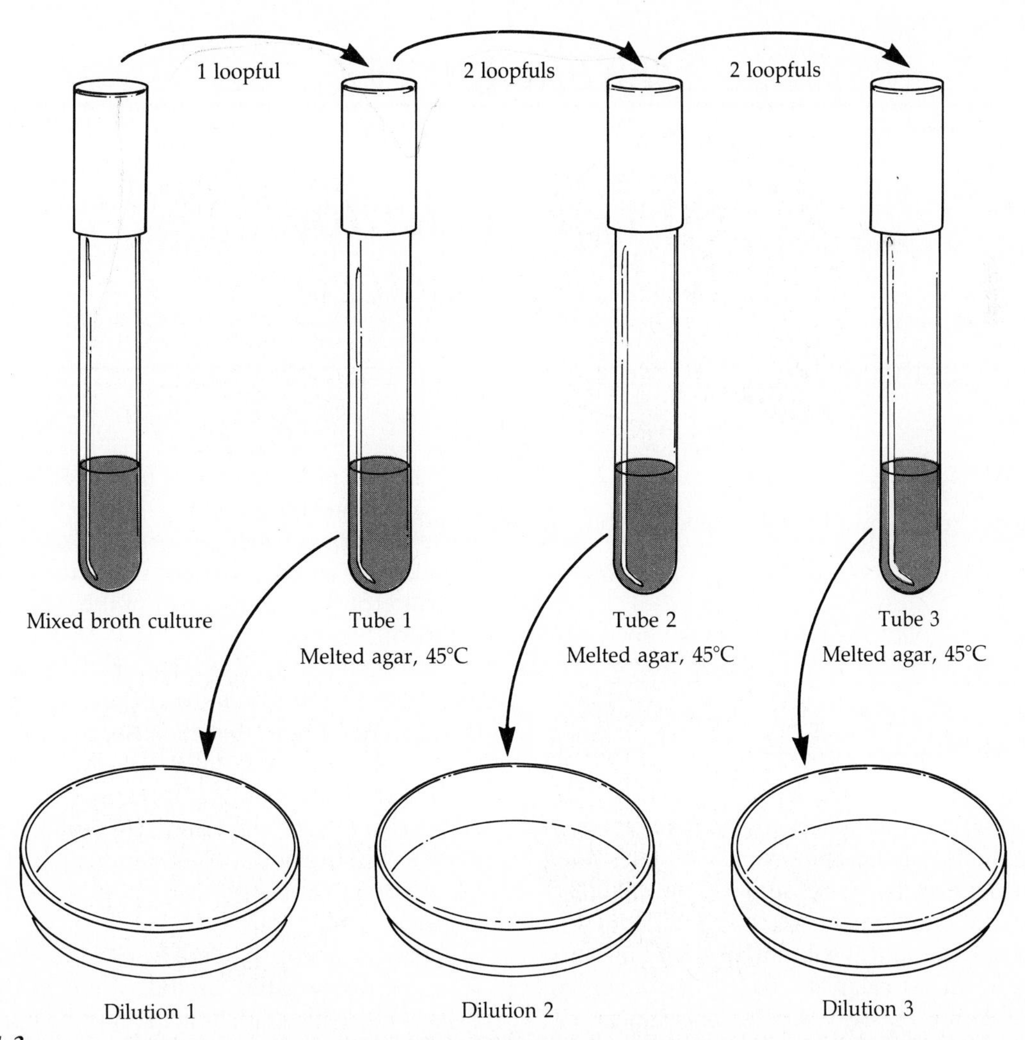

Figure 14-3
Pour plate technique. Bacteria are diluted through a series of tubes containing melted nutrient agar. The agar and bacteria are poured into sterile Petri plates. The bacteria will form colonies where they were trapped in the agar.

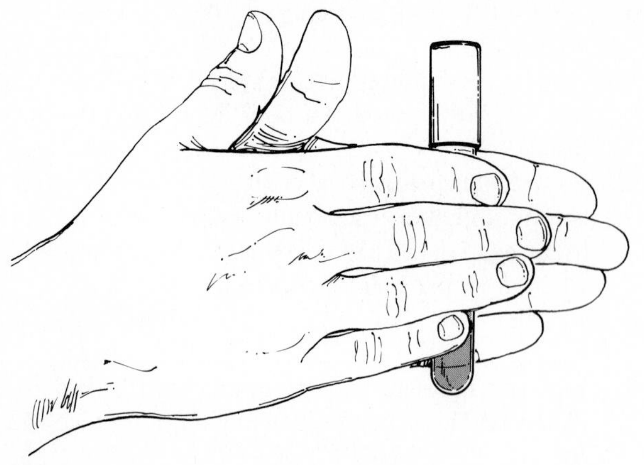

Figure 14-4

Mix the inoculum in a tube of melted agar by rolling the tube between your hands.

3. Select a mixed broth culture. Place the labeled plates on your workbench right side up.
4. Sterilize your loop and aseptically transfer 1 loopful of the broth to a tube of melted agar. Mix well as shown in Figure 14-4. Transfer 2 loopfuls to the second tube. Do you need to sterilize your loop between each transfer? No
   Aseptically pour the contents of the first tube into Petri plate "1."
5. Mix the second tube, transfer 2 loopfuls to the third tube. Pour the contents of the second tube into plate "2." Mix the third tube and pour into the remaining plate, "3."
6. Discard the tubes. Let the agar harden in the plates, then incubate at 35°C in an inverted position until growth is seen. *Suggestion:* When incubating multiple plates, use a rubber band to hold them together.
7. After incubation, describe the results.

Turn to the Laboratory Report for Exercise 14.

EXERCISE 15

Special Media for Isolating Bacteria

Objectives

After completing this exercise you should be able to

1. Differentiate between enrichment and selective media.
2. Provide an application for enrichment and selective media.

Background

One of the major limitations of dilution techniques used to isolate bacteria is that organisms present in limited amounts may be diluted out on plates filled with dominant bacteria. For example, if the culture to be isolated has 1 million of bacterium A and only 1 of bacterium B, bacterium B will probably be limited to the first sector in a streak plate. To help isolate organisms found in the minority, various enrichment and selective culturing methods are available that enhance the growth of some organisms and inhibit the growth of other organisms. **Selective media** contain chemicals that prevent the growth of unwanted bacteria without inhibiting the growth of the desired organism. Chemicals that enhance the growth of desired bacteria are added to an **enrichment medium.** Other bacteria will grow, but the growth of the desired bacteria will be increased.

Since multiple methods and media exist, a laboratory worker must match the correct procedure to the desired microbe. In this exercise, we will determine one criterion used to select a culture medium. For example, if bacterium B is salt-tolerant, salt could be added to the culture medium. Physical conditions can also be used to select for a bacterium. If bacterium B is heat-resistant, the specimen could be heated before isolation.

Another category of media useful in identifying bacteria are **differential media.** These media contain various nutrients that allow the investigator to distinguish one bacterium from another by how they metabolize or change the media. Differential media are not used in primary isolation to enhance growth because all bacteria generally grow well on them. Differential media will be used in later exercises (see, for example, Part 4).

Materials

Petri plate containing phenylethyl alcohol agar (1)

Gram staining reagents

Cultures (one of each)

Staphylococcus epidermidis

Escherichia coli

Mixed culture of *Escherichia* and *Staphylococcus*

Techniques Required

Exercises 7 and 13

Procedure

1. Using a marking pen, divide the phenylethyl alcohol agar plate into three sections by labeling the bottom. Label one section for each culture.

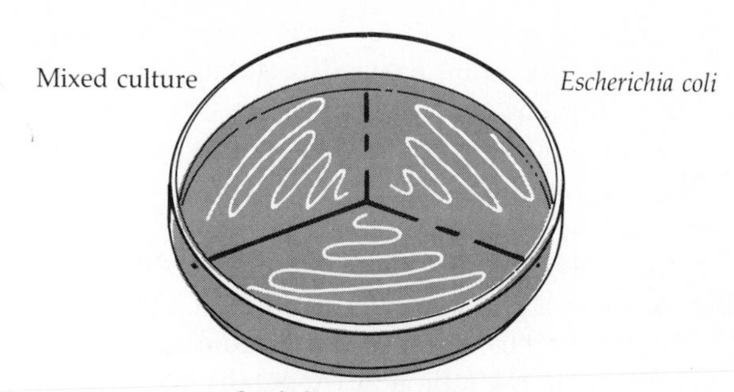

Figure 15-1

Divide a Petri plate into three sections by drawing lines on the bottom of the plate. Inoculate each section by streaking with an inoculating loop.

2. Streak each culture on the agar as in Figure 15-1.
3. Incubate the plates in an inverted position at 35°C. Record the results after 24 hours and after 48 to 72 hours of incubation. Gram stain different appearing colonies from each sector.

Turn to the Laboratory Report for Exercise 15.

EXERCISE 16

EXERCISE 16

Bacterial Nutrition

Recipe for Elephant Stew
One elephant
Brown gravy
Two rabbits (optional)
Cut elephant into bite-sized cubes. This takes about 2 months. Add enough brown gravy to cover. Simmer over a kerosene fire for four weeks at 465°. Yields 3,800 portions. If more guests are expected, add the rabbits only if necessary as most people do not like hare in their stew!

Objectives

After completing this exercise you should be able to

1. List the basic chemical requirements of a chemoheterotroph.
2. Perform and interpret the citrate test.

Background

Now that the elephant stew has *your* mouth watering, the nutritional needs of bacteria will be examined. Bacteria are isolated by the techniques introduced in Exercises 14 and 15 in order to study them further.

Nutritional requirements of bacteria are frequently studied in order to understand bacterial metabolism and to identify bacterial species. In this exercise, the nutritional requirements of two species of bacteria will be compared and contrasted. The composition of each medium is given in Table 16-1.

In one medium an indicator will be used. An **indicator** gives a quick visual indication of a change in the medium by changing color. Many different indicators are used in microbiology (Appendix G). The Simmons citrate agar used in this exercise contains the indicator bromthymol blue. Citric acid will be the only source of carbon, therefore only organisms capable of utilizing citric acid as a source of carbon will grow. When the citric acid is metabolized, an excess of sodium and ammonium ions results and the indicator turns from green to blue, indicating alkaline conditions.

Materials

1 tube of each medium listed in Table 16-1

Cultures (choose one)

Pseudomonas aeruginosa

Escherichia coli

Technique required

Exercise 13

Procedure

1. Label each tube of medium as you pick it up.
2. Work with a partner, and with another pair of students as a team. One pair should use *E. coli,* and the other, *P. aeruginosa.*
3. Aseptically inoculate the five tubes (G, GS, GSP, GSPM, and GSPMP) with 2 loopfuls of *P. aeruginosa* or *E. coli* (Exercise 13). Does the loop need to be sterilized each time when you are going from a sterile medium to a culture and back? ____

   ____
4. Inoculate one citrate slant by streaking with *P. aeruginosa* or *E. coli* (Figure 13-4).
5. Incubate all tubes for one week at 35°C.
6. Record amounts of growth in the broths: (−) = no growth; (+) = minimal growth; (2+) = moderate growth; (3+) = heavy growth; and (4+) = maximum growth. Is there any way you can determine differences in growth more easily?

   ____

   The citrate agar is interpreted by the color of the indicator. Green indicates citric acid was not used and is reported as a negative (−) test. Share your results with the other pair of your team.

Table 16-1
Media Used in This Exercise

Medium G
0.5% Glucose
Medium GS
0.5% Glucose 0.5% Sodium chloride
Medium GSP
0.5% Glucose 0.5% Sodium chloride 0.1% Ammonium dihydrogen phosphate + 0.1% dipotassium phosphate
Medium GSPM (Prepared in Exercise 11)
0.5% Glucose 0.5% Sodium chloride 0.1% Ammonium dihydrogen phosphate + 0.1% dipotassium phosphate 0.02% Magnesium sulfate
Medium GSPMP
0.5% Glucose 0.5% Sodium chloride 0.1% Ammonium dihydrogen phosphate + 0.1% dipotassium phosphate 0.02% Magnesium sulfate 0.5% Peptone
Simmons Citrate Agar
0.2% Sodium citrate 0.5% Sodium chloride 0.1% Monoammonium phosphate + 0.1% dipotassium phosphate 0.02% Magnesium sulfate 1.5% Agar 0.008% Bromthymol blue

Turn to the Laboratory Report for Exercise 16.

PART 4

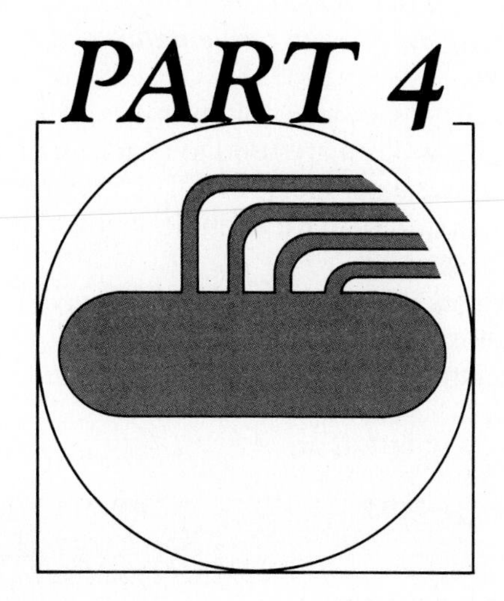

Metabolic Activities of Bacteria

EXERCISES

Leeuwenhoek saw microorganisms in wine, and Pasteur demonstrated that the microbes were living organisms. In 1872 Pasteur wrote, *"It is impossible that the organic matter of the newly formed ferments [microorganisms] contain a single carbon atom which has not been derived from the fermented substance."*

The chemical reactions observed by Pasteur and the chemical reactions that occur within all living organisms are referred to as **metabolism.** Metabolic reactions that release energy from the breakdown or degradation of complex organic molecules are **catabolic reactions.** Metabolic reactions that use energy to assemble smaller molecules into the large molecules that comprise a cell are **anabolic reactions.**

Metabolic processes involve **enzymes,** which are proteins that catalyze biologic reactions. Most enzymes function inside a cell; that is, they are

endoenzymes. A few enzymes, called **exoenzymes,** are released from the cell to catalyze reactions outside of the cell (see the figure).

Some bacteria use particular metabolic pathways in the presence of oxygen **(aerobic)** and other pathways in the absence of oxygen **(anaerobic).** *"I have demonstrated that although this ferment [microorganism] survived in the presence of some free oxygen, it lost its fermentative abilities in proportion to the concentration of this gas"* (Pasteur, 1872).

Since many bacteria share the same colony and cell morphology, additional characteristics such as metabolism are used to characterize and classify them. On the basis of which substrates a particular bacterium uses and which metabolic products it forms, laboratory tests have been designed to determine which enzymes the bacterium has.

The first seven exercises in Part 4 introduce concepts in microbial metabolism and laboratory tests used to detect various metabolic activities. In Exercises 24 and 25, unknown bacteria will be identified on the basis of metabolic characteristics.

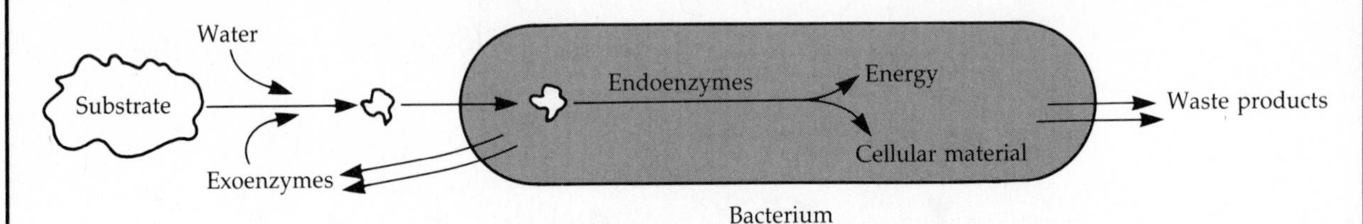

Large molecules are broken down outside of a cell by exoenzymes. Smaller molecules released by this reaction are taken into the cell and further degraded by endoenzymes.

EXERCISE 17

Carbohydrate Catabolism

The men who try to do something and fail are infinitely better than those who try nothing and succeed.

LLOYD JONES

Objectives

After completing this exercise you should be able to

1. Define the following terms: carbohydrate, catabolism, and exoenzyme.
2. Differentiate between oxidative and fermentative catabolism.
3. Perform and interpret microbial starch hydrolysis and OF tests.

Background

Chemical reactions that release energy from the decomposition of complex organic molecules are referred to as **catabolism.** Most bacteria catabolize carbohydrates for carbon and energy. **Carbohydrates** are organic molecules that contain carbon, hydrogen, and oxygen in the ratio $(CH_2O)_n$. Carbohydrates can be divided into three groups based on size: monosaccharides, disaccharides, and polysaccharides. **Monosaccharides** are simple sugars containing from three to seven carbon atoms, **disaccharides** are composed of two monosaccharide molecules, and **polysaccharides** consist of eight or more monosaccharide molecules.

Exoenzymes are mainly **hydrolytic enzymes** that break down, by the addition of water, large substrates into smaller components that can be transported into the cell. The exoenzyme amylase hydrolyzes the polysaccharide starch into smaller carbohydrates. Glucose, a monosaccharide, can be released by hydrolysis (Figure 17-1). In the laboratory, the presence of an exoenzyme is determined by looking for a change in the substrate outside of a bacterial colony.

Glucose can enter a cell and be catabolized; some bacteria catabolize glucose oxidatively and produce carbon dioxide and water. **Oxidative metabolism** requires the presence of molecular oxygen (O_2). Most bacteria, however, ferment glucose without using oxygen. **Fermentative metabolism** does not require oxygen but may occur in the presence of oxygen. The metabolic end products of fermentation are small organic molecules, usually organic acids. Some bacteria are both oxidative and fermentative; others are neither but obtain their carbon and energy by other means.

Whether an organism is oxidative or fermentative can be determined by using Hugh and Leifson's OF basal media with the desired carbohydrate added. **OF medium** is a nutrient semisolid agar deep containing a *high* concentration of carbohydrate and a *low* concentration of peptone. The peptone will support the growth of nonoxidative-nonfermentative bacteria. Two tubes are used: one open to the air and one sealed to keep air out. OF medium contains the indicator bromthymol blue, which turns yellow in the presence of acids, indicating catabolism of the carbohydrate. Alkaline conditions, due to the use of peptone and not the carbohydrate, are indicated by a dark blue color. If the carbohydrate is metabolized in both tubes, fermentation has occurred. Some bacteria produce gases from the fermentation of carbohydrates. Gases can be seen as pockets or cracks in the OF

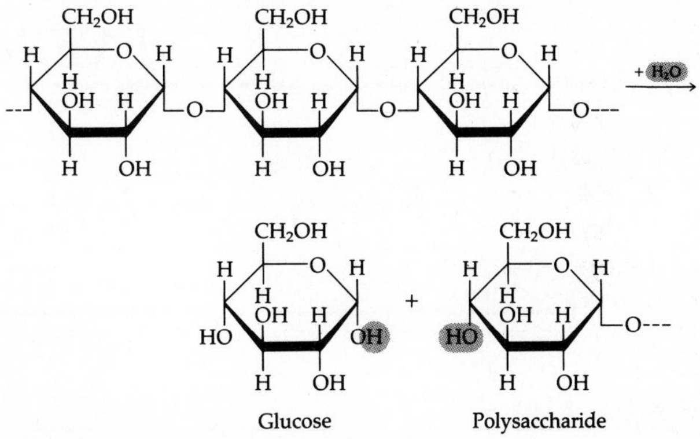

Figure 17-1

Starch hydrolysis. A molecule of water is used when starch is hydrolyzed.

medium. An organism that can only use the carbohydrate oxidatively will produce acid in the open tube only.

Carbohydrate metabolism will be introduced in this exercise. These differential tests will be very important in identifying bacteria in later exercises.

Materials

Petri plate containing nutrient starch agar (1)

OF-glucose deeps (2)

Vaspar (one-half paraffin and one-half petroleum jelly), melted

Gram's iodine

Cultures (as designated in each procedure)

Bacillus subtilis

Escherichia coli

Pseudomonas aeruginosa

Alcaligenes faecalis

Techniques Required

Exercise 13

Procedure

Starch Hydrolysis (Figure 17-2)

1. With a marking pen divide the starch agar into three sectors by labeling the bottom of the plate.
2. Streak a single line of *Bacillus, Escherichia,* and *Pseudomonas.*
3. Incubate, inverted, at 35°C until the next period. Record any bacterial growth, then flood the plate with Gram's iodine (Figure 17-2). Areas of starch hydrolysis will appear clear, while unchanged starch will stain dark blue. Record your results.

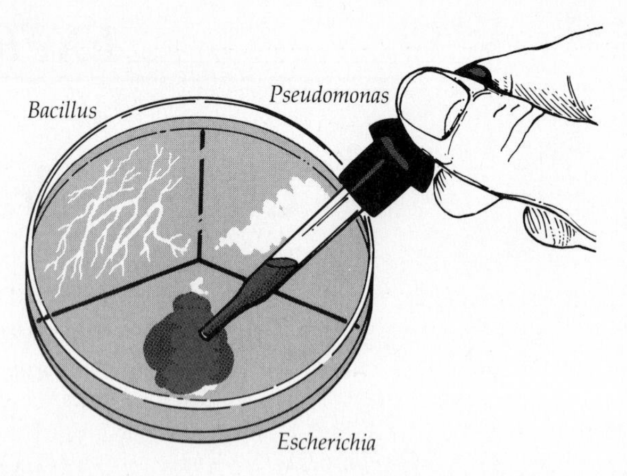

Figure 17-2

Starch hydrolysis test. After incubation, add iodine to the plate to detect the presence of starch.

OF-Glucose

1. Using an inoculating needle, inoculate two tubes of OF-glucose media with the assigned bacterial culture *(Escherichia, Pseudomonas,* or *Alcaligenes).*
2. Pour about 5 mm of melted vaspar into one of the tubes.
3. Incubate both tubes at 35°C until the next laboratory period. Observe the tubes and record the following: the presence of growth, whether glucose was utilized, and the type of metabolism. Motility can also be ascertained from the OF tubes. How? ______________

   Observe and record the results from the microorganisms you did not culture.

Turn to the Laboratory Report for Exercise 17.

EXERCISE 18

Fermentation of Carbohydrates

Objectives

After completing this exercise you should be able to

1. Define fermentation.
2. Perform and interpret carbohydrate fermentation tests.
3. Perform and interpret the MR and V–P tests.

Background

Once a bacterium has been determined to be fermentative by the OF test (Exercise 17), further tests can determine which carbohydrates are fermented; in some instances, the end products can also be determined. Many carbohydrates, including monosaccharides such as glucose, disaccharides like sucrose, and polysaccharides such as cellulose, can be fermented. Many bacteria produce organic acids (for example, lactic acid) and hydrogen and carbon dioxide gases from carbohydrate fermentation. A **fermentation tube** is used to detect acid and gas production from carbohydrates. The fermentation medium contains peptone, an acid–base indicator, an inverted tube to trap gas, and 0.5% to 1.0% of the desired carbohydrate. In Figure 18-1, phenol red indicator is red (neutral) in an uninoculated fermentation tube; fermentation that results in acid production will turn the indicator yellow (pH of 6.8 or below). When gas is produced during fermentation, some of the gas will be trapped in the inverted tube. Fermentation occurs with or without oxygen present; however, during prolonged incubation periods (greater than 48 hours) many bacteria will begin growing oxidatively on the peptone after exhausting the carbohydrate supplied, causing neutralization of the indicator.

Fermentation processes can produce a variety of end products, depending on the substrate, the incubation, and the organism. In some instances, large amounts of acid may be produced, while in others, a majority of neutral products may result (Figure 18-2*a*). The **MRVP test** is used to distinguish between organisms that produce large amounts of acid from glucose and those that produce the neutral product *acetoin*. MRVP medium is a glucose-supplemented nutrient broth used for the **methyl red (MR) test** and the **Voges–Proskauer (V–P) test.** If an organism produces a large amount of organic acid from glucose, the medium will remain red when methyl red is added, indicating the pH is below 4.4. If neutral products are produced, methyl red will turn yellow, indicating a pH above 6.0. The production of acetoin is detected by the addition of potassium hydroxide and α-naphthol. If acetoin is present, the upper part of the medium will turn red; a negative V–P test will turn the medium light brown. The chemical process is shown in Figure 18-2*b*. The production of acetoin is dependent on the length of incubation and the environmental conditions, as seen in Figure 18-3.

The fermentation process will be examined in this exercise.

Materials

Glucose fermentation tube (1)

Lactose fermentation tube (1)

Sucrose fermentation tube (1)

MRVP broth (4)

Methyl red

V–P reagent I, α-naphthol solution

V–P reagent II, potassium hydroxide (40 percent)

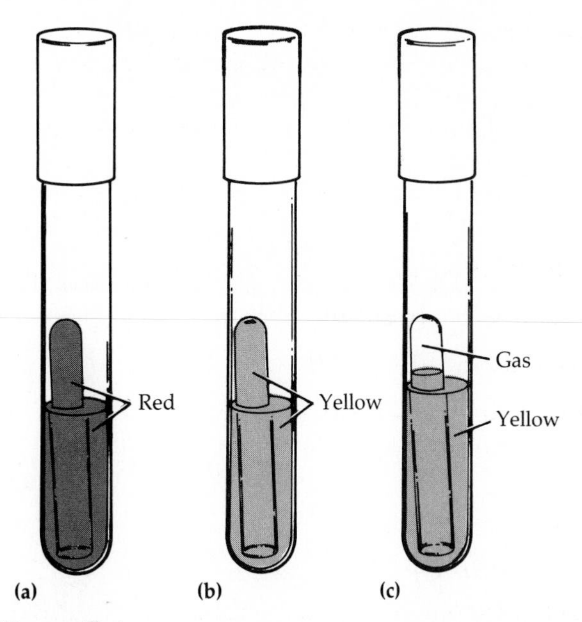

Figure 18-1

Carbohydrate fermentation tube. **(a)** The phenol red indicator is red in a neutral or alkaline solution. **(b)** Phenol red turns yellow in the presence of acids. **(c)** Gases are trapped in the inverted tube while the indicator shows the production of acid.

EXERCISE 18

Cultures (as assigned)

Escherichia coli

Enterobacter aerogenes

Alcaligenes faecalis

Proteus vulgaris

Techniques Required

Exercises 13 and 17

Procedure

Fermentation Tubes

1. Use your loop to inoculate the fermentation tubes with the assigned bacterial culture.
2. Incubate at 35°C. Examine at 24 and 48 hours for growth, acid, and gas. Why is it important to record the presence of growth? ______

______

Record your results with this culture, as well as your results for the other species tested.

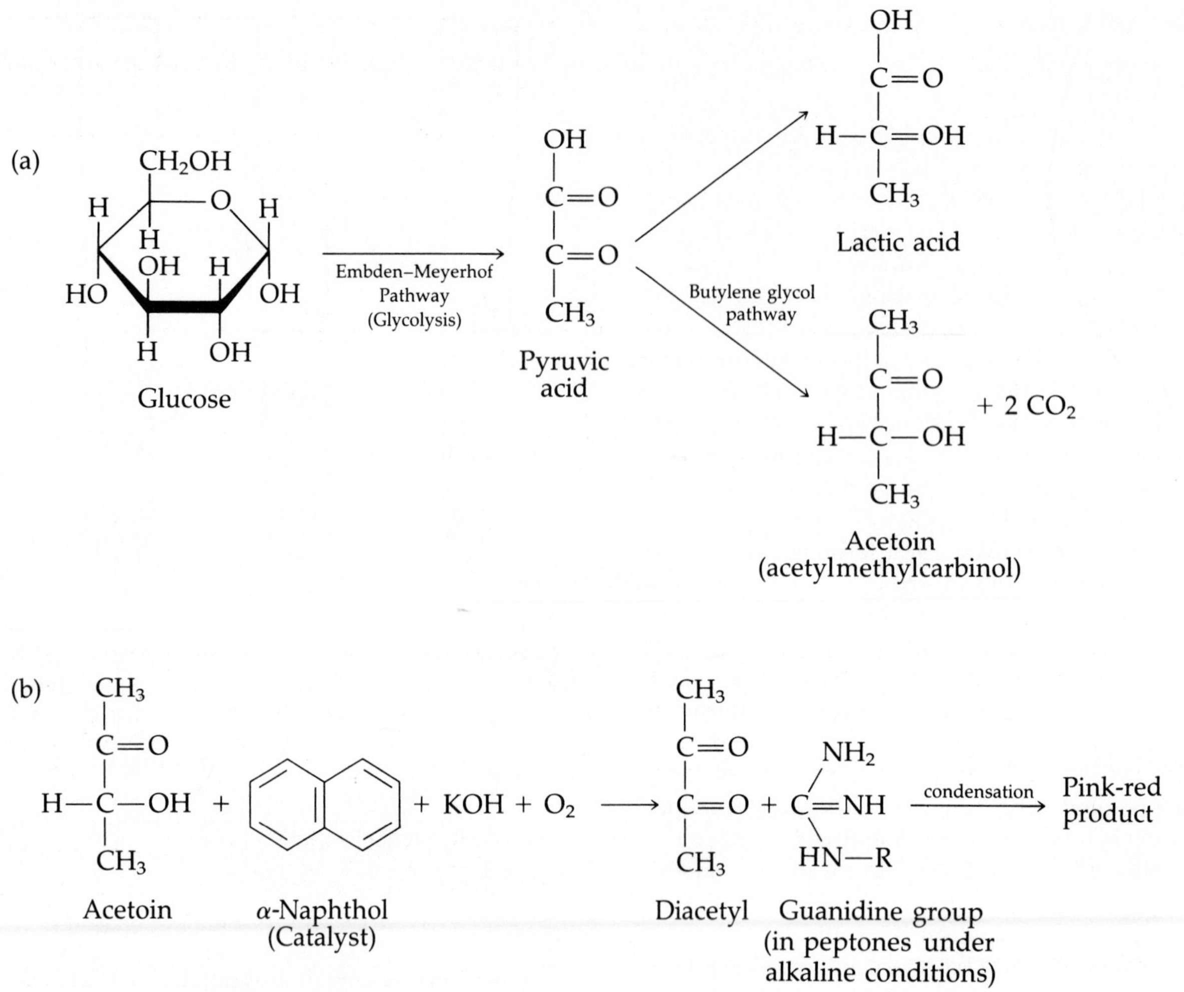

(a)
Glucose
Embden–Meyerhof Pathway (Glycolysis)
Pyruvic acid
Lactic acid
Butylene glycol pathway
+ 2 CO_2
Acetoin (acetylmethylcarbinol)
(b)
Acetoin
α-Naphthol (Catalyst)
+ KOH + O_2
Diacetyl
Guanidine group (in peptones under alkaline conditions)
condensation
Pink-red product

Figure 18-2

Fermentation of glucose. **(a)** Organic acids such as lactic acid or neutral products such as acetoin may be produced from fermentation. **(b)** Potassium hydroxide (KOH) and α-naphthol are used to detect acetoin.

pH of medium
Conversion of acid to neutral products (acetoin)
Color changes of methyl red
6.8
5.1
4.4
Yellow
Yellowish-red
Red
E. aerogenes
E. coli
24
48
72
Hours of incubation

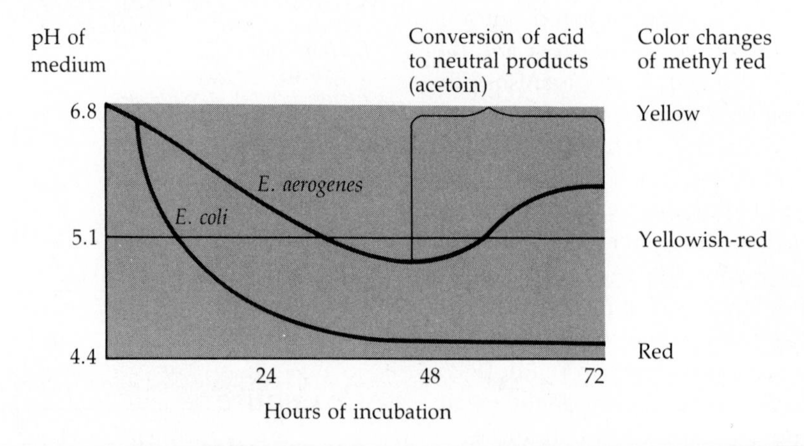

Figure 18-3

The production of acetoin is dependent on incubation time and pH.

MRVP Tests

1. Using your loop, inoculate two MRVP tubes with *Escherichia* and two with *Enterobacter.*
2. Incubate at 35°C for 48 hours or longer. Why is time of incubation important? ____________________

   ____
3. To one tube of each set *(Escherichia* and *Enterobacter),* add 5 drops of methyl red. Record the resulting color. Red is a positive methyl red test.
4. To the other set of two tubes add 0.6 ml (12 drops) of V–P reagent I and 0.2 ml (2 or 3 drops) V–P reagent II.

5. Gently shake the tubes with the caps off to expose the media to oxygen in order to oxidize the acetoin.
6. Allow the tubes to stand for 15 to 30 minutes. A positive V–P test will develop pink to red color. Record your results.

Turn to the Laboratory Report for Exercise 18.

EXERCISE 19

Lipid Hydrolysis

Only a mediocre person is always at his best.

W. SOMERSET MAUGHAM

Objective

After completing this exercise you should be able to

1. Determine lipid hydrolysis.

Background

Many bacteria use exoenzymes such as lipase to hydrolytically decompose lipids. The **lipid** molecule is broken down into *glycerol* and *three fatty acids*. Some bacteria can ferment the glycerol; others oxidize the fatty acids. When lipid hydrolysis occurs in decomposition of foods such as butter, it results in rancid flavor and aroma due to the fatty acids (such as butyric acid). Sewage systems have a great deal of difficulty decomposing lipids because the process is slow and requires large amounts of oxygen.

In this exercise the hydrolysis of tributyrin (Figure 19-1) to glycerol and butyric acid will be examined. Whether an organism hydrolyzes lipids can be useful in identifying and characterizing the organism.

Materials

Petri plate containing tributyrin nutrient agar (1)

Cultures (one of each)

Pseudomonas aeruginosa

Bacillus subtilis

Escherichia coli

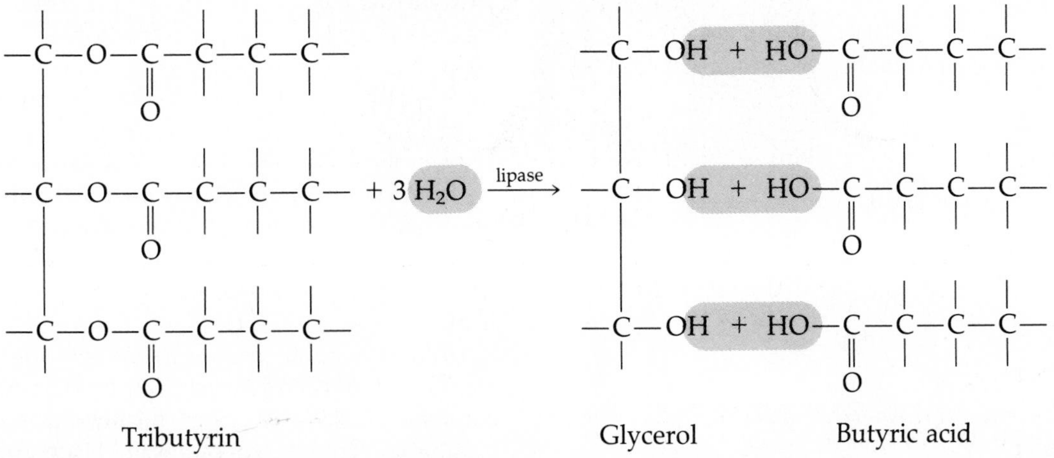

Tributyrin + 3 H_2O —lipase→ Glycerol + Butyric acid

Figure 19-1
Tributyrin hydrolysis. Lipids are composed of a glycerol and three fatty acid molecules.

Techniques Required

Exercises 13 and 17

Procedure

1. Divide the plate into three sections and streak each organism on the surface of a sector.
2. Incubate at 35°C for 2 to 3 days. Examine. Hydrolysis of the lipid should result in the opaque emulsified tributyrin becoming soluble fatty acids and glycerol. What will a positive test for lipid hydrolysis look like? ______

Record your observations in the Laboratory Report.

Turn to the Laboratory Report for Exercise 19.

EXERCISE 20

Protein Catabolism, Part 1

Objectives

After completing this exercise you should be able to

1. Determine a bacterium's ability to hydrolyze gelatin.
2. Test for the presence of urease.
3. Perform and interpret a litmus milk test.

Background

Proteins are organic molecules that contain carbon, hydrogen, oxygen, and nitrogen. Additionally, some proteins contain sulfur. The subunits that make up a protein are **amino acids** (Figure 20-1). Amino acids bond together by **peptide bonds** (Figure 20-2) to form a small chain (a **peptide**) or a larger molecule **(polypeptide).**

Bacteria can use a variety of proteins and amino acids as carbon and energy sources when carbohydrates are not available. However, amino acids are primarily used in anabolic reactions (Exercise 23).

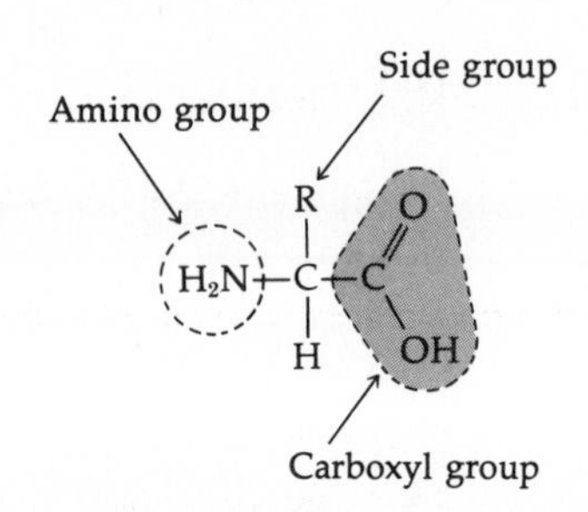

Figure 20-1

General structural formula for an amino acid. The letter R stands for any of a number of groups of atoms. Different amino acids have different R groups.

Large protein molecules such as gelatin are hydrolyzed by exoenzymes, and the smaller products of hydrolysis are transported into the cell. Hydrolysis of gelatin can be demonstrated by growing bacteria in nutrient gelatin. Nutrient gelatin dissolves in warm water (50°C), solidifies **(gels)** when cooled below 25°C, and liquefies **(sols)** when heated to about 25°C. When an exoenzyme hydrolyzes gelatin, it liquefies and does not solidify even when cooled below 20°C.

Bacteria can hydrolyze the protein in milk called **casein.** Casein hydrolysis can be detected in litmus milk. Litmus milk consists of skim milk and the indicator litmus. The medium is opaque due to casein in colloidal suspension and the litmus is blue. After **peptonization** (hydrolysis of the milk proteins), the medium becomes clear due to hydrolysis of casein to soluble amino acids and peptide fragments. Litmus milk is also used to detect **lactose fermentation,** since litmus turns pink in the presence of acid. Excessive amounts of acid will cause **coagulation** (curd formation) of the milk. **Catabolism of amino acids** will result in an alkaline (purple) reaction. Additionally, some bacteria can **reduce litmus** (Exercise 22), causing the litmus indicator to turn white in the bottom of the tube.

Urea is a waste product of protein digestion in most vertebrates and is excreted in the urine. Presence of the enzyme **urease,** which liberates ammonia from urea (Figure 20-3), is a useful diagnostic test for identifying bacteria. **Urea agar** contains peptone, glucose, urea, and phenol red. The pH of the prepared medium is 6.8 (phenol red turns yellow). During in-

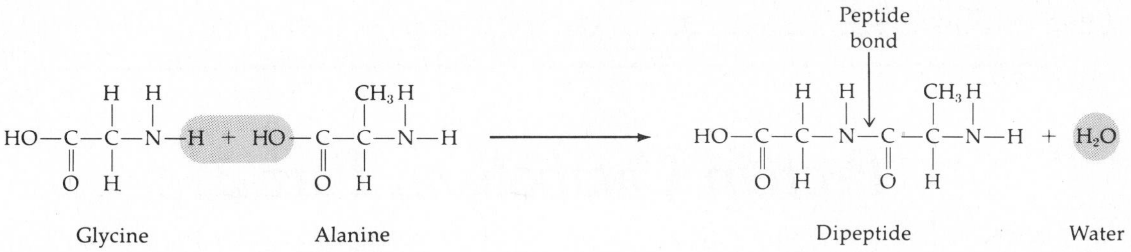

Peptide bond

$$HO{-}C(=O){-}C(H)(H){-}N{-}H + HO{-}C(=O){-}C(H)(CH_3){-}N{-}H \longrightarrow HO{-}C(=O){-}C(H)(H){-}N{-}C(=O){-}C(H)(CH_3){-}N{-}H + H_2O$$

Glycine Alanine Dipeptide Water

Figure 20-2

Peptide bond formation. The amino acids glycine and alanine combine to form a dipeptide. The newly formed bond between the nitrogen atom of glycine and the carbon atom of alanine is called a *peptide bond.*

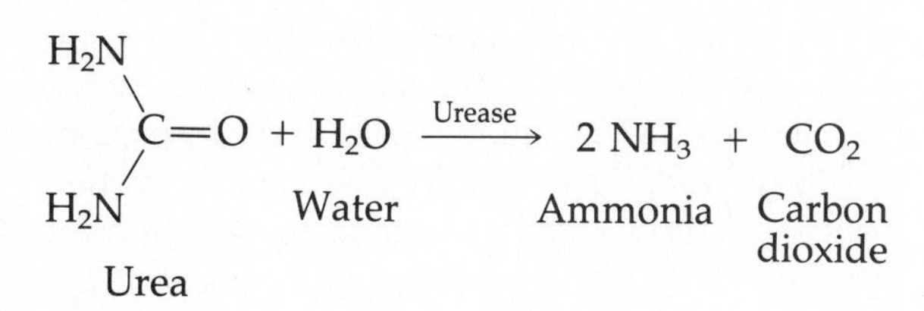

Figure 20-3

Urea hydrolysis.

cubation, bacteria possessing urease will produce ammonia that raises the pH of the medium, turning the indicator fuchsia (hot pink) at pH 8.4.

Bacterial action on nutrient gelatin, litmus milk, and urea agar will be investigated in this exercise. Amino acid metabolism will be studied in Exercise 21, "Protein Catabolism, Part 2."

Materials

Tubes containing nutrient gelatin (2)

Tubes containing litmus milk (2)

Tubes containing urea agar (2)

Cultures (one of each)

Pseudomonas aeruginosa

Proteus vulgaris

Techniques Required

Exercise 13

Procedure

1. Label one tube of each medium *"Pseudomonas"* and the other tube *"Proteus."*
2. Gelatin hydrolysis. Examine the nutrient gelatin: Is it solid or liquid? ______
   What is the temperature of the laboratory? ______
   If the gelatin is solid, what would you need to do to liquefy it? ______
   To resolidify it? ______
   a. Inoculate one tube with *Pseudomonas* and one with *Proteus.*
   b. Incubate at room temperature and record your observations at 2 to 4 days and at 4 to 7 days. Do not agitate the tube when the gelatin is liquid. Why? ______
   c. If the gelatin has liquefied, place the tube in a beaker of crushed ice for a few minutes. Is the gelatin still liquefied? ______
   Record your results. Liquefication or hydrolysis is recorded as (+).
3. Litmus milk. Describe the appearance of litmus milk. ______
   a. Inoculate one tube with *Pseudomonas* and one with *Proteus.*
   b. Incubate the litmus milk at 35°C for 24 to 48 hours and record the results. Litmus is pink under acidic conditions, purple in alkaline conditions, and white when reduced.
4. Urease test.
   a. Inoculate one urea agar slant with *Pseudomonas* and one with *Proteus.*
   b. Incubate for 24 to 48 hours at 35°C. Record your results: (+) for the presence of urease (red) and (−) for no urease. What color is phenol red at pH 6.8 or below? ______
   At pH 8.4 or above? ______

Turn to the Laboratory Report for Exercise 20.

EXERCISE 21

Protein Catabolism, Part 2

Objectives

After completing this exercise, you should be able to

1. Define the following terms: deamination and decarboxylation.
2. Explain the derivation of H_2S in decomposition.
3. Perform and interpret an indole test.

Background

Once amino acids are taken into a bacterial cell, various metabolic processes can occur. Before an amino acid can be used as a carbon and energy source, the amino group must be removed. The removal of an amino group is called **deamination.** The amino group is converted to ammonia that can be excreted from the cell. Deamination results in the formation of an organic acid. Deamination of the amino acid phenylalanine can be detected by forming a colored ferric ion complex with the resulting acid (Figure 21-1). Deamination can also be ascertained by testing for the presence of ammonia using **Nessler's reagent,** which turns deep yellow in the presence of ammonia.

Various amino acids may be decarboxylated. **Decarboxylation** is the removal of carbon dioxide from an amino acid. The presence of a specific decarboxylase

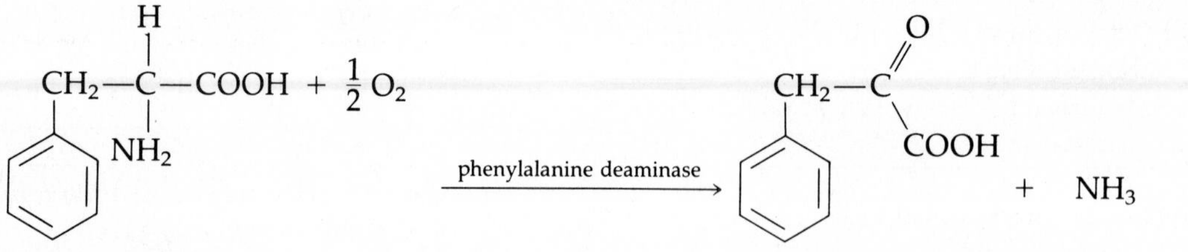

$$C_6H_5-CH_2-CH(NH_2)-COOH + \frac{1}{2}O_2 \xrightarrow{\text{phenylalanine deaminase}} C_6H_5-CH_2-C(=O)-COOH + NH_3$$

L-phenylalanine → Phenylpyruvic acid + Ammonia

Phenylpyruvic acid + Ferric ion (Fe^{3+}) ⟶ Green complex

Figure 21-1

Phenylpyruvic acid produced as a result of the deamination of phenylalanine is detected by the addition of ferric ion (Fe^{3+}).

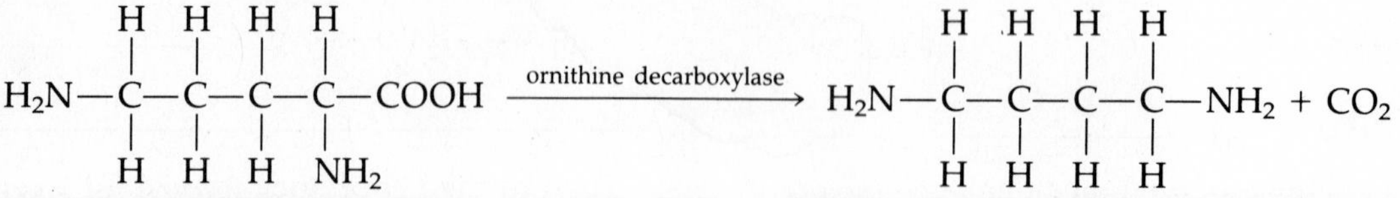

$$H_2N-CH_2-CH_2-CH_2-CH(NH_2)-COOH \xrightarrow{\text{ornithine decarboxylase}} H_2N-CH_2-CH_2-CH_2-CH_2-NH_2 + CO_2$$

Ornithine → Putrescine

Bromcresol purple (yellow) → Bromcresol purple (lavender-purple)

Figure 21-2

Decarboxylation of the amino acid ornithine causes a change in the bromcresol purple indicator.

enzyme results in the breakdown of the amino acid with the formation of the corresponding amine, liberation of carbon dioxide, and a shift in pH to alkaline. Media for decarboxylase reactions consist of glucose, nutrient broth, a pH indicator, and the desired amino acid. In Figure 21-2, bromcresol purple is used as a pH indicator and a positive decarboxylase test yielding excess amines is indicated by purple. (Bromcresol purple is yellow in acidic conditions.) The names given to some of the amines, such as putrescine, indicate how foul smelling they are. Cadaverine was the name given to the foul-smelling diamine derived from decarboxylation of lysine in decomposing bodies on a battlefield.

Some bacteria liberate **hydrogen sulfide (H_2S)** from the sulfur-containing amino acids: cystine, cysteine, and methionine. H_2S can also be produced from the reduction of inorganic compounds (Exercise 22) such as thiosulfate ($S_2O_3^{2-}$). H_2S is commonly called "rotten egg" gas because of the copious amounts liberated when eggs decompose. To detect H_2S production, a heavy metal salt containing ferrous ion (Fe^{2+}) or lead ion (Pb^{2+}) is added to a nutrient culture medium. When H_2S is produced, the sulfide (S^{2-}) reacts with the metal salt to produce a visible black precipitate. The production of hydrogen sulfide from cysteine is shown in Figure 21-3.

The ability of some bacteria to convert the amino acid tryptophan to indole is a useful diagnostic tool (Figure 21-4). The **indole test** is performed by inoculating a bacterium into tryptone broth and detecting indole by the addition of dimethylaminobenzaldehyde *(Kovacs reagent)*.

$$\underset{\text{Cysteine}}{HSCH_2CH(NH_2)COOH} \xrightarrow{\text{cysteine desulfhydrase}} H_2S + NH_3 + \underset{\text{Pyruvic acid}}{CH_3COCOOH}$$

$$\underset{\text{Hydrogen sulfide}}{H_2S} + \underset{\text{Ferrous sulfate}}{FeSO_4} \longrightarrow \underset{\text{Ferrous sulfide}}{FeS} + \underset{\text{Sulfuric acid}}{H_2SO_4}$$

Figure 21-3

The release of H_2S from the amino acid cysteine is detected by the formation of ferrous sulfide (black precipitate).

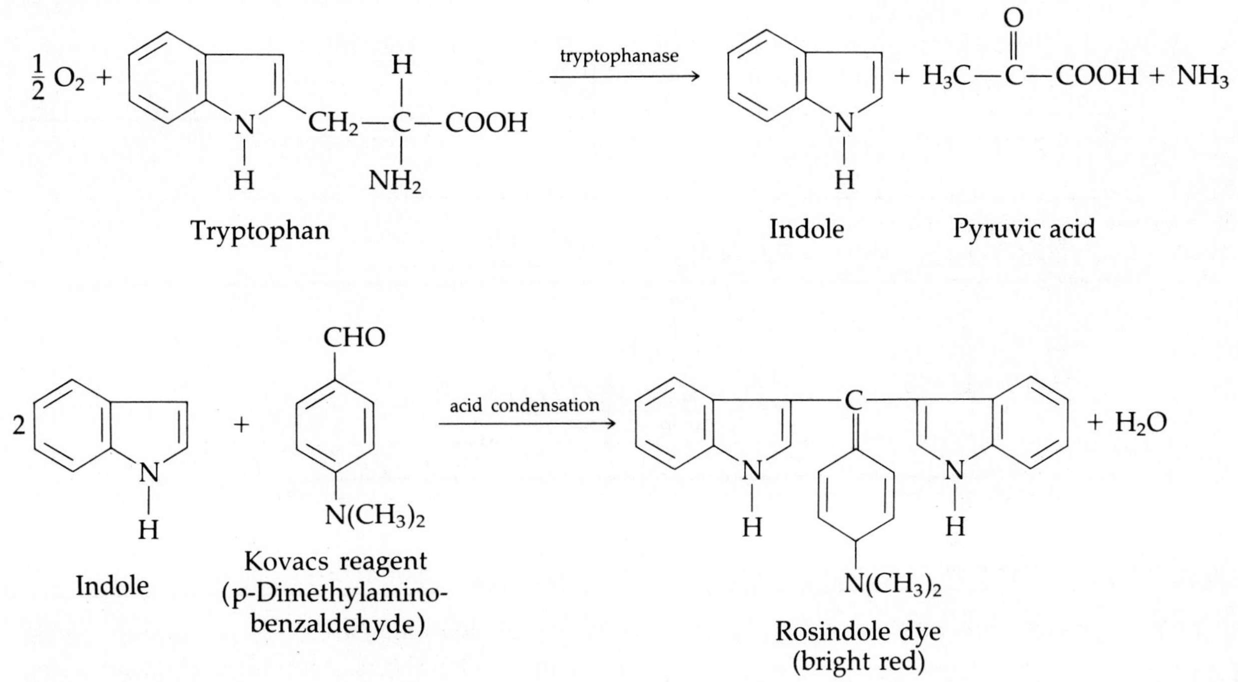

$\frac{1}{2}O_2$ + Tryptophan $\xrightarrow{\text{tryptophanase}}$ Indole + $H_3C{-}C({=}O){-}COOH$ (Pyruvic acid) + NH_3

2 Indole + Kovacs reagent (p-Dimethylamino-benzaldehyde) $\xrightarrow{\text{acid condensation}}$ Rosindole dye (bright red) + H_2O

Figure 21-4

Many bacteria produce indole from the amino acid tryptophan.

Materials

Tubes containing ornithine broth (2)

Phenylalanine slants (2)

Tubes containing peptone iron deeps (2)

Tubes containing 1% tryptone broth (2)

Vaspar

Ferric chloride reagent

Kovacs reagent

Cultures (as designated in the procedure)

Escherichia coli

Pseudomonas aeruginosa

Proteus vulgaris

Enterobacter aerogenes

Techniques Required

Exercise 13

Procedure

Ornithine Decarboxylation

1. Note the color of the ornithine broth.
2. Inoculate one tube with *Enterobacter aerogenes* and one with *Proteus vulgaris.* Add approximately 5 mm of vaspar to the surface of each tube.
3. Incubate at 35°C for 1 to 2 days.
4. Observe for the presence of growth. A positive test is purple whereas a negative test is yellow. Why? ______

Phenylalanine Deamination

1. Streak one phenylalanine slant heavily with *Proteus vulgaris* and the other with one of the remaining bacterial cultures.
2. Incubate for 1 to 2 days at 35°C. Observe for the presence of growth.
3. Add 4 or 5 drops of ferric chloride reagent to the top of the slant, allowing the reagent to run through the growth on the slant. A positive test gives a dark green color.

Hydrogen Sulfide Production

TSI

1. Stab one peptone iron deep with *Escherichia coli* and the other with *Proteus vulgaris.*
2. Incubate at 35°C for up to 7 days. Observe initially at 24 or 48 hours.
3. Observe for the presence of growth. Blackening in the butt of the tube indicates a positive test.

Indole Production

1. Inoculate one tryptone broth tube with *Escherichia coli* and the other with *Enterobacter aerogenes.*
2. Incubate at 35°C for 24 to 48 hours. Add 5 to 10 drops of Kovacs reagent. Agitate gently. A cherry red color in the top layer of the tube indicates a positive test.

Results

Record your results for each test.

Turn to the Laboratory Report for Exercise 21.

EXERCISE 22

Respiration

Objectives

After completing this exercise you should be able to

1. Define reduction.
2. Compare and contrast the following terms: aerobic respiration, anaerobic respiration, and fermentation.
3. Perform nitrate reduction, catalase, and oxidase tests.

Background

Molecules that combine with electrons liberated during metabolic processes are called **electron acceptors.** Electron acceptors become **reduced** when they gain electrons. Electrons are formed from the ionization of a hydrogen atom as shown here:

$$\underset{\text{Hydrogen atom}}{H} \longrightarrow \underset{\text{Hydrogen ion}}{H^+} + \underset{\text{Electron}}{e^-}$$

When an electron acceptor picks up an electron, it becomes negatively charged and combines with the positively charged hydrogen ion. **Reduction,** then, is a gain of electrons or hydrogen atoms.

Organic molecules act as electron acceptors in fermentative metabolism. Inorganic molecules serve as electron acceptors in oxidative metabolism or **respiration.** Molecular oxygen (O_2) is the final electron acceptor in **aerobic respiration.** In the process **anaerobic respiration,** inorganic compounds other than O_2 act as final electron acceptors. (A few bacteria use organic electron acceptors in anaerobic respiration.)

In some bacteria, electrons are transferred to **cytochrome oxidase** before being transferred to oxygen in aerobic respiration. The presence of cytochrome oxidase is used to identify these bacteria.

During aerobic respiration in the presence of oxygen, hydrogen atoms may be combined with oxygen, forming hydrogen peroxide (H_2O_2), which is lethal to the cell. Most organisms have the enzyme **catalase,** which breaks down hydrogen peroxide to water and oxygen as shown:

$$\underset{\text{Hydrogen peroxide}}{2H_2O_2} \xrightarrow{\text{catalase}} \underset{\text{Water}}{2H_2O} + \underset{\text{Oxygen}}{O_2\updownarrow}$$

The reduction of some inorganic compounds in respiration is shown in Figure 22-1. In this exercise we will examine the reduction of inorganic, nitrogen-containing compounds. During anaerobic respiration some bacteria reduce nitrates to nitrites; others further reduce nitrates to nitrous oxide or nitrogen gas. Other bacteria reduce nitrates to nitrites and then ammonia.

Nitrate broth (nutrient broth plus 0.1 percent potassium nitrate) is used to determine a bacterium's ability to reduce nitrates. Nitrites are detected by the addition of dimethyl-α-naphthylamine and sulfanilic acid to nitrate broth. A negative test (no nitrites) is further checked for the presence of nitrate in the broth by the addition of zinc. If nitrates are present, reduction has not taken place.

Materials

Tubes containing nitrate broth (3)

Petri plates containing trypticase soy agar (2)

Nitrate reagent A (dimethyl-α-naphthylamine)

Nitrate reagent B (sulfanilic acid)

Hydrogen peroxide, 3%

Oxidase reagent (p-aminodimethyl-aniline monohydrochloride)

Zinc dust

Cultures (one of each)

Bacillus subtilis

Bacillus megaterium

Pseudomonas aeruginosa

Streptococcus lactis

Techniques Required

Exercise 13 and 17

Procedure

Nitrate Reduction Test

Label three tubes of nitrate broth and inoculate one with *Bacillus megaterium,* one with *Pseudomonas aeruginosa,* and one with *B. subtilis.*

1. Incubate at 35°C for 2 days.

Aerobic respiration

$$\underset{\text{Oxygen}}{\tfrac{1}{2}O_2} + 2\,H^+ + 2\,e^- \longrightarrow \underset{\text{Water}}{2\,H_2O}$$

Anaerobic respiration

$$\underset{\text{Sulfate ion}}{SO_4^{2-}} + 10\,H^+ + 10\,e^- \longrightarrow \underset{\text{Hydrogen sulfide}}{H_2S} + 4\,H_2O$$

$$\underset{\text{Carbonate ion}}{CO_3^{2-}} + 10\,H^+ + 10\,e^- \longrightarrow \underset{\text{Methane}}{CH_4} + 3\,H_2O$$

$$\underset{\text{Nitrate ion}}{NO_3^-} + 2\,H^+ + 2\,e^- \longrightarrow \underset{\text{Nitrite ion}}{NO_2^-} + H_2O \xrightarrow{6\,H^+ + 6\,e^-} \underset{\text{Nitrous oxide}}{N_2O} \xrightarrow{2\,H^+ + 2\,e^-} \underset{\text{Nitrogen gas}}{N_2}$$

Figure 22-1
Chemical equations showing the reduction of some electron acceptors in bacterial respiration.

2. Add 5 drops of nitrate A and 5 drops of nitrate B to each tube. Shake gently.
3. A red color within 30 seconds is a positive test. If red, what compound is present? ______________

____
4. If it does not turn red, add a small pinch of zinc dust and if it turns red now, the test is negative. If not, it is positive for nitrate reduction. Why? ______

____
5. Record your results.

Oxidase Test

Divide one plate in half. Label one-half "*B. megaterium*" and the other "*P. aeruginosa*." Streak each organism on the appropriate half.

1. Incubate the plate inverted at 35°C for 24 to 48 hours.
2. Drop oxidase reagent on the colonies and observe for a color change to pink within a minute, then blue to black. Oxidase negative colonies will not change color.

Catalase Test

Divide the other plate in half and inoculate it with *S. lactis* and *B. subtilis*.

1. Incubate the plate inverted at 35°C for 24 to 48 hours.
2. Drop hydrogen peroxide on the colonies and observe for bubbles (catalase positive). What gas is in the bubbles? ______________

Turn to the Laboratory Report for Exercise 22.

EXERCISE 23

Anabolic Activity

People are lonely because they build walls instead of bridges.

J. F. NEWTON

Objectives

After completing this exercise you should be able to

1. Differentiate between anabolic and catabolic reactions.
2. Define the terms: common intermediate and secondary metabolite.
3. Provide several examples of anabolism.

Background

Catabolic reactions were examined in Exercises 17 through 22. The energy produced during degradation of substances is used in anabolism.

About 40% of the energy released from oxidation reactions in catabolism is stored in high energy bonds in triphosphate compounds such as **ATP** (the remaining 60% is lost as heat). Energy storage compounds serve as the link or **common intermediate** between catabolism and anabolism. In the following reaction, ATP is produced from ADP (a diphosphate), inorganic phosphate (P_i), and energy liberated from the reaction:

$$A \xrightarrow{ADP + P_i \;\curvearrowright\; ATP} B + C$$

The ATP is then used to power another reaction as shown:

$$X + Y \xrightarrow{ATP \;\curvearrowright\; ADP + P_i} Z$$

ATP can be regenerated again from the ADP + P_i in a catabolic reaction.

The anabolic activity of bacteria can vary from heat production to motility to a wide range of biosynthetic reactions. Biosynthetic reactions involve activities such as cell wall formation, protein synthesis, enzyme formation, and nucleic acid synthesis. Some bacteria can secrete products such as capsular polysaccharides (Ex-

(a)

CH_2OH … Glucose → Repeating cellobiose unit of cellulose

(b)

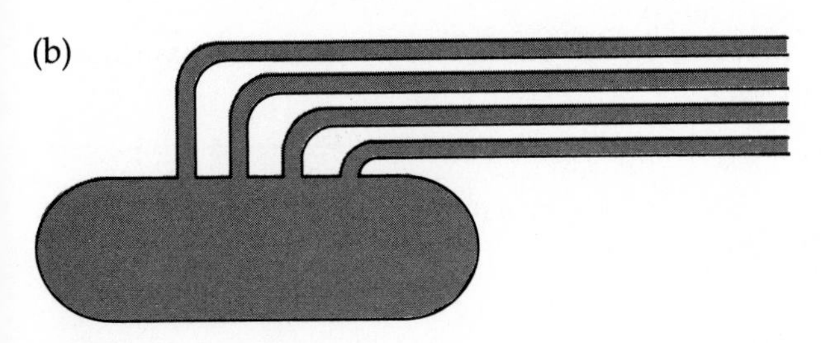

Figure 23-1

Cellulose is synthesized from glucose. **(a)** Cellobiose subunits are synthesized first. **(b)** Cellulose strands are produced which crystallize into multistranded polymers of cellulose.

ercise 9), extracellular enzymes (Exercise 17), and flagella (Exercise 9). Bacterial flagella are composed of helical molecules of the protein flagellin. They are anchored to the cell wall and cell membrane with a series of rings. The helical filament is attached to a hook inserted in the rings. The filaments rotate and push the cell. The rotary motion requires energy although the nature of the energy is unknown. A common intermediate *other* than ATP is needed, however, since bacteria are also motile when their ATP has been destroyed. *Proteus* has a large number of flagella and can move on the surface of solid media. The resultant spreading of the colony across an agar plate is described as *swarming*.

Some bacteria produce compounds that have no apparent function. The red pigment *prodigiosin* is synthesized from amino acids after the cell's needs are met. The function, if any, of such **secondary metabolites** is not known. Another distinguishing factor of secondary metabolites is that their synthesis depends on environmental factors such as oxygen and temperature.

The cellulose strands secreted by *Acetobacter aceti* subspecies *xylinum* are an unusual product. The cellulose is synthesized from glucose into a ribbon 0.05 to 0.1 μm in diameter, at a rate of 2 μm per minute. The ribbons of cellulose appear to polymerize and crystallize into larger strands (Figure 23-1). The cellulose may provide a floating mat or pellicle that provides the aerobic *Acetobacter* with a mechanism for growing at the surface of a liquid medium.

In this exercise several anabolic activities will be examined.

Materials

Schramm and Hestrin's glucose medium (SH) (1)

Semisolid agar deeps (2)

Nutrient agar slants (2)

Sterile Petri plate (1)

Glass rod

Paper towels

Cultures (one of each)

Acetobacter aceti subspecies *xylinum*

Proteus vulgaris

Micrococcus luteus

Serratia marcescens

Techniques Required

Exercise 13

Procedure

Cellulose Production

1. Inoculate the SH medium with 2 loopfuls of *Acetobacter aceti* subspecies *xylinum*. Aseptically pour the SH medium into a sterile Petri plate.
2. Incubate at 28°C for 5 to 7 days. *Do not* invert.
3. After 5 to 7 days, or when a pellicle has formed, observe the pellicle. Carefully discard the culture medium by pouring into a beaker containing disinfectant.
4. Wash the pellicle several times by adding water to the Petri plate.
5. Describe the appearance and texture of the pellicle. Lift the washed pellicle out of the Petri plate with a glass rod and spread out on a paper towel. Test its wet strength by pulling on it with your finger. Let dry and test its dry strength.

Motility

1. With your needle, inoculate one semisolid agar deep with *Proteus vulgaris* and one with *Micrococcus luteus.* Plunge the needle straight down the middle of the deep (Exercise 13).
2. Incubate at 35°C until the next period.
3. Draw the patterns seen in the semisolid agar. Motile organisms will give growth patterns similar to an inverted Christmas tree while nonmotile organisms will grow in a line limited by penetration of the needle. If an organism is extremely motile it may spread throughout the entire deep, giving a cloudy appearance. Use an uninoculated tube for comparison. Why does growth only on the top surface of the deep not indicate motility?

Pigment Production

1. Inoculate the nutrient agar slants with *Micrococcus luteus* and *Serratia marcescens* (Exercise 13).
2. Incubate at 28°C until good growth appears. (Some students should incubate the *Serratia* slants at 35°C.)
3. After incubation, describe the growth patterns on the agar slants. Describe the pigment seen.

Turn to the Laboratory Report for Exercise 23.

EXERCISE 24

Read

Unknown Identification

The strategy of discovery lies in determining the sequence of choice of problems to solve. Now it is in fact very much more difficult to see a problem than to find a solution to it. The former requires imagination, the latter only ingenuity.

BERNAL, 1971

Objectives

After completing this exercise you should be able to

1. Explain how bacteria are characterized and classified.
2. Use *Bergey's Manual.*
3. Identify an unknown bacterium.

Background

In microbiology, a system of classification must be available to allow the microbiologist to categorize and classify organisms. Communication among scientists would be very limited if no universal system of classification existed. The **taxonomy** (grouping) of bacteria is difficult because few definite anatomical or visual differences exist and no clear evolutionary relationships have been found. With these limitations, most bacteria are characterized by evaluation of primary characteristics such as morphology and growth patterns and secondary characteristics such as metabolism and serology (Exercise 74). A characteristic that is critical for distinguishing one bacterial group from another may be irrelevant for characterization of other bacteria. The most important reference for bacterial taxonomy is *Bergey's Manual of Determinative Bacteriology.** In *Bergey's Manual,* bacteria are grouped into

*Ř. D. Buchanan and N. E. Gibbons, eds. 1974. *Bergey's Manual of Determinative Bacteriology,* 8th ed. Baltimore: Williams and Wilkins; see also J. G. Holt, 1977. *The Shorter Bergey's Manual of Determinative Bacteriology, Eighth Edition.* Baltimore: Williams and Wilkins.

19 numbered **parts,** according to Gram stain reaction, cell shape, cell arrangements, oxygen requirements, motility, and nutritional and metabolic properties. Although the characteristics of a given group are relatively constant, through repeated laboratory culture, atypical bacteria will be found. This variability, however, only heightens the fun of classifying bacteria.

You will be given an unknown heterotrophic bacterium to characterize and identify. By using careful deduction and by systematically compiling and analyzing data you should be able to identify the bacterium.

A key to species of bacteria has been provided in this exercise (Table 24-1). The key is an example of a dichotomous classification system; that is, a population is repeatedly divided into two parts until a description identifies a single member. Such a key is sometimes called an "artificial key" because there is no single correct way to write one. You may want to check your conclusion with the species description given in *Bergey's Manual.*

To begin your identification, ascertain the purity of the culture you have been given and prepare stock and working cultures. Avoid contamination of your unknown. Note growth characteristics and Gram stain appearance to give you a clue as to how to proceed. After culturing and staining the unknown, many bacte-

Table 24-1

Key to the Identification of Selected Heterotrophic Bacteria

I. Cells are cocci	
A. Gram-positive	
1. Catalase produced	
a. Cells arranged in tetrads or glucose not fermented	
(1) Yellow pigment	*Micrococcus luteus*
(2) Red pigment	*M. roseus*
b. Cells not in tetrads or glucose fermented	
(1) Acid from mannitol	*Staphylococcus aureus*
(2) No acid from mannitol	
(a) Acid from fructose	*S. epidermidis*
(b) No acid from fructose	*S. saprophyticus*
2. Catalase not produced	
a. Growth in 6.5% NaCl or at pH 9.6	*Streptococcus faecalis*
b. No growth in 6.5% NaCl or at pH 9.6	
(1) Acid from glycerol	*S. equisimilis*
(2) No acid from glycerol	
(a) Acid from inulin	
aa. Acid from raffinose	*S. salivarius*
bb. No acid from raffinose	*S. sanguis*
(b) No acid from inulin	*S. mitis*
B. Gram-negative	
1. Glucose fermented	
a. Nitrate reduced	*Neisseria mucosa*
b. Nitrate not reduced	*N. sicca*
2. Glucose not fermented	*N. flavescens*

Continued

Table 24-1 (continued)
Key to the Identification of Selected Heterotrophic Bacteria

II. Cells are rods

- A. Gram-positive
  - 1. Endospores present; catalase-positive
    - a. Acid from mannitol
      - (1) V–P-positive — *Bacillus subtilis*
      - (2) V–P-negative — *B. megaterium*
    - b. No acid from mannitol — *B. cereus*
  - 2. No endospores; catalase-negative
    - a. Acid and gas from glucose — *Lactobacillus fermentum*
    - b. Acid, not gas, from glucose
      - (1) Acid from mannitol — *L. casei*
      - (2) No acid from mannitol — *L. acidophilus*
- B. Gram-negative
  - 1. Glucose not fermented
    - a. Oxidase-positive; glucose oxidized
      - (1) Litmus milk coagulated, peptonized, reduced — *Pseudomonas aeruginosa*
      - (2) Litmus milk alkaline — *P. fluorescens*
    - b. Oxidase-negative
      - (1) Gelatin hydrolyzed
        - (a) White colonies — *Alcaligenes faecalis*
        - (b) Yellow colonies — *A. paradoxus*
      - (2) Gelatin not hydrolyzed — *A. aquamarinus*
  - 2. Glucose fermented; oxidase-negative
    - a. Acid and gas from lactose
      - (1) Citrate utilized
        - (a) MR-negative, V–P-positive — *Enterobacter aerogenes*
        - (b) MR-positive, V–P-negative — *Citrobacter freundii*
      - (2) Citrate not utilized — *Escherichia coli*
    - b. Lactose not fermented
      - (1) Red pigment produced at 25°C — *Serratia marcescens*
      - (2) No red pigment
        - (a) Indole produced
          - aa. H_2S produced — *Proteus vulgaris*
          - bb. H_2S not produced — *Morganella morganii*
        - (b) Indole not produced — *P. mirabilis*

rial groups can be eliminated. Final determination of your unknown will depend on careful selection of the relevant biochemical tests and weighing the value of one test over another in case of contradictions. Enjoy!

Materials

Petri plates containing trypticase soy agar (2)

Trypticase soy agar slant (2)

All stains, reagents, and media previously used

Culture

Unknown bacterium # ________

Techniques Required

Exercises 1, 2, 5–9, 13–23

Procedure

1. Streak your unknown onto the agar plates for isolation. Incubate one plate at 35°C and the other at room temperature for 24 to 48 hours. Note the growth characteristics and the temperature at which it grew best.
2. Aseptically inoculate two trypticase soy agar slants. Incubate. Describe the resulting growth. Keep one slant culture in the refrigerator as your stock culture; the other is your working culture. Subculture onto another slant from your stock culture when your working culture is contaminated or not viable.
3. Use your working culture for all identification procedures. When a new slant is made and its purity demonstrated, the old working culture should be discarded. What should you do if you think your culture is contaminated? ________
4. Read the key in Table 24-1 to develop ideas on how to proceed. Perhaps determining staining characteristics might be a good place to start. What shape is it? ________ What can be eliminated? ________
5. After determining its staining and morphologic characteristics, determine which biochemical tests you will need. Do not be wasteful. Inoculate *only* what is needed. It is not necessary to repeat a test—do it once accurately. Do not perform unnecessary tests.
6. If you come across a new test—one not previously done in the course—ask if it is essential. Can you circumvent it? ________ If not, consult your instructor.

Turn to the Laboratory Report for Exercise 24.

EXERCISE 25

EXERCISE 25

Rapid Identification Methods

Objectives

After completing this exercise you should be able to

1. Evaluate three methods of identifying enterics.
2. Name three advantages of the "systems approach" over conventional tube methods.

Background

The clinical microbiology laboratory must identify bacteria quickly and accurately. Accuracy is improved by using standardized tests. The IMViC tests were developed as a means of separating enterics, particularly the coliforms*, using a standard combination of four tests. Each capital letter in **IMViC** represents a test; the *i* is added for easier pronunciation. The tests are:

I for indole production from tryptophan (Exercise 21)

M, methyl red test for acid production from glucose (Exercise 18)

*Coliforms (Exercise 49) are aerobic or facultatively anaerobic, gram-negative, non-endospore-forming, rod-shaped bacteria that ferment lactose with acid and gas formation within 48 hours at 35°C.

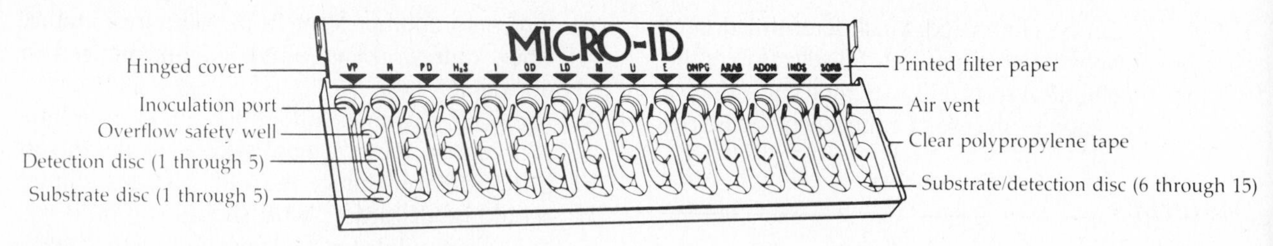

Figure 25-1
Micro-ID® System. The first five reaction wells contain a substrate disk and a detection disk. The remaining ten wells contain a single substrate/detection disk.

Table 25-1
IMViC Reactions for Selected Species of Enterics

Species	Indole	Methyl red	Voges–Pros-kauer	Citrate
Escherichia coli	+(v)	+	−	−
Citrobacter freundii	−	+	−	+
Enterobacter aerogenes	−	−	+	+
Enterobacter cloacae	−	−	+	+
Serratia marcescens	−	+ or −*	+	+
Proteus vulgaris	+	−	−	−(v)
Proteus mirabilis	−	+	− or +**	+(v)

v = variable
* majority of strains give + results
** majority of strains give − results

V is the Voges–Proskauer test for production of acetoin from glucose (Exercise 18)

C for the utilization of citrate as the sole carbon source (Exercise 16)

Although variation among strains does exist, IMViC reactions for selected species of enterics are given in Table 25-1.

The IMViC tests are four tests of equal value that require 24 to 48 hours of incubation. In the last few years, however, **rapid identification methods** have been developed. An example is the Micro-ID®* system (Figure 25-1) for identifying oxidase-negative, gram-negative bacteria belonging to the family Entero-

*General Diagnostics, Division of Warner-Lambert Company, Morris Plains, NJ 07950.

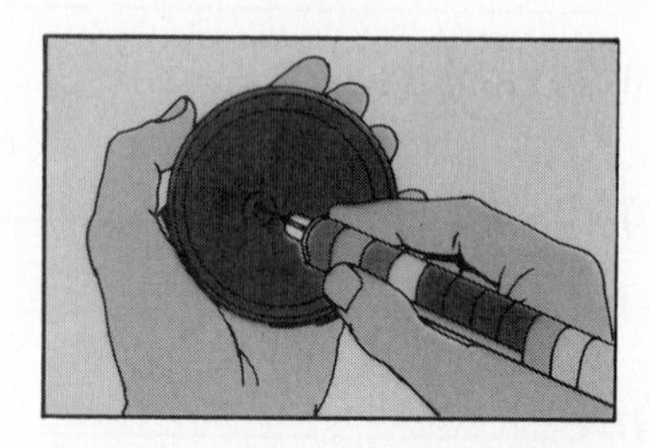

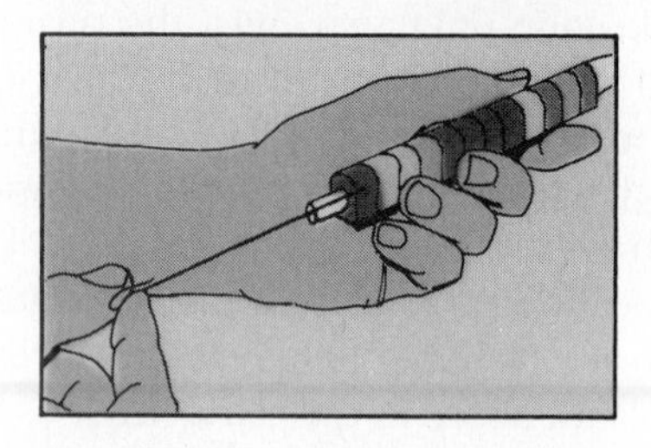

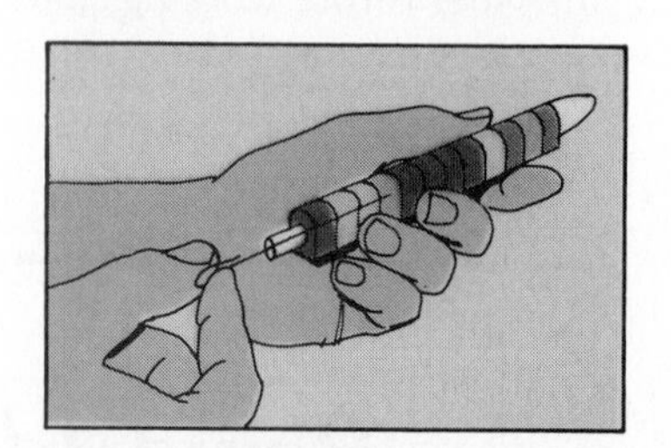

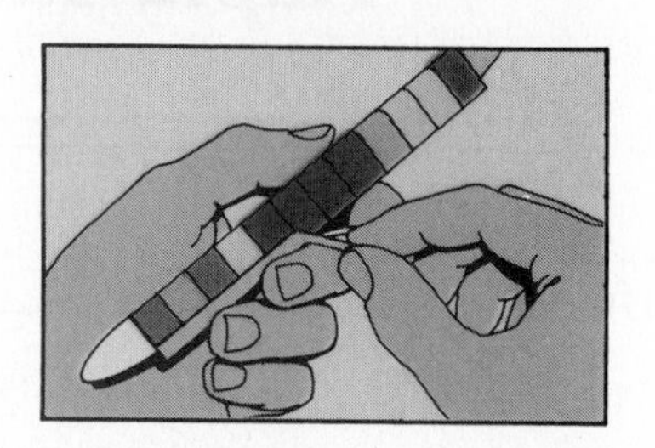

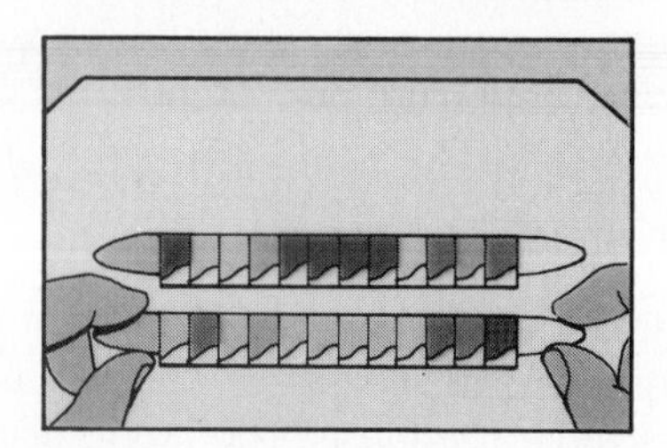

Figure 25-2
Inoculating an Enterotube®

bacteriaceae. Micro-ID® consists of 15 paper disks impregnated with reagents that detect the presence of specific enzymes or metabolic end products. These reagents include a substrate to be acted on by a bacterial enzyme and an indicator system that reacts with the metabolic end product to yield a readily identifiable color change (Table 25-2). No culturing beyond the initial isolation is necessary and results can be obtained in 4 hours. Comparisons between these rapid identification methods and conventional culture methods show between 92% and 98% agreement.

Computerized analysis of test results can increase accuracy because each test is given a point value. Tests that are more important than others get more points. The Enterotube®* (Figure 25-2) employs this type of data analysis. Although the Enterotube® requires a 24-hour incubation period, one tube performs 15 tests (Table 25-3). Identification of enterics using the Enterotube® has been shown to be more accurate than conventional tube methods.

At the present time, six commercial identification systems are available. As the systems are perfected,

*Roche Diagnostics, Division of Hoffman-LaRoche Inc., Nutley, NJ 07110.

Table 25-2
Micro-ID® Biochemical Reactions

Test	Comments	Indicator Color		Refer to Exercise
		+Reaction	−Reaction	
V–P	Addition of KOH detects the presence of acetoin	Pink to red	Yellow	18
N	To test for nitrate reduction	Red	Colorless to pink	22
PD	Deamination of phenylalanine is indicated by a color change	Green	Yellow	21
H_2S	Blackening indicates reduction of thiosulfate	Brown to black	White	21
I	Tryptophan is metabolized to indole	Pink to red	Yellow to orange	21
OD	Decarboxylation of ornithine produces putrescine	Purple to red-purple	Amber to yellow	21
LD	Decarboxylation of lysine liberates cadaverine	Purple to red-purple	Amber to yellow	21
M	Malonate can be used as a carbon source resulting in alkaline end products	Green to blue	Yellow	16
U	Urea is hydrolyzed by the enzyme urease	Orange to red-purple	Yellow	20
E	Esculetin from esculin hydrolysis causes a black precipitate	Brown to black	Beige	17
ONPG	O-nitrophenyl- β-D-galactopyranoside is hydrolyzed by the enzyme that hydrolyzes lactose	Yellow	Colorless	18
ARAB	Fermentation of arabinose	Yellow to amber	Purple	18
ADON	Fermentation of adonitol	Yellow to amber	Red-purple to purple	18
INOS	Fermentation of inositol	Yellow to amber	Red-purple to purple	18
SORB	Fermentation of sorbitol	Yellow to amber	Red-purple to purple	18

Source: General Diagnostics, Division of Warner- Lambert Company.

they can provide greater standardization in identification because they overcome the limitations of hunting through a key (Exercise 24), differences in media preparation, and evaluation of tests within a laboratory or between different laboratories. They are also time-, cost-, and labor-saving.

Materials

Petri plate containing nutrient agar (1)

IMViC tests and reagents

Micro-ID® tray and reagents

Enterotube® and reagents

Serological tube containing 3.5 ml saline (1)

1-ml pipette

Oxidase reagent

Culture

Unknown enteric # __________

Techniques Required

Exercises 13, 16, 17, 18, 20, 21, and 22

Procedure

Isolation

1. Streak the nutrient agar plate with your unknown for isolation and to determine purity of the culture. Incubate inverted at 35°C for 24 to 48 hours. Record the appearance of the colonies.
2. Determine the oxidase reaction of one of the colonies remaining on the plate. Why? __________
   How will you determine the oxidase reaction? __________

IMViC Tests

1. Inoculate tubes of tryptone broth, MRVP broth, and Simmons citrate agar with your unknown.
2. Incubate the tubes at 35°C for 24 to 48 hours; perform the appropriate tests, and record your results.

Table 25-3
Biochemical Reactions in the Enterotube®

Test	Comments	Indicator Changed		Refer to Exercise
		From	To	
GLU	Acid from glucose	Red	Yellow	18
GAS	Gas produced from fermentation of glucose trapped in this compartment, causing separation of the wax			
LYS	Lysine decarboxylase	Yellow	Purple	21
ORN	Ornithine decarboxylase	Yellow	Purple	21
H_2S	Ferrous ion reacts with sulfide ions forming a black precipitate			21
IND	Kovacs reagent is added to the H_2S/IND compartment to detect indole	Beige	Red	21
LAC	Lactose fermentation	Red	Yellow	18
ARAB	Arabinose fermentation	Red	Yellow	18
SORB	Sorbitol fermentation	Red	Yellow	18
V–P	Voges–Proskauer reagents detect acetoin	Beige	Red	18
PA	Pyruvic acid released from phenylalanine after its deamination combines with iron salts to form a black precipitate			21
UREA	Ammonia changes the pH of the medium	Yellow	Pink	20
CIT	Citric acid used as a carbon source	Green	Blue	16

Source: Roche Diagnostics, Division of Hoffman-LaRoche Inc.

Micro-ID®

1. Prepare a suspension of identical, isolated (18 to 24 hours old) colonies in 3.5 ml saline.
2. Open the sealed, moisture-proof foil package and remove the Micro-ID® unit. *Do not remove the clear plastic tape covering test wells.*
3. Open the cover and allow the Micro-ID® unit to lie flat on your workbench.
4. Pipette approximately 0.2 ml of the bacterial suspension into each inoculation well at the top of the Micro-ID® unit. Why isn't asepsis necessary? ____________________
5. Close the cover and stand the Micro-ID® tray upright in the support rack. *Make sure that the organism suspension is in contact with all substrate disks. Do not moisten detection disks.*
6. Incubate for 4 hours at 35°C.
7. After incubation, place unit flat on workbench, open the lid, and add 0.1 ml (2 drops) of 20% KOH to the inoculation well of the V–P test *only. Do not add KOH to any other inoculation well.*
8. Close the lid and hold tray upright. *Be certain that the KOH flows down into the V–P test solution.*
9. Rotate the Micro-ID® unit clockwise about 90 degrees so that the upper disks in the first five wells become wet. Hold the tray upright and tap gently on the laboratory bench to dislodge any suspension trapped under the upper disks. *Be certain that each upper disk in reaction chambers 1 through 5 is moistened by this procedure.*
10. Read all reactions immediately, except the V–P test, as positive or negative according to the color changes listed in Table 25-2.
11. Allow color to develop in the V–P well for approximately 10 minutes, then read.
12. Read the color of the *upper disk* for the first five tests, and the color of the *organism suspension* for the remaining ten tests.
13. Record results in the Laboratory Report. Each positive test receives a point value. Add the numbers in each group of three tests to get a five-digit number. What is your number? ________
    Find the five-digit number in the *Micro-ID® Identification Manual.**
14. Dispose of used Micro-ID® trays and saline suspensions by placing in the autoclave basket.

Enterotube® (Figure 25-2)

1. Remove both caps from the Enterotube®. One end of the wire is straight and is used to pick up the inoculum; the bent end of the wire is the handle. Holding the Enterotube®, pick a well-isolated colony with the inoculating end of the wire. Avoid touching the agar with the needle.

*General Diagnostics, Division of Warner-Lambert Company, Morris Plains, NJ 07950.

2. Inoculate the Enterotube® by holding the bent end of the wire and twisting; then withdraw the needle through all 12 compartments using a turning motion.
3. Reinsert the needle into the Enterotube®, using a turning motion, through all 12 compartments. Then withdraw to the H_2S/indole compartment. Break the needle at the notch by bending, discard the handle in disinfectant, and replace caps loosely on both ends of the tube. The portion of the needle remaining in the tube maintains anaerobic conditions necessary for fermentation, production of gas, and decarboxylation. The part of the wire in the H_2S/indole compartment will not interfere with these tests.
4. Strip off the blue tape after inoculation but before incubation, to provide aerobic conditions. Slide the clear band over the dextrose (glucose) compartment to contain any small amount of sterile wax that may escape due to excessive gas production by some bacteria.
5. Incubate the tube lying on its flat surface at 35°C for 24 hours.
6. Interpret and record all reactions (see Table 25-3) in the Laboratory Report. Read all the other tests before the indole and V–P tests, which follow:

Indole test. Melt a small hole in the plastic film covering the H_2S/indole compartment using a warm inoculating loop. Add 1 to 2 drops of Kovacs reagent, and allow the reagent to contact the agar surface. A positive test is indicated by a red color within 10 seconds.

V–P test. Add 2 drops of 40% KOH containing 5% α-naphthol to the V–P compartment. A positive test is indicated by development of a red color within 20 minutes.

7. Indicate each positive reaction by circling the number appearing below the appropriate compartment of the Enterotube® outlined in the Laboratory Report. Add the circled numbers only within each bracketed section and enter this sum in the space provided below the arrow. Read the five numbers in these spaces across as a five-digit number in the *Computer Coding and Identification System.**
8. Dispose of the Enterotube® by placing in the autoclave basket.

*Roche Diagnostics, Division of Hoffman-LaRoche Inc., Nutley, NJ 07110.

Turn to the Laboratory Report for Exercise 25.

PART 5

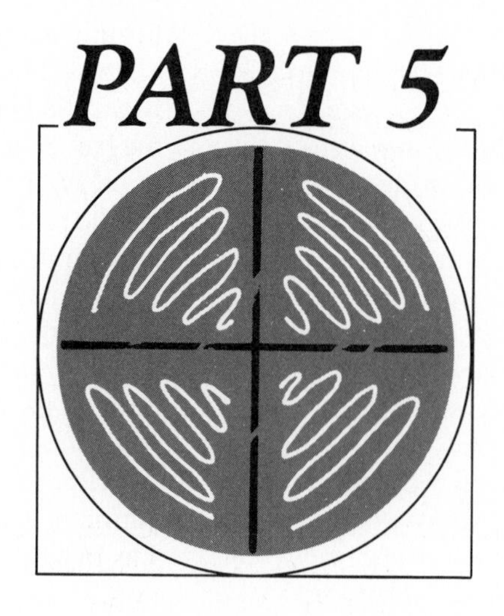

Bacterial Genetics

EXERCISES

All of the characteristics of bacteria including growth patterns, metabolic activities, pathogenicity, and chemical composition are inherited. These traits are transmitted from parent cell to offspring through genes. **Genetics** is the study of what genes are, how they carry information, how they are replicated and passed to the next generation or to another organism, and how their information is expressed within an organism to determine the particular characteristics of that organism.

The information stored in genes is called the **genotype.** All of the genes may not be expressed. **Phenotype** refers to the actual expressed characteristics, such as the ability to perform certain biochemical reactions (for example, synthesis of a capsule or pigment).

Bacteria are invaluable to the study of genetics because large numbers can be cultured inexpensively, and their relatively simple genetic composition facilitates studying the structure and function of genes.

Bacteria undergo genetic change due to mutation or recombination. These genetic changes result in a change in the genotype. When a change occurs, in most instances we can expect that the product coded by certain genes will be changed. An enzyme coded by a changed gene may become inactive. This might be disadvantageous or even lethal if the cell loses a phenotypic trait it needs. Some genetic changes may be beneficial. For instance, if an altered enzyme coded by a changed gene has new enzymatic activity, the cell may be able to grow in new environments.

Genetic change in microbial populations is relatively easy to observe. In 1949, H. B. Newcombe wrote:

> *Numerous bacterial variants are known which will grow in environments unfavorable to the parent strain, and to explain their occurrence two conflicting hypotheses have been advanced. The first assumes that the particular environment produces the observed change in some bacteria exposed to it, whereas the second assumes that the variants arose spontaneously during growth under normal conditions.*

Luria and Delbruck confirmed the latter **spontaneous mutation** hypothesis.

Genetic changes can be detected by using culturing methods that select for bacteria that have altered phenotypes. In Exercise 26, bacteria that are incapable of synthesizing essential amino acids will be selected.

Genetic recombination is the rearrangement of genes to form new combinations. Usually groups of genes from two different members of a species recombine or reshuffle, producing sequences containing nucleic acid from both participants and resulting in a new genetic variety (see the figure). Genetic recombination is an infrequent event but has given rise to progeny with important changes such as antibiotic resistance or enhanced pathogenicity.

In bacteria, genetic recombination can result from transformation (Exercise 27), conjugation (Exercise 28), or transduction (Exercise 29).

Cancer is thought to be the result of *permanent changes* in the nucleotide sequence of chromosomes. A test using principles of microbial genetics for possible cancer-inducing substances will be performed in Exercise 30.

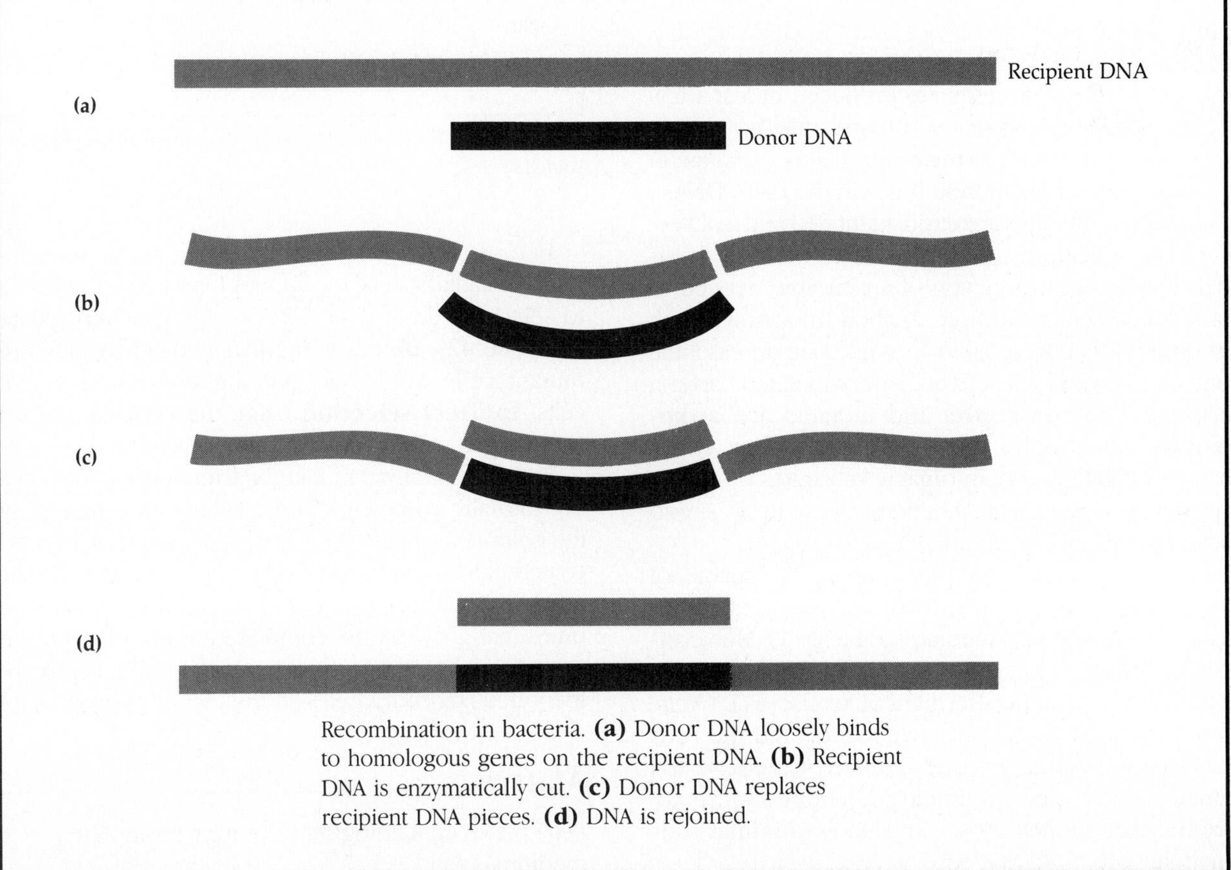

Recombination in bacteria. **(a)** Donor DNA loosely binds to homologous genes on the recipient DNA. **(b)** Recipient DNA is enzymatically cut. **(c)** Donor DNA replaces recipient DNA pieces. **(d)** DNA is rejoined.

EXERCISE 26

Isolation of Bacterial Mutants

Though coli *may bother and vex us,*
It's hard to believe they outsex us.
They accomplish seduction
by viral transduction
Foregoing the joys of amplexus.
S. GILBERT

Objectives

After completing this exercise you should be able to

1. Define the following terms: prototroph, auxotroph, and mutation.
2. Differentiate between direct and indirect selection.
3. Isolate bacterial mutants by replica plating.

Background

For practical purposes, genes and the characteristics for which they code are stable. However, when millions of bacterial progeny are produced in just a few hours of incubation, genetic variants may be seen. A variant is the result of a **mutation,** that is, a change in the sequence of nucleotide bases in the cell's DNA.

Metabolic mutants are easily identified and isolated. Catabolic mutations could result in a bacterium that is deficient in the production of an enzyme needed to utilize a particular substrate. Anabolic mutations result in bacteria that are unable to synthesize an essential organic chemical. "Wild type" or nonmutated bacteria are called **prototrophs** and mutants are **auxotrophs.** Auxotrophs have been isolated that either cannot catabolize certain organic substrates or cannot synthesize certain organic chemicals such as amino acids, purine or pyrimidine bases, or sugars.

The culture used in this exercise is capable of synthesizing all of its growth requirements from glucose-minimal salts medium (Table 26-1). Mutations will be induced by exposing the cells to the mutagenic wavelengths of ultraviolet light (Exercise 44). (*Note:* ultraviolet light is not very penetrating and the Petri plate cover will have to be removed to expose the culture.) After exposure to ultraviolet radiation, auxotrophs that cannot grow on glucose-minimal salts medium will be identified.

Table 26-1
Composition of Minimal Agar*

Component	Amount
Dipotassium phosphate	0.7%
Monopotassium phosphate	0.2%
Sodium citrate	0.05%
Magnesium sulfate	0.01%
Ammonium sulfate	0.1%
Agar	1.5%
Water	

*Glucose (0.1%) is commonly added as a carbon and energy source.

Because of the low rate of appearance of recognizable mutants, special techniques have been developed to select for desired mutants. In the **gradient plate** (Exercise 42), **direct selection** is used to pick out mutant cells while rejecting the unmutated parent cells. **Indirect selection** using the **replica-plating technique** will be used in this exercise.

In replica plating, mutated bacteria are grown on a nutritionally complete solid medium. An imprint of the colonies is made on velveteen-covered or rubber-coated blocks and transferred to glucose-minimal salts solid medium and to a complete solid medium. Colonies that grow on the complete medium but not on the minimal medium are auxotrophs. Auxotrophs are identified *indirectly* because they will not grow.

Materials

Petri plates containing nutrient agar (complete medium) (4)

Petri plate containing glucose-minimal salts agar (minimal medium) (1)

99-ml water dilution blanks (3)

Sterile 1-ml pipettes (4)

Ultraviolet lamp, 260 nm

Glass spreader ("hockey stick")

Sterile velveteen

Sterile replica-plating block

Rubber band

Alcohol

Cultures (one of the following)

Serratia marcescens

Escherichia coli

Techniques Required

Exercises 13 and 16 and Appendices A and B

*Procedure**

*Adapted from C. W. Brady. "Replica plate isolation of an auxotroph." Unpublished, University of Wisconsin Whitewater, Whitewater, Wisconsin 53190.

First Period

1. Label the nutrient agar plates "A," "B," and "C" and the dilution blanks "1," "2," and "3."
2. Aseptically pipette 1 ml of the assigned broth culture to dilution blank 1 and mix well (Appendix A).
3. Using another pipette, transfer 1 ml from dilution blank 1 to dilution blank 2 as shown in Figure 26-1, and mix well. What dilution is in bottle 2? (Refer to Appendix B.) ____________

   ____
4. Using another pipette, transfer 1 ml from dilution blank 2 to bottle 3 and 1 ml to the surface of plate A.
5. Mix bottle 3, and transfer 1 ml to the surface of plate B with a sterile pipette and 0.1 ml to the surface of plate C.
6. Disinfect a glass spreader ("hockey stick") by dipping in alcohol, quickly igniting the alcohol in a Bunsen burner flame, and letting the alcohol burn off. While it is burning, hold the hockey stick pointed *down*. Why? ____________

   ____________ Let cool.
7. Spread the liquid on the surface of plate C over the entire surface. Do the same with plates B and A. See Figure 26-2.
8. Disinfect the hockey stick and return it. Why is it not necessary to disinfect the hockey stick between each plate when proceeding from plate C

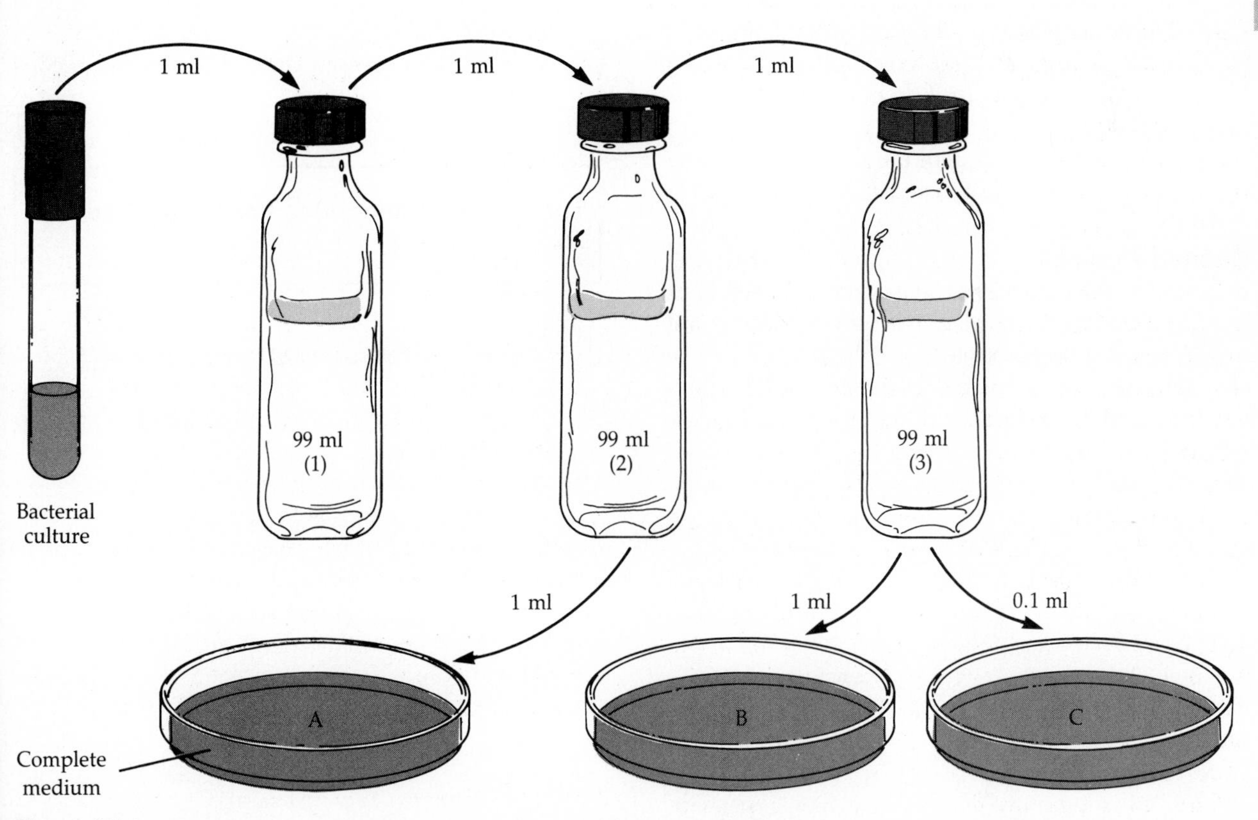

Figure 26-1
Dilution procedure.

Figure 26-2
Inoculation using a glass spreading rod ("hockey stick").

to B to A? ______________________

____

9. Expose each plate to the ultraviolet lamp for 30 to 60 seconds with the cover off. *Do not look directly at the light and do not leave your hand exposed to it.*
10. Incubate the plates until the next period: *Escherichia coli* at 35°C; *Serratia marcescens* at 25°C.

Second Period

1. Select a plate with 25 to 50 isolated colonies. Mark the bottom of the plate with a reference point. This is the master plate.
2. Mark the uninoculated complete and minimal media with a reference point on the bottom end of each plate (Figure 26-3*a*).
3. If your replica-plating block is already assembled, proceed to step 4. If not, follow these instructions: Carefully open the package of velveteen. Place the replicator block on the center of the velveteen. Pick up the four corners of the cloth and secure tightly on the handle with a rubber band.
4. Inoculate the sterile media by either (a) or (b), as follows:
   a. Hold the replica-plating block by resting the handle on the table with the rough surface or velveteen up (Figure 26-3*b*). Invert the master plate selected in step 1 on the block and allow the master plate agar to lightly touch the block. Remove the cover from the minimal medium, align the reference marks, and touch the uninoculated minimal agar with the inoculated replica-plating block. Replace the cover. Remove the cover from the uninoculated complete medium and inoculate with the replica-plating block, keeping the reference marks the same.
   b. Replica plating can also be done by placing the master plate and the uninoculated plates on the table. Place the reference marks in a 12 o'clock position and remove the covers. Touch the replica-plating block to the master plate, then, without altering its orientation, gently touch it to the minimal medium, then the complete medium (Figure 26-3*c*).
5. Replace covers and incubate as before. Refrigerate the master plate.
6. After incubation, compare the plates and record your results.

Turn to the Laboratory Report for Exercise 26.

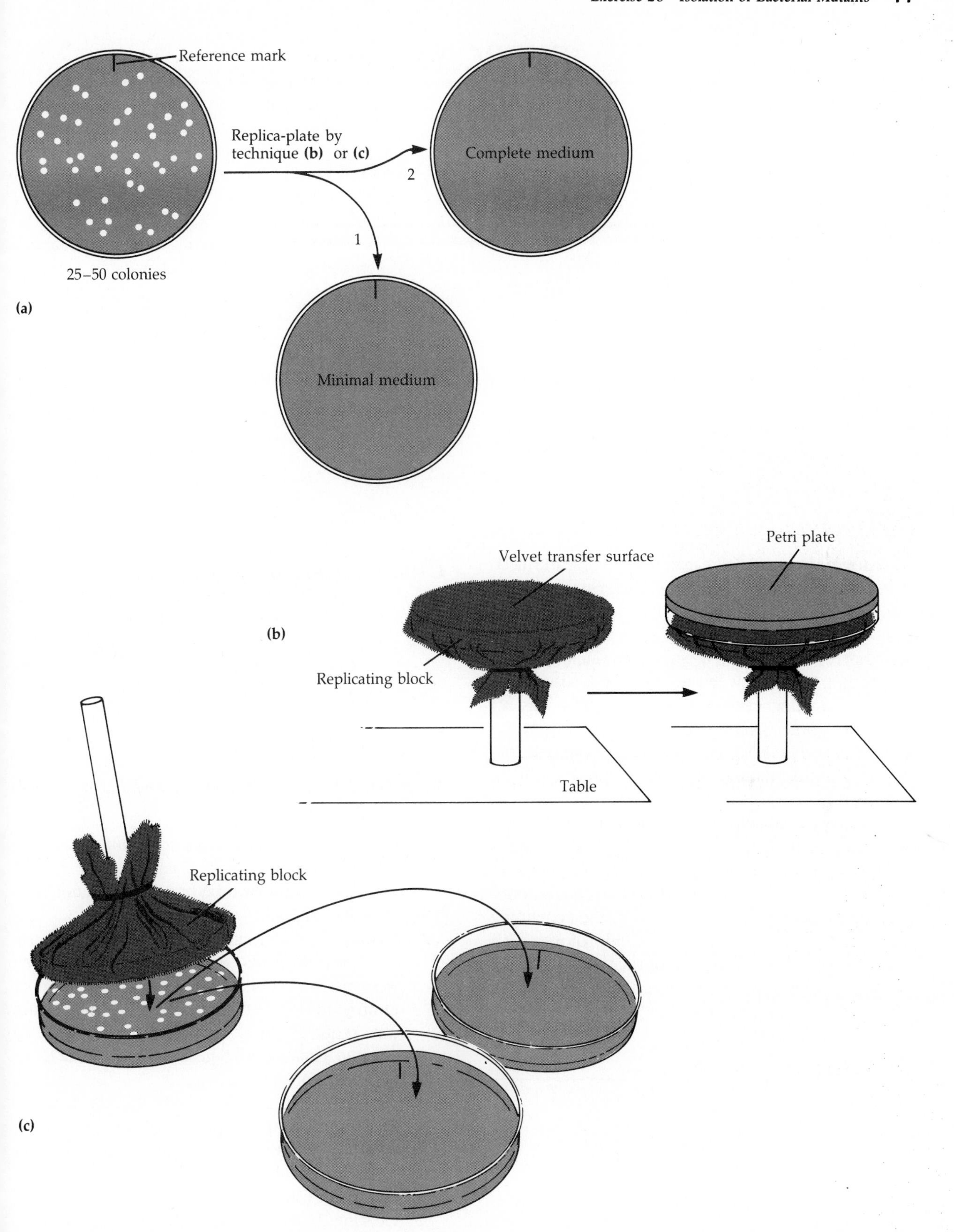

Figure 26-3

Replica plating. **(a)** Schematic diagram. **(b)** Transfer with stationary replicating block. **(c)** Transfer by moving the replicating block from plate to plate.

EXERCISE 27

Transformation

Objectives

After completing this exercise you should be able to

1. Explain the process of isolating DNA and then isolate DNA.
2. Define transformation.
3. Accomplish genetic change through transformation.

Background

Transformation is a rare event involving the accidental acquisition by a *recipient* bacterium of small pieces of DNA released from a dead bacterium *(donor)*. The acquired pieces of DNA can recombine with the recipient's DNA if sufficient nucleotide sequence homology between the two DNA strands exists. The resulting **recombinant** bacterium has acquired new genes from the donor.

In this exercise DNA will be isolated from one species, and transformation of another species will be attempted.

Donor bacterial cells will be concentrated by centrifugation resulting in a wet cell pellet. The cell walls will be enzymatically destroyed with lysozyme to release the nucleic acid and other cellular components. The nucleic acid can be isolated from the remaining cellular debris by the addition of chloroform. This DNA will be added to a culture of recipient cells. The recipient cells will be observed for phenotypic changes due to transformation.

Materials

Tubes containing nutrient broth (3)

Tube containing distilled water + EDTA (ethylenediaminetetraacetate) (1)

Tube containing 0.015*M* NaC1 + 0.0015*M* sodium citrate (1)

Petri plates containing sodium citrate agar (2)

Lysozyme (crystalline)

Chloroform

95% ethyl alcohol

Sterile 10-ml pipette (1)

Glass rod (6 mm × 200 mm)

10-ml graduated cylinder

Vortex

Balance

Centrifuge

Test tube (1)

Cultures (one of each)

Bacillus megaterium

Escherichia coli

Techniques Required

Exercises 7, 13, and 16 and Appendices A and C

Procedure* (see Figure 27-1)

First Period

1. Centrifuge the *Bacillus megaterium* culture for 10 minutes at 3000 rpm. Balance the centrifuge by placing a similar tube or a tube containing 10 ml water opposite your culture tube.
2. Discard the supernatant (liquid portion) by carefully pouring it into a container of disinfectant. Resuspend the pellet in 4 ml distilled water containing EDTA. EDTA chelates the magnesium ions required for DNAase activity. DNAase is an enzyme capable of digesting DNA. What does chelate mean? ______________________
____
3. Add 5 mg lysozyme, mix gently, and incubate at 35°C for 1 hour. The resulting protoplasts are lysed in this hypotonic environment and DNA is released, resulting in an increased viscosity and clearing. Why? ______________________
4. Add 4 ml chloroform to the solution under a fume hood. *All flames should be out.* Vortex for a few minutes to remove protein contaminants, then centrifuge for 10 minutes.

*Adapted from Conrad Wieler. 1975. "Experiment in Genetic Transformation," *American Biology Teacher 37*: 537–538.

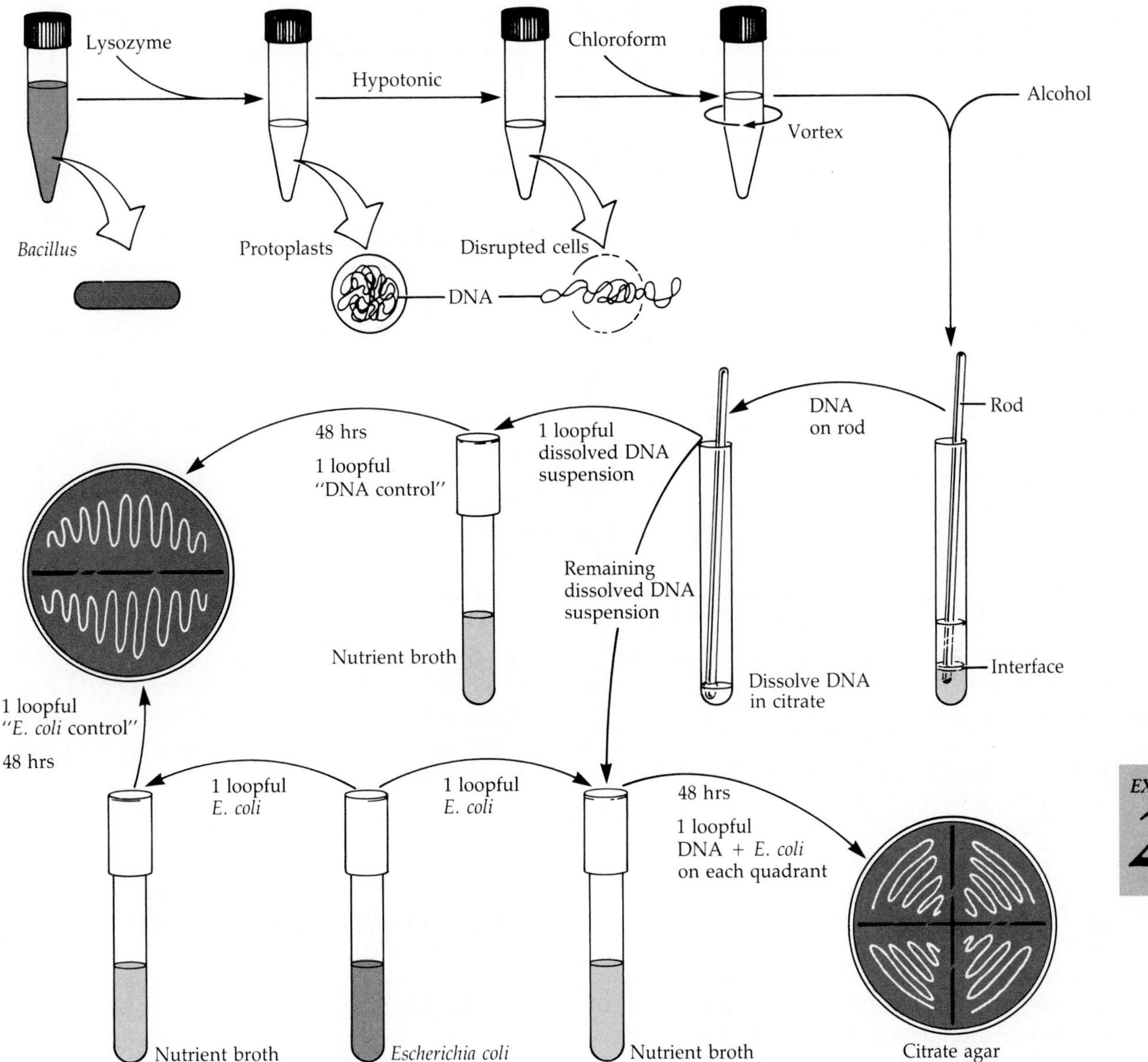

Figure 27-1
Procedure for isolation of DNA and transformation.

5. Pour the supernatant into a large test tube. Layer 4 ml 95% ethyl alcohol over the turbid aqueous layer. Collect DNA strands that appear at the interface by winding them gently onto a glass rod as the alcohol is added.
6. Dissolve the DNA in 0.5 ml sterile saline-citrate solution.
7. For a sterility check, inoculate nutrient broth (labeled "DNA control") with a loopful of the DNA suspension. How will this determine sterility? ______
8. Add the DNA suspension to a tube of nutrient broth and inoculate with a loopful of *Escherichia coli.* Mix gently.
9. Inoculate a tube of nutrient broth with *E. coli* (label "*E. coli* control").
10. Incubate all tubes for 48 hours at 35°C.

Second Period

1. Obtain two plates of citrate agar.
2. Divide one citrate agar plate in half. Streak one half with a loopful of the DNA control and the other with *E. coli* control. Does *E. coli* wild type grow on citrate agar? ______
3. Divide another citrate agar plate into four quadrants and inoculate each quadrant with a loopful from the DNA/*E. coli* tube.

4. Incubate the plates inverted at 35°C and observe after 24 to 48 hours. Record your results. Gram stain any colonies to ensure that they are *E. coli.* How is *E. coli* differentiated from *B. megaterium* in a Gram stain? ______________

Turn to the Laboratory Report for Exercise 27.

EXERCISE 28

Conjugation

Where all men think alike,
no one thinks very much.

WALTER LIPPMAN

Objectives

After completing this exercise you should be able to

1. Define the following terms: F plasmid, Hfr, and F^-.
2. Perform a conjugation experiment.
3. Construct a gene map.

Background

Conjugation is the transfer of genetic information (DNA) from a live, donor bacterium to a recipient cell. The donor bacterium is designated $\mathbf{F^+}$ because it contains a **fertility** or **F plasmid.** The recipient or $\mathbf{F^-}$ cell lacks an F plasmid. A **plasmid** is an extrachromosomal, circular piece of DNA.

F^+ cells contain a gene that codes for the synthesis of *sex pili.* During conjugation between an F^+ and F^-, the F plasmid is replicated and one copy is transferred to the F^- cell through a pilus bridge between the cells. The F^- cell is converted to an F^+.

The F plasmid can become integrated into the chromosomal DNA. When the F plasmid is integrated into the chromosome, the cell is called a **high frequency recombinant (Hfr).** Recombination can result from conjugation between an Hfr and F^-. The chromosome of the Hfr and its integrated plasmid replicate, and one copy of the entire chromosome can be transferred to an F^- cell.

In this exercise, Hfr and F^- cells with different genotypes will be mixed together. Recombinants will be F^- cells with some phenotypic traits of the Hfr.

Chromosome maps showing the locations of specific genes can be constructed from conjugation experiments. If conjugation is interrupted at intervals, the order in which genes were transferred can be determined. The chromosome map begins at zero minutes and ends at 90 minutes or the length of time for transfer of the entire chromosome.

The Hfr culture used in this exercise is capable of fermenting lactose and synthesizing the amino acids threonine and leucine. Additionally, it cannot grow in the presence of the antibiotic streptomycin; that is, it is sensitive to streptomycin. Its genotype is designated lac^+, thr^+, leu^+, str^s. The F^- culture cannot ferment lactose, cannot synthesize threonine and leucine, and is resistant to streptomycin. The genotype of the F^- is lac^-, thr^-, leu^-, str^r.

As a preliminary exercise for your experiment, fill in the following table with the *expected* results. Indicate whether growth should occur on each medium (+) or there should be no growth (−).

Medium: Minimal salts (see Table 26-1) plus:	Growth?	
	Hfr	F^-
Glucose + streptomycin + threonine		
Glucose + streptomycin + leucine		
Lactose + streptomycin + threonine + leucine		

Materials

Petri plate containing glucose-minimal salts agar + streptomycin and threonine (1)

Petri plate containing glucose-minimal salts agar + streptomycin and leucine (1)

Petri plate containing lactose-minimal salts agar + streptomycin, threonine, leucine (1)

Sterile cotton swabs (3)

Sterile 1-ml pipettes (2)

Small sterile test tube (1)

Cultures (one of each)

Escherichia coli Hfr (*lac*$^+$ *thr*$^+$ *leu*$^+$ *str*^{s})

Escherichia coli F$^-$ (*lac*$^-$ *thr*$^-$ *leu*$^-$ *str*^{r})

Techniques Required

Exercises 13 and 16 and Appendix A

Procedure*

1. Divide each minimal agar plate into thirds. Which medium is testing a catabolic gene? ______________________
2. Aseptically place 0.5 ml Hfr culture into the empty test tube and swab one of the sectors on each of the Petri plates with the Hfr culture. Label "Hfr."
3. Add 0.5 ml F$^-$ bacteria to the test tube and swab one of the sectors on each of the Petri plates with the F$^-$ bacteria. Label "F$^-$." Gently shake the Hfr/F$^-$ mixture to mix.
4. Leave the Hfr/F$^-$ mixture at room temperature for 10 to 45 minutes as assigned.
5. Swab the Hfr/F$^-$ mixture on the remaining sectors on each plate and label "Hfr/F$^-$."
6. Incubate all plates at 35°C until the next period.
7. Examine for growth and record your results. Qualitatively measure growth and record amounts as follows: (−) = no growth, (+) = minimal growth, (2+) = moderate growth, (3+) = heavy growth, and (4+) = maximum growth.

Turn to the Laboratory Report for Exercise 28.

EXERCISE 29

Transduction

Objectives

After completing this exercise you should be able to

1. Explain the mechanism of transduction.
2. Differentiate between generalized, specialized, and abortive transduction.
3. Demonstrate transduction experimentally.

Background

Genetic change due to a filterable agent was first observed in bacteria by Joshua Lederberg in 1952. The filterable agent was bacteriophage P22 and the bacterium was *Salmonella typhimurium*. Recombination that is mediated by a bacteriophage is called **transduction.**

One type of transduction is called **generalized transduction.** In the process of infection (Exercise 35), fragments of the bacterial chromosome may occasionally be accidentally incorporated into the virus. The virus particle then carries bacterial DNA in addition to parts of the viral DNA. If the virus particle infects a new bacterial host, the new host could receive several bacterial genes. Transduction involving the bacterial DNA can lead to recombination (see the figure on page 73) between this DNA and the DNA of the new host cell.

In generalized transduction, all genes are equally likely to be picked up in a phage coat and transferred. In **specialized transduction,** only certain bacterial genes can be transferred. Specialized transduction is mediated by a lysogenic phage, and only genes near

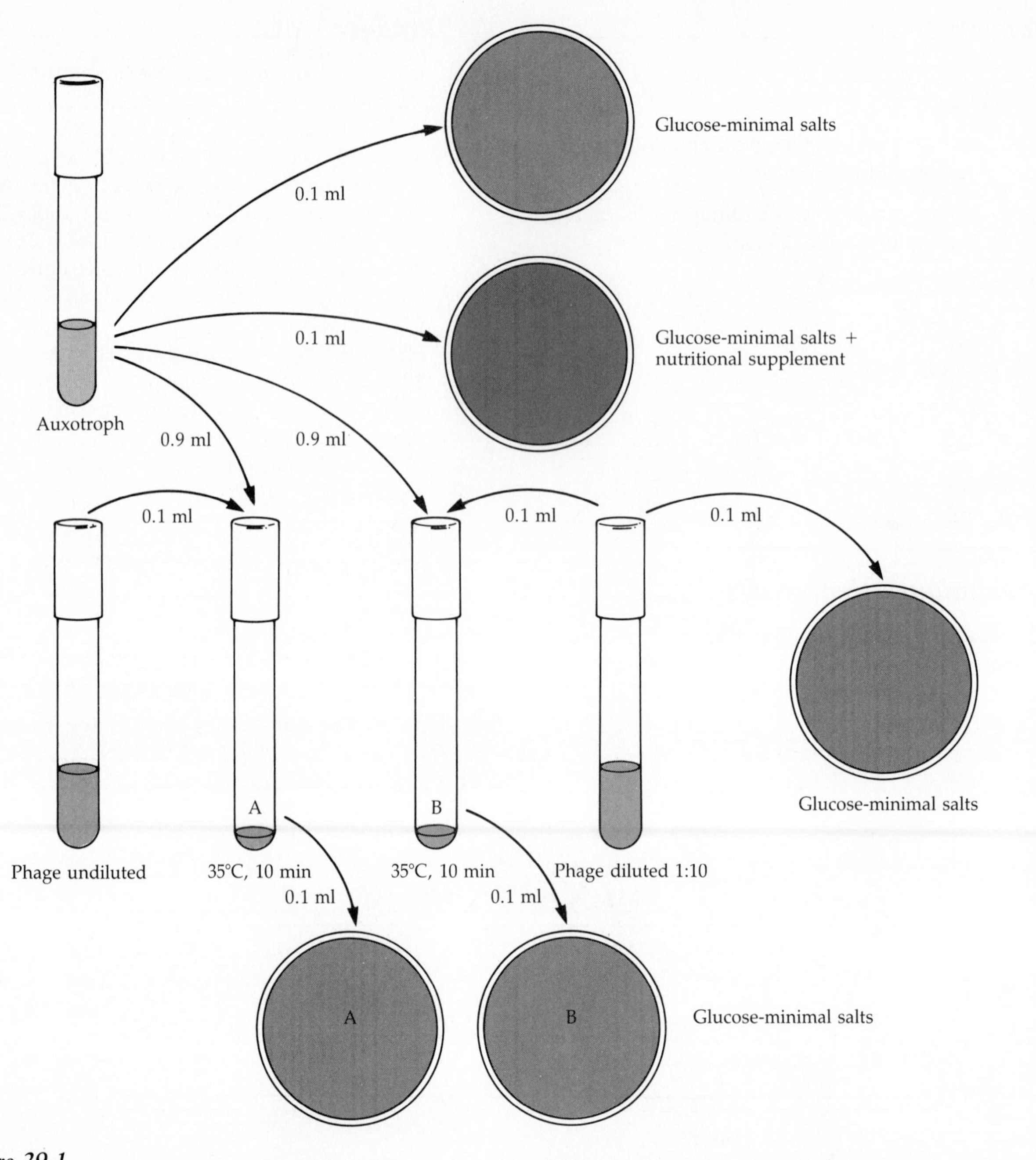

Figure 29-1

Procedure used to demonstrate transduction.

the prophage can be transferred. In **abortive transduction** the newly transferred fragment of DNA does not become integrated into the bacterial chromosome, but functions independently. When the cell divides, the DNA fragment is passed to only one offspring. Cells carrying genes acquired by abortive transduction form minute (small) colonies.

In this experiment, you will be provided with a phage grown on an *E. coli* prototroph *(wild type)* and an auxotrophic strain of *E. coli.* The auxotroph is a nutritional mutant that is incapable of synthesizing an amino acid. The genotype is noted by the name of the amino acid and a minus; for example, *E. coli* his^- cannot synthesize histidine. This auxotroph will grow on glucose-minimal salts plus histidine but will not grow on glucose-minimal salts (see Table 26-1).

Materials

Petri plates containing glucose-minimal salts agar (5)

Petri plate containing glucose-minimal salts agar + nutritional supplement (1)

Sterile 1-ml pipettes (4)

Sterile serological tubes (2)

Glass spreader ("hockey stick")

Alcohol

35°C water bath

Cultures (one of each)

Bacteriophage P1 grown on *E. coli* prototroph

Bacteriophage P1, diluted 1:10

E. coli auxotroph

E. coli K12 prototroph

Techniques Required

Exercises 13 and Appendix A. Reading Exercise 35 will be helpful.

Procedure

1. Label one glucose-minimal salts plate "Prototroph" and aseptically inoculate it with 0.1 ml prototroph culture. Disinfect a hockey stick by dipping it into alcohol and burning off the alcohol. Spread the 0.1 ml inoculum over the surface of the agar (see Figure 26-2).
2. Label two tubes: "A" and "B." Aseptically pipette 0.9 ml of the auxotroph into each tube. Using the same pipette, place 0.1 ml auxotroph on glucose-minimal salts agar and 0.1 ml on glucose-minimal salts + nutritional supplement. Spread the inoculum on each plate using a hockey stick, as described in step 1 and label each plate.
3. Aseptically pipette 0.1 ml of the 1:10 phage culture into tube "B," 0.1 ml onto another glucose-minimal salts agar, and 0.1 of the undiluted phage culture into tube "A." See Figure 29-1. Why can the same pipette be used for all three inoculations? ______________________
   Spread the inoculum on the agar as before and label the plate. Incubate the tubes at 35°C for 10 minutes.
4. Label two glucose-minimal salts plates "A" and "B" and inoculate plate "B" with 0.1 ml from tube "B." Spread with a disinfected hockey stick. Repeat with plate "A" and tube "A."
5. Incubate all plates at 35°C until the next lab. period and record your results.

Turn to the Laboratory Report for Exercise 29.

EXERCISE 30

Ames Test for Detecting Possible Chemical Carcinogens

The requirements of health can be stated simply. Those fortunate enough to be born free of significant congenital disease or disability will remain well if three basic needs are met: they must be adequately fed; they must be protected from a wide range of hazards in the environment; and they must not depart radically from the pattern of personal behavior under which man evolved, for example, by smoking, overeating, or sedentary living.

THOMAS McKEOWN

Objectives

After completing this exercise you should be able to

1. Differentiate between the following terms: mutagenic and carcinogenic.
2. Provide the rationale for the Ames Test.
3. Perform the Ames Test.

Background

Every day we are exposed to a variety of chemicals, some of which are **carcinogens;** that is, they can induce cancer. Historically, animal models have been used to evaluate the carcinogenic potential of a chemical, but the procedures are very costly and time-consuming, and result in the inadequate testing of some chemicals. Many chemical carcinogens induce cancer because they are **mutagens** that alter the nucleotide base sequence of DNA.

Bruce Ames and his co-workers at the University of California at Berkeley have developed a fast, inexpensive assay for mutagenesis using a *Salmonella* auxotroph. Most chemicals that have been shown to cause cancer in animals have proven mutagenic in the **Ames Test.** Since not all mutagens induce cancer, the Ames Test is a screening technique that can be used to identify high-risk compounds that must then be tested for carcinogenic potential. The Ames Test utilizes auxotrophic strains of *Salmonella typhimurium,* which cannot synthesize the amino acid histidine (*his*$^-$). The strains are also defective in *dark excision* repair of mutations ***(uvrB),*** and an ***rfa*** mutation eliminates a portion of the lipopolysaccharide capsule that coats the bacterial surface. The *rfa* mutation prevents the *Salmonella* from growing in the presence of sodium desoxycholate or crystal violet and increases the cell wall permeability so more mutagens enter the cell. The *uvrB* mutation minimizes repair of mutations and as a result the bacteria are much more sensitive to mutations. To grow the auxotroph, histidine and biotin (due to the *uvrB*) must be added to the culture media.

In the Ames Test, a small sample of the test chemical (a suspected mutagen) is either added to a melted soft agar (containing histidine) suspension of the *Salmonella* that is overlayed on minimal (lacking histidine) media or is placed in the center of minimal agar seeded with a lawn of the *Salmonella* auxotroph. (A small amount of histidine allows all the cells to go through a few divisions.) If certain mutations occur, the bacteria may *revert* to a wild type or prototroph and grow to form a colony. In theory, the number of colonies is proportional to the mutagenicity of the chemical.

In nature many chemicals are neither carcinogenic nor mutagenic, but they are metabolically converted to mutagens by liver enzymes. The original Ames Test could not detect the mutagenic potential of these in vivo (in the body) conversions. Suspect chemicals can be incubated with a suspension of rat liver enzyme preparation in an oxygenated environment to be "activated." *Salmonella* is exposed to the "activated" chemicals to test for mutagenicity (Figure 30-1). In this laboratory exercise the chemicals to be evaluated are mutagenic and will not require activation.

Be very careful doing this exercise. The chemicals are toxic and potentially dangerous. *Salmonella typhimurium* is a potential pathogen.

Materials

Petri plates containing glucose-minimal salts agar (see Table 26-1) (2)

Tubes containing 2 ml soft agar (glucose-minimal salts + 0.05 m*M* histidine and 0.05 m*M* biotin) (2)

Sterile filter paper disks

Sterile 1-ml pipette

Forceps and alcohol

Suspected mutagens:
- benzo (α) pyrene
- nitrous acid
- 2-aminofluorene
- cigarette smoke condensates
- saccharin
- household compound (your choice)

Culture

Salmonella typhimurium his$^-$, *uvrB, rfa*

Techniques Required

Exercises 13 and 26

Procedure*

1. Label one minimal-salts agar plate "Control" and the other "Test."
2. Aseptically pipette 0.1 ml *Salmonella* to one of the soft agar tubes, and quickly pour over the surface of the control plate. Tilt the plate back and forth gently to spread the agar evenly. Let harden in the dark. Why the dark? ____________
3. Obtain another soft agar tube and repeat step 2 for the test plate. Why does the soft agar contain histidine? ____________

____

*Adapted from Deborah J. Pringnitz. 1975. "The Effects of Ascorbic Acid on the Ability of Dimethylnitrosamine to Cause Reversion in a *Salmonella typhimurium* Mutant." Master's thesis, Mankato State University, Mankato, Minnesota.

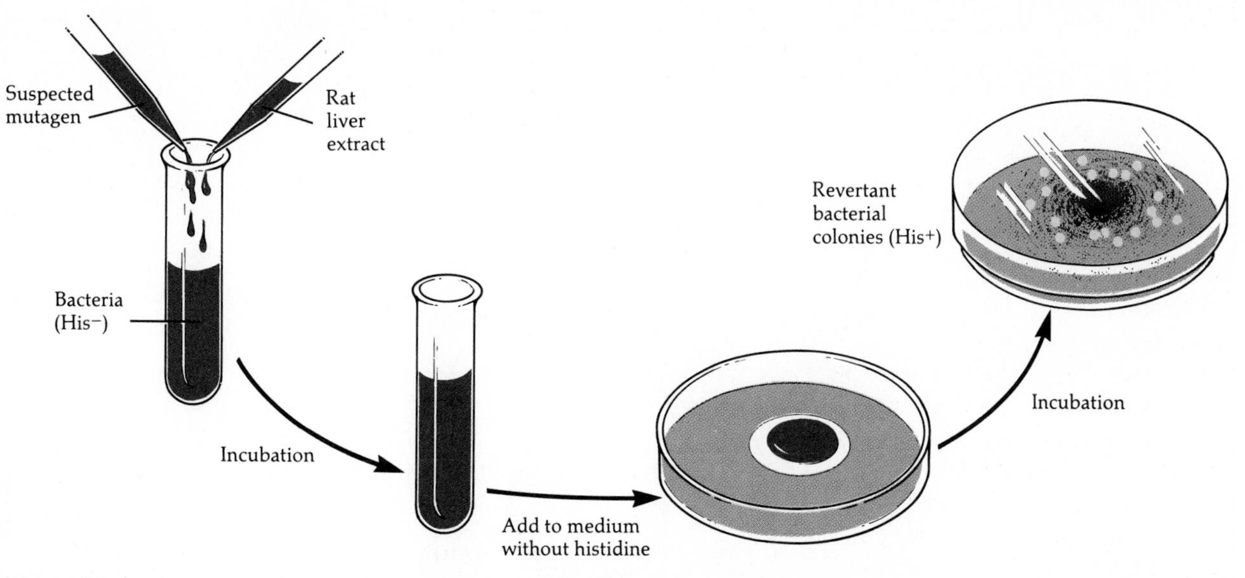

Figure 30-1

Ames Test. Mutant *Salmonella* unable to synthesize histidine (*his*$^-$) are mixed with a suspected mutagen and rat liver extract and inoculated onto a medium lacking histidine. Only bacteria that have further mutated (reverted) to *his*$^+$ (able to synthesize histidine) will grow into colonies. The liver extract "activates" the suspected mutagen.

4. Dip forceps in alcohol and, with the tip pointed down, burn off the alcohol. Aseptically place 3 to 5 sterile white disks on the surface of the test plate. Label the bottom of the Petri plate with the chemical to be applied to each disk.
   a. Add a few drops of the mutagen to the corresponding disk. *Be very careful.* These are potentially dangerous compounds.
   b. If the chemical is crystalline, place a few crystals directly on the soft agar overlay.
5. Incubate both plates right side up at 35°C for 48 hours.
6. Describe your results.

Turn to the Laboratory Report for Exercise 30.

PART 6

Fungi, Protozoans, and Algae

EXERCISES

In Part 6, we examine some representative free-living **eucaryotic organisms.** Organisms with eucaryotic cells include algae, protozoans, fungi, and higher plants and animals. The eucaryotic cell is typically larger and structurally more complex than the procaryotic cell. The DNA of a eucaryotic cell is enclosed within a membrane-bounded **nucleus.** In addition, eucaryotic cells contain membrane-bounded **organelles,** which have specialized structures and perform specific functions (see the figure).

The eucaryotic microorganisms studied in this Part are all free-living chemoheterotrophs (Exercises 31, 32, and 34) except for the algae (Exercise 33), which are photoautotrophs. Disease-causing fungi and protozoans will be studied in later exercises (Exercises 65 and 66).

Yeasts are possibly the best known microorganisms. They are widely used in commercial processes and can be purchased in the supermarket for baking. Yeasts are unicellular fungi (Exercise 31).

Leeuwenhoek was the first to observe the yeast responsible for fermentation in beer:

> *I have made divers observations of the yeast from which beer is made and I have generally seen that it is composed of globules floating in a clear medium (which I judged to be the beer itself).*

Many algae are only visible through the microscope. Algae are important producers of oxygen and food for protozoans and other organisms. A few unicellular algae, such as the agents of "red tides," *Gonyaulax catanella* and related species, are toxic to animals, including humans, when ingested in large numbers.

Protozoans, originally called "infusoria," were of interest to early investigators. In 1778, von Gleichen studied food vacuoles by feeding red dye to his infusoria. A refinement of this experiment will be done in Exercise 34.

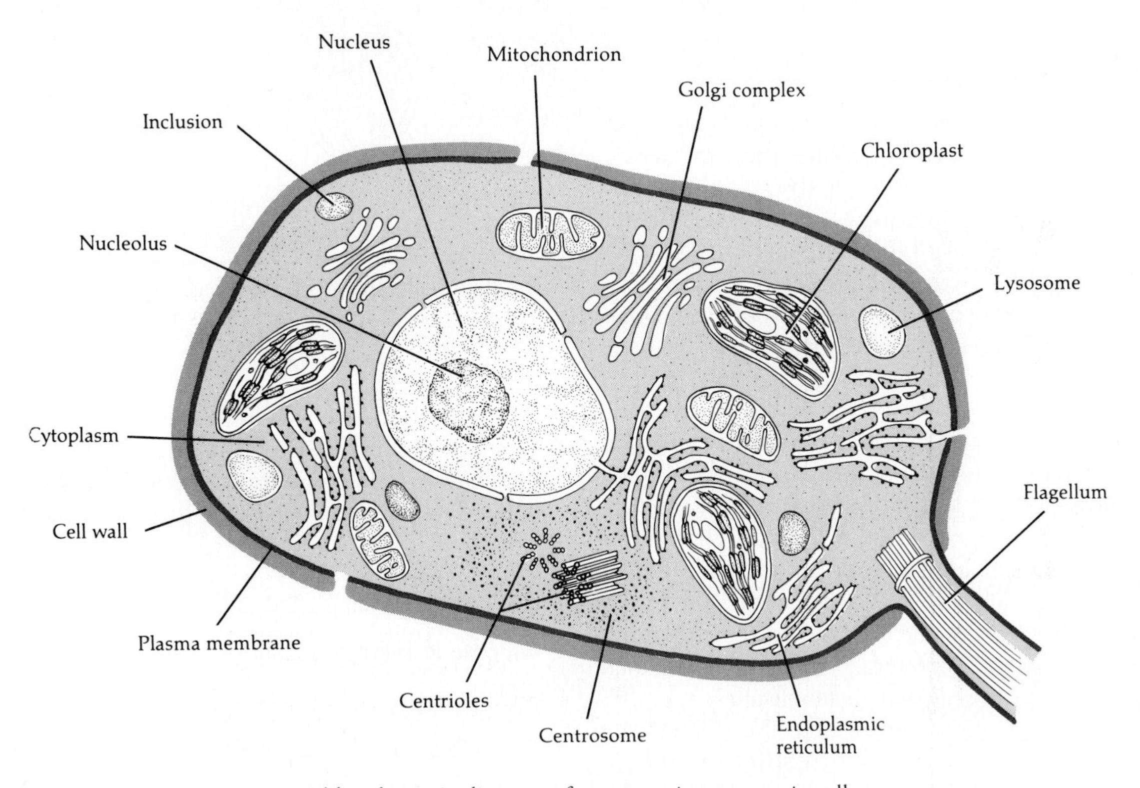

Highly schematic diagram of a composite eucaryotic cell.

EXERCISE 31

Fungi: Yeasts

Objectives

After completing this exercise you should be able to

1. Culture and identify yeasts.
2. Differentiate between yeasts and bacteria.

Background

Fungi possess eucaryotic cells and can exist as unicellular or multicellular organisms. They are heterotrophic and obtain nutrients by absorbing dissolved organic material through their cell walls and plasma membranes. Fungi (with the exception of yeasts) are aerobic. Unicellular yeasts, multicellular molds (Exercise 32), and macroscopic species such as mushrooms are included in the kingdom Fungi. Many of the techniques useful in working with bacteria can be applied to fungi. Fungi generally prefer more acidic conditions and tolerate higher osmotic pressure and lower moisture than bacteria. They are larger than bacteria, with more cellular and morphologic detail. In contrast to bacterial characterization, primary characteristics such as morphology and cellular detail are used to classify fungi, with little attention given to secondary characteristics such as metabolism and antigenic composition. Fungi are structurally more complex than bacteria but are less diverse metabolically.

Yeasts are nonfilamentous, unicellular fungi that are typically spherical or oval in shape. Yeasts are widely distributed in nature, frequently found on fruits and leaves as a white, powdery coating. Yeasts reproduce asexually by budding, a process in which a new cell forms as a protuberance (bud) from the parent cell (Figure 31-1*a*). In some instances, when buds fail to detach themselves, a short chain of cells called a **pseudohypha** forms (Figure 31-1*b*). When yeasts reproduce sexually, they may produce one of several types of sexual spores. The type of sexual spore produced by a species of yeast is used to classify the yeast as to phylum. (Phyla of fungi are described in Exercise 32.)

Yeasts are facultative anaerobes (Exercise 40). Their metabolic activities are used in many industrial fermentation processes. Yeasts are used to prepare many foods (such as bread) and beverages such as wine and beer (refer to Exercise 55). Metabolic activities are also used to identify yeasts.

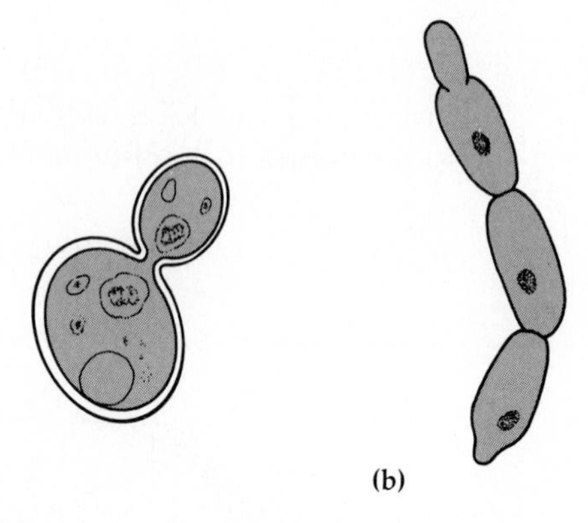

Figure 31-1.
Budding yeasts. **(a)** A bud forming from a parent cell. **(b)** Pseudohyphae are short chains of cells formed by some yeasts.

In the laboratory, **Sabouraud agar,** a selective medium, is commonly used to isolate yeast. Sabouraud agar has very simple nutrients (glucose and peptone) and a low pH that inhibits the growth of most other organisms.

Materials

Glucose fermentation tubes (2)

Sucrose fermentation tubes (2)

Petri plates containing Sabouraud agar (2)

Bottle containing glucose–yeast extract broth (1)

Sterile cotton swab (1)

Cover slip

Test tube (1)

Methylene blue

Balloon

Fruit or leaves

Cultures (as assigned)

Baker's yeast

Rhodotorula rubra

Candida albicans

Saccharomyces cerevisiae

Techniques Required

Exercises 2, 14, 15, and 18

Procedure

Yeasts

1. Gently suspend a pinch of baker's yeast in a small amount of lukewarm water in a test tube to give a milky solution.
2. Each pair of students will use one of the yeast cultures and the suspension of baker's yeast.
   a. Divide one Sabouraud agar plate in half and streak one-half with a known yeast culture and the other half with the baker's yeast suspension.
   b. Inoculate each organism into a glucose fermentation tube and a sucrose fermentation tube.
3. Incubate all media at 35°C until growth is seen.
4. Make a wet mount (Exercise 2) of each culture by using a small drop of methylene blue. Record your observations.
5. After the yeast have grown, smell the plates. Record your results. Examine cultures of the yeasts you did not culture and record pertinent results.

Yeast Isolation

1. Cut the fruit or leaves into small pieces. Place them in the bottle of glucose–yeast extract broth. Cover the mouth of the bottle with a balloon. Incubate the bottle at room temperature until growth has occurred. Record the appearance of the broth after incubation. Was gas produced? ___
2. Divide a Sabouraud agar plate in half. Each partner inoculates half of the medium with either *a* or *b,* as follows.
   a. Swab the surface of your tongue with a sterile swab. Inoculate one-half of the agar surface with the swab. Why will few bacteria grow on this medium? ___
   b. Using a sterile inoculating loop, streak one-half of the agar surface with a loopful of broth from the bottle just prepared in step 1.
3. Incubate the plate inverted at room temperature until growth has occurred. Smell the plate. Prepare wet mounts with methylene blue from different appearing colonies. Record your results.

Turn to the Laboratory Report for Exercise 31.

EXERCISE 32

Fungi: Molds

And what is weed? A plant whose virtues have not been discovered.

RALPH WALDO EMERSON

Objectives

After completing this exercise you should be able to

1. Characterize and classify fungi.
2. Compare and contrast fungi and bacteria.
3. Identify common saprophytic molds.
4. Cultivate mushrooms.

Background

The multicellular filamentous fungi are called **molds.** There is wide diversity in mold morphology—therefore, morphology is very useful in classification of these fungi.

The macroscopic mold colony is called a **thallus** and is composed of a mass of strands called **mycelia.** Each strand is a **hypha,** with the **vegetative hyphae** growing in or on the surface of the growth medium. Aerial hyphae, called **reproductive hyphae,** originate from the vegetative hyphae and produce a variety of asexual reproductive **spores** (Figure 32-1). The hyphal strand of most molds is composed of individual cells separated by a crosswall or **septum.** These are called **septate hyphae.** A few fungi have hyphae that lack septa and are a continuous mass of cytoplasm with multiple nuclei. These are called **coenocytic hyphae.**

Fungi are characterized and classified by the appearance of their colony (color, size, and so on), hyphal organization (septate or coenocytic), and the structure and organization of reproductive spores. Because of the importance of colony appearance and organization, culture techniques and microscopic examination of fungi are very important.

Members of the phylum **Oomycota** are saprophytes or plant parasites. A common example is *Saprolegnia.* It is found growing on organic matter in aquaria. The Oomycota have coenocytic hyphae and are similar to the Zygomycota (below) except that their sporangiospores are flagellated. Of what advantage would this be? ______________________________

____

Members of the phylum **Zygomycota** are saprophytic molds that have coenocytic hyphae. A common example is *Rhizopus* (bread mold). The asexual spores are formed inside a **sporangium** or spore sac and are called **sporangiospores** (Figure 32-1*a*). Sexual spores called **zygospores** are formed by the fusion of two cells.

The **Ascomycota** include molds with septate hyphae and some yeasts. They are called sac fungi because their sexual spores, called **ascospores,** are produced in a sac or **ascus.** The saprophytic molds usually produce **conidiospores** asexually (Figure 32-1*b*). The arrangements of the conidiospores are used to identify these fungi. Common examples are *Penicillium* and *Aspergillus.*

The **Basidiomycota** also possess septate hyphae. The sexual spores, called **basiodiospores,** are produced by a club-shaped structure called a **basidium.** In mushrooms, the basidia are found along the gills or pores on the underside of the cap.

A classification scheme for saprophytic fungi is shown in Table 32-1.

Fungi, especially molds, are important clinically (Exercise 65) and industrially. Spores in the air are the most common source of contamination in the laboratory. Sabouraud agar is a selective medium that is commonly used to isolate fungi (see Exercise 31).

In this laboratory exercise we will examine different molds and attempt to culture mushrooms.

MOLDS

Materials

Melted Sabouraud agar

Petri plates containing Sabouraud agar (2)

Vaspar (one-half petroleum jelly and one-half paraffin)

Transparent tape

Petri plate

Cover slips

Bent glass rod

Pasteur pipette

Cultures (as assigned)

Rhizopus stolonifer

Aspergillis niger

Penicillium notatum

Prepared slides of the following:
- Zygospores
- Ascospores
- Basidiospores

Techniques Required

Exercises 2, 13, and 31, and Appendix E

Table 32-1
Characteristics of Common Saprophytic Fungi

Phylum	Growth Characteristics	Asexual Reproduction	Sexual Reproduction
Oomycota "Water molds"	Coenocytic hyphae	Motile sporangiospores called zoospores	Zygospores
Zygomycota "Conjugation fungi"	Coenocytic hyphae	Sporangiospores	Zygospores
Ascomycota "Sac fungi"	Septate hyphae Yeastlike	Conidiospores Budding	Ascospores
Basidiomycota "Club fungi"	Septate hyphae includes fleshy fungi (mushrooms)	Fragmentation	Basidiospores

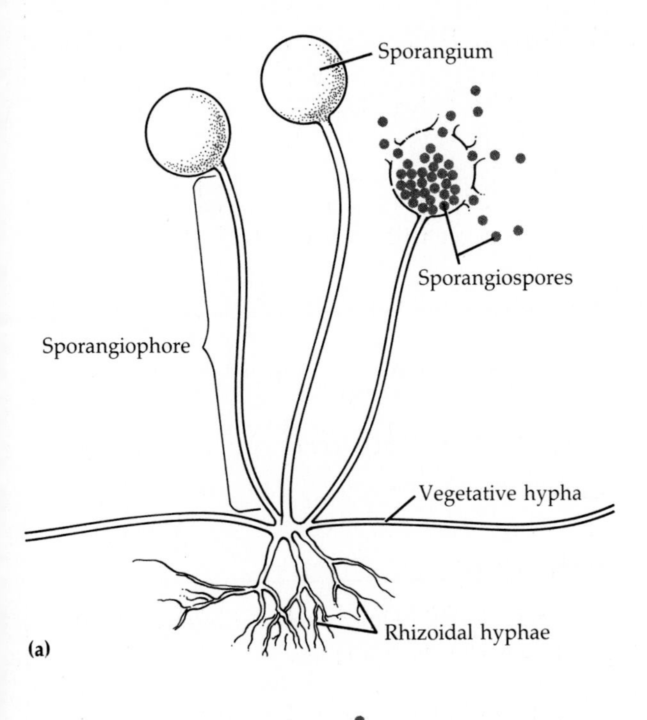

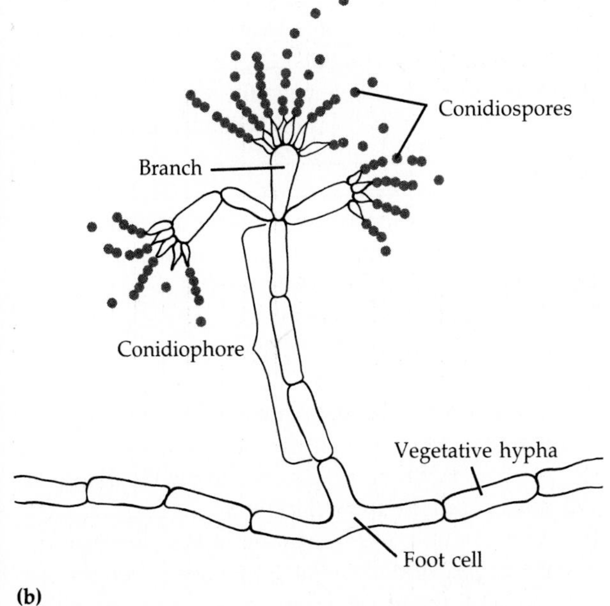

Figure 32-1.
Asexual spores are produced by aerial hyphae.
(a) Sporangiospores are formed within a sporangium.
(b) Conidiospores are formed in chains. One possible arrangement is shown here.

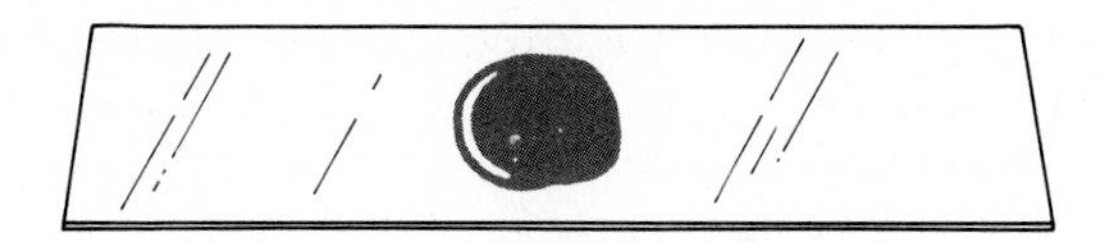

(a) Place agar medium on the slide.

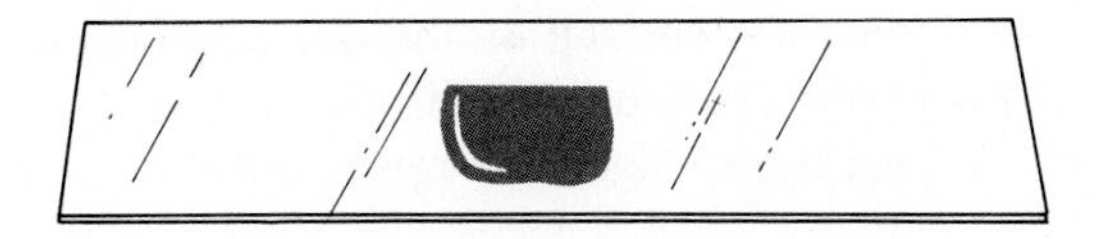

(b) Cut a straight edge on one side of the solidified agar.

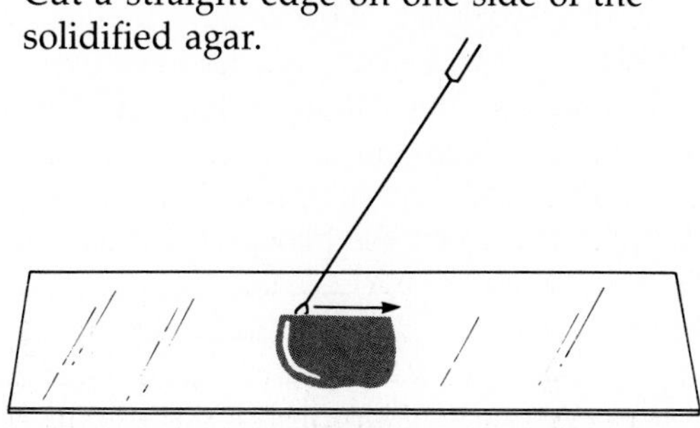

(c) Inoculate the mold onto the straight edge.

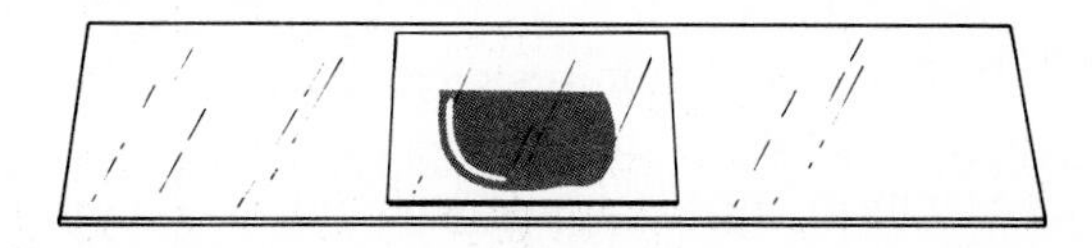

(d) Place a cover slip on the agar.

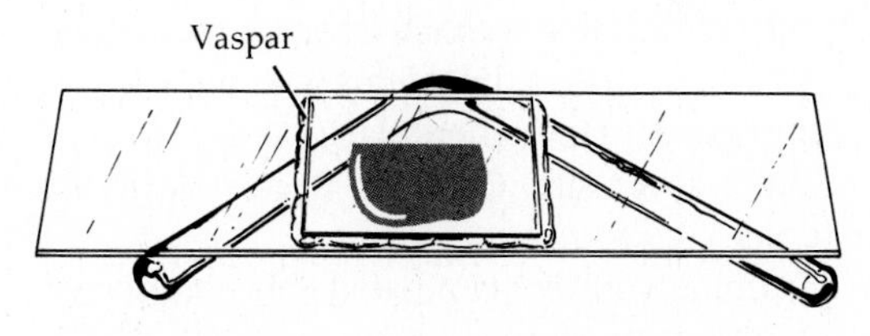

(e) Seal the three uninoculated edges with vaspar and place the slide on a glass rod in a Petri dish.

Figure 32-2.
Preparation of a slide culture.

Procedure

First Period

1. Contaminate one Sabouraud agar plate in any manner you desire. Expose it to the air (outside, hall, lab, or wherever) for 15 to 30 minutes, or touch it. Incubate the plate inverted at room temperature for 5 to 7 days.
2. Inoculate the other Sabouraud agar plate with the mold culture assigned to you. Make one line in the center of the plate. Why don't you streak it?

   ____

   Incubate the plate inverted at room temperature for 5 to 7 days.
3. Set up a slide culture of your assigned mold (Figure 32-2).
   a. Clean a slide and dry it.
   b. Place a drop of Sabouraud agar on the slide,

flatten it to the size of a dime, and let it solidify.

c. Using a sterile, hot loop, carefully scrape off half of the circle of agar, leaving a smooth edge.

d. Gently shake the mold culture to resuspend it and inoculate the straight edge of the agar with a loopful of mold.

e. Place a cover slip on the inoculated agar.

f. Using a Pasteur pipette, add melted vaspar to seal three edges. Do not seal the inoculated edge.

g. Put a piece of wet paper towel in the bottom of a Petri plate. Why? ______________

____

Place the slide culture on the glass rod on top of the towel and close the Petri plate. Incubate it at room temperature for 2 to 5 days. Do *not* invert the plate. What is the purpose of the glass rod? ______________

____

____

4. Observe the prepared slides showing sexual spores. Carefully diagram each of the spore formations in the space provided in the Laboratory Report.

Second Period

1. Examine plate cultures of *each* mold (without a microscope), and describe their color and appearance. Then examine with a dissecting microscope (Appendix E). Look at the top and the underside.
2. Prepare a wet mount of *each* mold by the following method. Put a dilute drop of methylene blue on a slide. Touch the thallus with the sticky side of transparent tape. Then place on the drop of stain like a cover slip. Observe. Record your observations.
3. Examine your contaminated Sabouraud plate and describe the results.
4. Examine slide cultures of each mold using a dissecting microscope. Record your observations.

MUSHROOM GROWING*

Materials

Vermiculite

Growth medium

1-liter beaker

1-liter graduated cylinder

Brown paper

Plastic wrap

Rubber band

Needed for later: fine-textured soil

Culture

Agaricus campestris (table mushroom) spawn

Techniques Required

Exercise 13

Procedure

1. Place 600 ml vermiculite in a beaker and add 450 ml growth medium.
2. Place a paper towel over the beaker, then a square of brown paper. Secure with a rubber band.
3. Sterilize as instructed. How should this culture medium be sterilized? ______________
4. Allow to cool. The vermiculite will have expanded to fill most of the beaker. What caused the expansion? ______________

   ____
5. Aseptically inoculate with a bit of the spawn about 2.5 cm below the vermiculite surface.
6. Cover *loosely* with plastic wrap and place in the 18° to 20°C incubator.
7. After 2 weeks, white mycelia should be present. At this time, cover the vermiculite with about 2.5 cm of fine-textured soil. Sprinkle with water so that it is moist. Keep the soil moist by adding water every 2 to 3 days. Keep covered in the incubator. Do mushrooms require darkness? ______________
8. After about 1 week, fruiting stages develop: first the aggregated mycelial "pinheads," then the button, and finally, the spore stage. They *are* edible!

*Adapted from D. J. Holden and S. J. Wallner. 1971. "Experimental Mushroom Growing." *American Biology Teacher 33: 91–93.*

Turn to the Laboratory Report for Exercise 32.

EXERCISE 33

Algae

Objectives

After completing this exercise you should be able to

1. List criteria used to classify algae.
2. List general requirements for algal growth.

Background

Algae are photosynthetic eucaryotes. Algae exhibit a wide range of shapes: from the giant brown algae or kelp and delicate marine red algae to unicellular green algae. Algae are classified according to pigments, storage products, chemical composition of their cell walls, and flagella. The diversity of algae is illustrated by their classification: Algal groups are found in three kingdoms.

Most freshwater algae belong to the groups listed in Table 33-1. Of primary interest to microbiologists are the blue-green algae or **cyanobacteria.** Cyanobacteria have procaryotic cells and belong to the kingdom Monera along with bacteria.

While algal growth is essential in providing oxygen and food for other organisms, some filamentous algae, such as *Spirogyra,* are a nuisance to humans because they clog filters in water systems. Algae can be used to determine the quality of water. Polluted waters containing excessive nutrients from sewage or other sources have more cyanobacteria and fewer diatoms than clean waters. Additionally, the *number* of algal cells indicates water quality. More than 1000 algal cells per milliliter indicates that excessive nutrients are present.

Materials

Pond water samples. (Distilled water should be added to the pond water periodically during incubation to replace water lost by evaporation.)

A. Incubated in the light for 4 weeks.
B. Incubated in the dark for 4 weeks.
C. Nitrates and phosphates added; incubated in the light for 4 weeks.

Table 33-1
Some Characteristics of Major Groups of Algae Found in Fresh Water

Characteristics	Algae classified as			
	Monera	Algal Protists		Plants
	Cyanobacteria	Euglenoids	Diatoms	Green algae
Color	Blue-green	Green	Yellow-brown	Green
Cell wall	Bacteria-like	Lacking	Readily visible with regular markings	Visible
Cell type	Procaryote	Eucaryote	Eucaryote	Eucaryote
Flagella	Absent	Present	Absent	Present in some
Cell arrangement	Unicellular or filamentous	Unicellular	Unicellular or colonial	Unicellular, colonial, or filamentous
Nutrition	Autotrophic	Facultatively heterotrophic	Autotrophic	Autotrophic
Produce O_2	Yes; some use bacterial photosynthesis	Yes	Yes	Yes

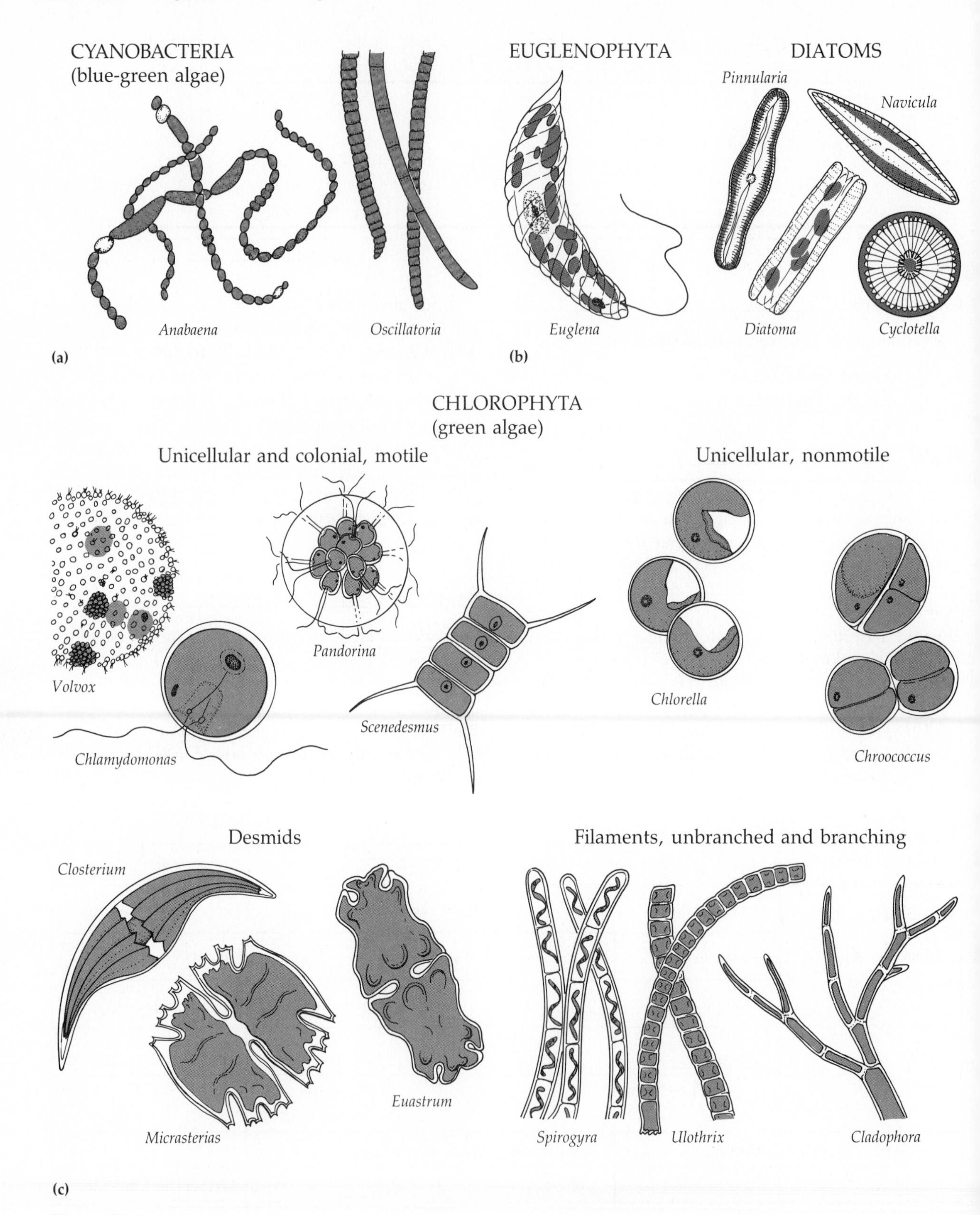

Figure 33-1.

Representative freshwater algae. **(a)** Representatives of the kingdom Monera (Procaryota) found in fresh water. **(b)** Some members of the kingdom Protista found in fresh water. **(c)** Examples of the kingdom Plantae found in fresh water.

D. Copper sulfate added; incubated in the light for 4 weeks.

Techniques Required

Exercise 2

Procedure

1. Prepare a hanging-drop slide from a sample of pond water **A.** Take your drop from the bottom of the container. Why? ______________________
____
2. Examine the slide using the low and high-dry objectives. Identify the algae present in the pond water. Use Figure 33-1 for identification. Draw those algae that you cannot identify. Record the relative amounts of each algae from 4+ (most abundant) to + (one representative seen).
3. Repeat the observation and data collection for the remaining pond water samples.

Turn to the Laboratory Report for Exercise 33.

EXERCISE 34

Protozoans

Objectives

After completing this exercise you should be able to

1. List three characteristics of protozoans.
2. Explain how protozoans are classified.

Background

Protozoans are unicellular eucaryotic organisms that belong to the kingdom Protista. They are found in areas with a large water supply. Many protozoans live in soil and water and some are normal flora in animals. A few species of protozoans are parasites. These will be examined in Exercise 66.

Protozoans are aerobic heterotrophs. They feed on other microorganisms and on small particulate matter. Protozoans lack cell walls; in some (flagellates and ciliates), the outer covering is a thick, elastic membrane called a **pellicle.** Cells with a pellicle require specialized structures to take in food. The **contractile vacuole** may be visible in some specimens. This organelle fills with fresh water and then contracts to eliminate excess fresh water from the cell, allowing the organism to live in low solute environments. Would you expect more contractile vacuoles in freshwater or marine protozoans? ______________

Protozoans may be classified on the basis of their method of motility. In this exercise, we will examine free-living members of three phyla of protists. The **Sarcodina,** or **amoebas** (Figure 34-1*a*), move by extending lobelike projections of cytoplasm called **pseudopods.** As pseudopods flow from one end of the cell, the rest of the cell flows toward the pseudopods.

The **Mastigophora,** or **flagellates** (Figure 34-1*b*), have one or more flagella. Although flagellates are heterotrophs, the organism used in this exercise is a facultative heterotroph. It grows photosynthetically in the presence of light and heterotrophically in the dark.

The **Ciliata** (Figure 34-1*c*) have many cilia extending from the cell. In some ciliates, the cilia occur in rows over the entire surface of the cell. In ciliates that live attached to solid surfaces, the cilia occur only around the oral groove. Why only around the oral groove? ______________________
Food is taken into the **oral groove** through the cytostome (mouth), and into the cytopharynx where a **food vacuole** forms.

Materials

Methylcellulose, 1.5%

Congo red–yeast suspension

Acetic acid, 5%

Pasteur pipettes

Toothpicks

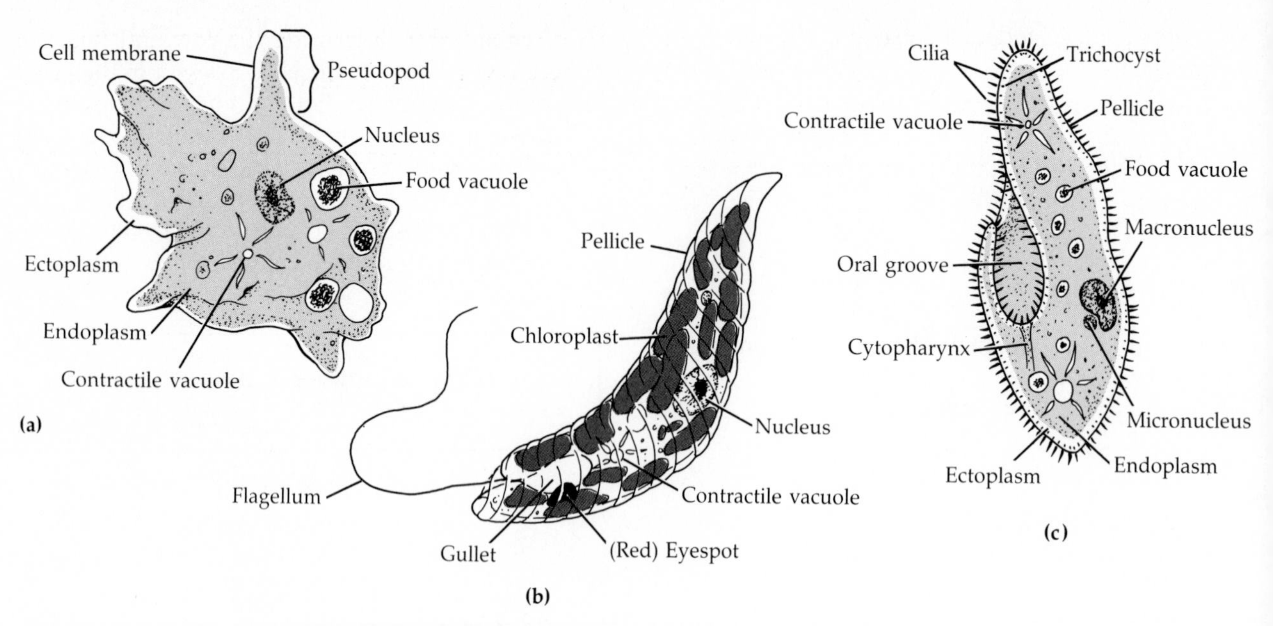

Figure 34-1.
Protozoans. **(a)** *Amoeba* moves by extending pseudopods. **(b)** *Euglena* has a whiplike flagellum. **(c)** *Paramecium* has cilia over its surface.

Cultures

Amoeba

Paramecium

Euglena

Techniques Required

Exercise 2

Procedure

1. Prepare a wet mount of *Amoeba*. Place a drop from the bottom of the *Amoeba* culture on a slide. Place one edge of the cover slip into the drop and let the fluid run along the cover slip (Figure 34-2). Gently lay the cover slip over the drop. Observe the movement of *Amoeba* and diagram amoeboid movement. Which region of the cytoplasm has more granules: the ectoplasm or the endoplasm (refer to Figure 34-1*a*)? ________

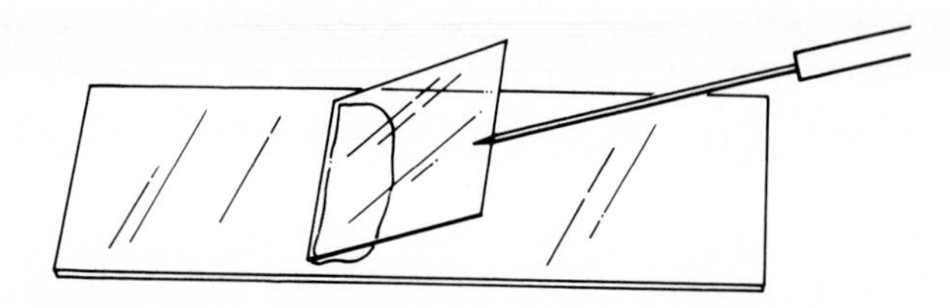

Figure 34-2.
Gently lower the cover slip.

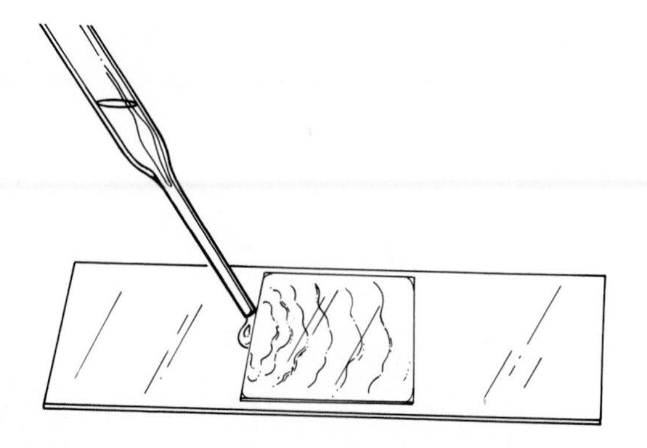

Figure 34-3.
Add a drop of acetic acid at one edge of the cover slip. Allow it to diffuse into the wet mount.

2. Prepare a wet mount of *Euglena*. Follow one individual and diagram its movement. Can you see *Euglena's* orange "eyespot"? ________
Why do you suppose it is present in photosynthetic strains and not in nonphotosynthetic strains? ________
Allow a drop of acetic acid to seep under the cover slip (Figure 34-3). How does *Euglena* respond? ________
3. Prepare a wet mount of *Paramecium* and observe its movement. Why do you suppose it rolls and *Amoeba* does not? ________
4. Place a drop of methylcellulose on a slide. With a toothpick, stir some Congo red–yeast mixture into the methylcellulose. Make a wet mount of *Paramecium* in this mixture. The *Paramecium* will move more slowly in the viscous methylcellu-

lose. Observe the ingestion of the red-stained yeast cells by *Paramecium*. Congo red is a pH indicator. As the contents of food vacuoles are digested, the indicator will turn blue (pH 3). What metabolic products would produce acidic conditions in the vacuoles? ______________________

Turn to the Laboratory Report for Exercise 34.

PART 7

Viruses

EXERCISES

Viruses are fundamentally different from the bacteria studied in previous exercises as well as from all other organisms. W. M. Stanley, who was awarded the Nobel prize (1946) for his work on tobacco mosaic virus (see Exercise 38), described the virus as *"one of the great riddles of biology. We do not know whether it is alive or dead, because it seems to occupy a place midway between the inert chemical molecule and the living organism."*

Viruses are submicroscopic, filterable, infectious agents (see the figure). That is, viruses are too small to be seen with a light microscope and are visualized with an electron microscope. Filtration is frequently used to separate viruses from other microorganisms. A suspension containing viruses and bacteria is filtered through a membrane filter (Appendix F) with a small pore size (0.45 μm) that retains the bacteria but that allows viruses to pass through.

Viruses are *obligate intracellular parasites* and have a simple structure and organization. Viruses contain DNA or RNA enclosed in a protein coat. Viruses exhibit quite strict specificity for host cells. Once inside the host cell, the virus may use the host cell's synthetic machinery and material to cause synthesis of virus particles that can infect new cells.

Viruses are widely distributed in nature and have been isolated from virtually every eucaryotic and procaryotic organism. Based on their host specificity, viruses are divided into general categories such as **bacterial viruses (bacteriophages), plant viruses,** and **animal viruses.** A virus's host cells must be grown in order to culture viruses in a laboratory.

In this part, isolation of bacteriophages and plant viruses will be studied. Animal viruses are examined in Exercise 64.

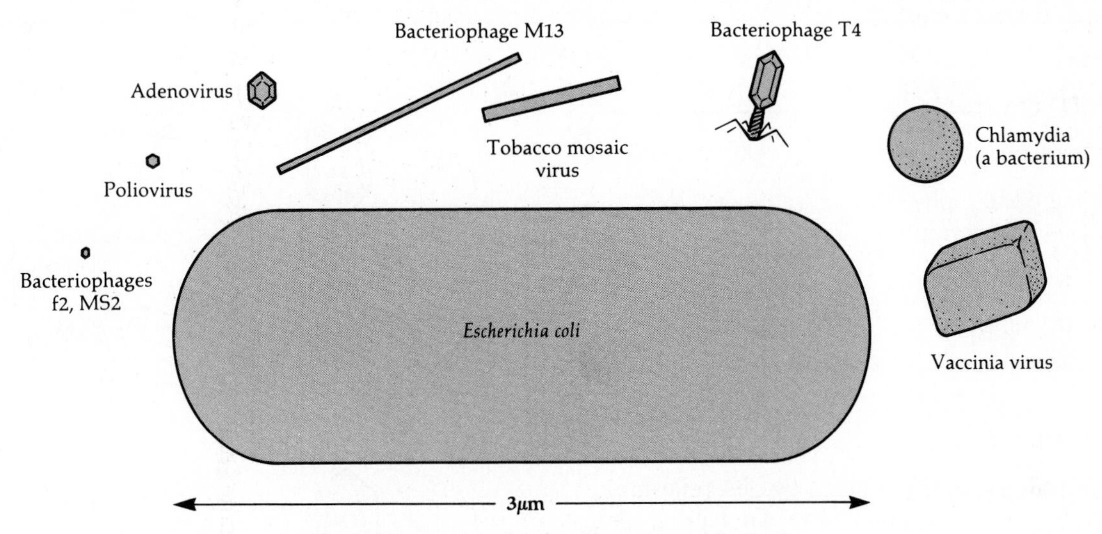

Comparative sizes of several viruses and bacteria.

EXERCISE 35

Isolation of Bacteriophages

Objectives

After completing this exercise you should be able to

1. Define bacteriophage.
2. Isolate a bacteriophage from a natural environment.
3. Describe the cultivation of bacteriophages.
4. List the steps in the multiplication of a bacteriophage.

Background

Bacteriophages, a term coined by d'Herelle meaning "bacteria eater," parasitize most, if not all, bacteria. Some bacteriophages, or **phages,** such as T-even bacteriophages, have a **complex structure** (Figure 35-1*a*). The protein coat consists of a polyhedral head and a helical tail to which other structures are attached. The head contains the nucleic acid. To initiate an infection, the bacteriophage **adsorbs** onto the surface of a bacterial cell by means of its tail fibers and base plate. The bacteriophage injects its DNA into the bacterium during **penetration** (Figure 35-1*b*). The tail sheath contracts, driving the core through the cell wall and injecting the DNA into the bacterium.

Bacteriophages can be grown in liquid or solid cultures of bacteria. The use of solid media makes location of the bacteriophage possible by the **plaque method.** Host bacteria and bacteriophages are mixed together in melted agar. The agar is then poured into a Petri plate containing hardened nutrient agar. Each bacteriophage that infects a bacterium, multiplies, and releases several hundred new viruses. The new viruses infect other bacteria, and more new viruses are produced. All of the bacteria in the area surrounding the original virus are destroyed, leaving a clear area or **plaque** against a confluent "lawn" of bacteria. The "lawn" of bacteria is produced by the growth of uninfected bacterial cells.

In this exercise a bacteriophage will be isolated from host cells in a natural environment (that is, in sewage and houseflies). Since the numbers of phages in a natural source are low, the desired host bacteria and additional nutrients are added as an **enrichment procedure** (Exercise 15). After incubation, the bacteriophage can be isolated by centrifugation of the enrichment and membrane filtration. **Filtration** has been used for removing microbes from liquids for purposes of sterilization since 1884, when (Pasteur's associate) Chamberland designed the first filter candle. Today most filtration of viruses is done using membrane filters with pore sizes (usually 0.45 μm) that physically exclude bacteria from the filtrate (Appendix F).

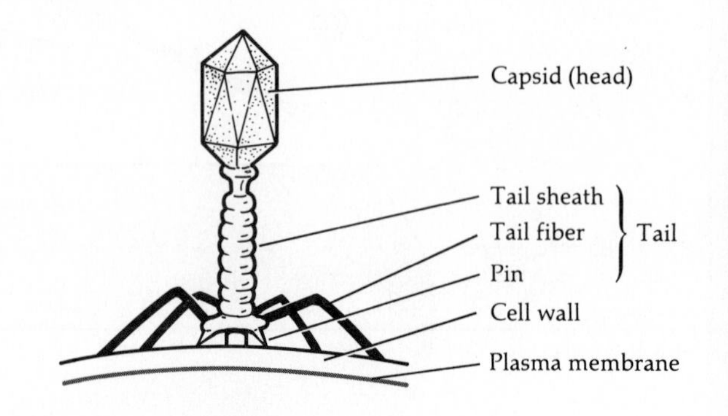

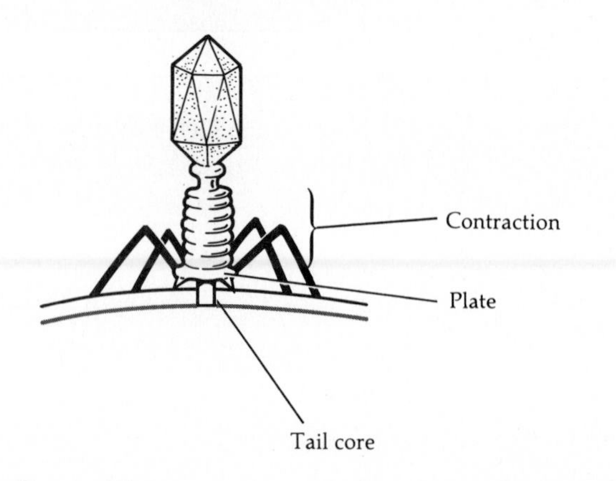

Figure 35-1.
T-even bacteriophage. **(a)** Diagram of a bacteriophage showing its component parts and adsorption onto a host cell. **(b)** Penetration of a host cell by the bacteriophage.

Materials

Use flies or sewage.

Flies as source:
- Houseflies (fresh) (20–25); bring your own (Figure 35-2)
- Trypticase soy broth (20 ml)

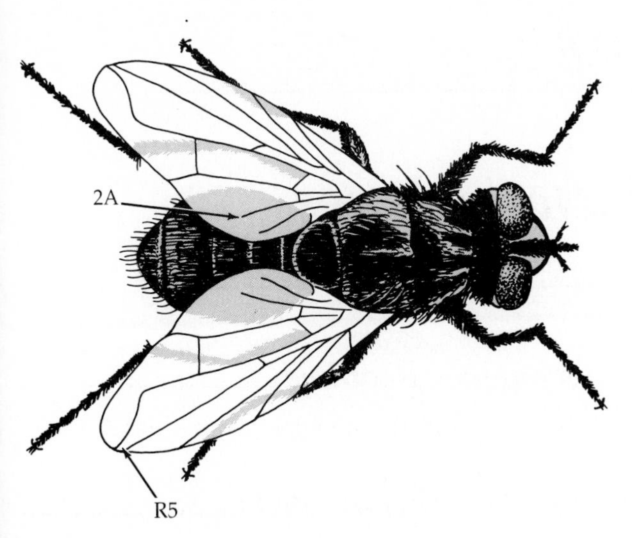

Figure 35-2.
Houseflies may harbor bacteriophages. To identify a housefly, observe that the 2A vein doesn't reach the wing margin, and the R5 cell narrows.

Mortar and pestle
Escherichia coli (2 ml)

Sewage as source:
Raw sewage (45 ml)
10× nutrient broth (5 ml)
50-ml graduated cylinder
Escherichia coli (5 ml)

Sterile 125-ml Erlenmeyer flask

Centrifuge tubes

Petri plate containing trypticase soy agar

Sterile membrane filter (0.45 μm)

Sterile membrane filter apparatus

Sterile cotton swab

Culture

Escherichia coli broth

Techniques Required

Exercises 13 and 14, and Appendix F

Procedure

Use enrichment procedure a or b.

1. **a.** Add the flies to half of the trypticase soy broth in the mortar, and grind with the pestle to a fine pulp. Transfer this mixture to a sterile flask, and add the remainder of the broth and 2 ml *E. coli.* Wash the mortar and pestle.
   b. Add 45 ml sewage to 5 ml 10× nutrient broth in a sterile flask. Add 5 ml *E. coli* broth. Mix gently. Why is the broth ten times more concentrated than normal? ____________
2. Incubate the enrichment for 24 hours at 35°C.
3. Decant 10 ml of the enrichment into a centrifuge tube. Place the tube in a centrifuge and balance the centrifuge with a similar tube or a tube containing 10 ml water. Centrifuge at 2500 rpm for 10 minutes to remove most bacteria and solid materials.
4. Filter the supernatant through a membrane filter. (Refer to Appendix F.) Decant the clear liquid into a screw-capped tube. How does filtration separate viruses from bacteria? ____________
   ____
5. Swab a plate containing trypticase soy agar twice with the fresh *E. coli* to ensure that a confluent lawn will develop. Mark three circles on the bottom of the plate and label 1, 2, and 3.
6. Aseptically add 1 loopful of your phage filtrate to circle 1. Add 3 loopfuls to circle 2, and 5 loopfuls to circle 3. What is the purpose of using three different amounts? ____________
   ____
7. Incubate for 24 hours at 35°C and observe for lysis.

Turn to the Laboratory Report for Exercise 35.

EXERCISE 36

Titration of Bacteriophages

Objectives

After completing this exercise you should be able to

1. Define titer.
2. Determine the titer of a bacteriophage sample using the broth-clearing and plaque-forming methods.

Background

When studying bacteriophages, it is often necessary to *measure* the viral activity in a preparation. The viral activity can be determined quantitatively by performing sequential dilutions of the viral preparation and assaying for the presence of viruses. In the **broth-clearing assay,** the **end point** is the highest dilution (smallest amount of virus) producing lysis of bacteria and clearing of the broth. The **titer,** or concentration, that results in a recognizable effect is the reciprocal of the end point. In the **plaque-forming assay,** the titer is determined by counting plaques. Each plaque theoretically corresponds to a single infective virus in the initial suspension. Some plaques may arise from more than one virus particle and some virus particles may not be infectious. Therefore, the titer is determined by counting the number of **plaque-forming units (p.f.u.).** The titer, plaque-forming units per milliliter, is determined by counting the number of plaques and multiplying times the reciprocal of the dilution. For example, 32 p.f.u. in a $1{:}10^4$ dilution is equal to 32×10^4 or 3.2×10^5 p.f.u.

In this exercise we will determine the viral activity by a broth-clearing assay and a plaque-forming assay.

Materials

Tubes containing trypticase soy broth (6)

Petri plates containing nutrient agar (6)

Tubes containing 3 ml melted soft trypticase soy agar (0.7% agar) (6)

1-ml sterile pipettes (7)

Cultures (one of each)

Escherichia coli broth

T-even phage suspension

Techniques Required

Exercises 13 and 14, and Appendices A and B

Procedure (Figure 36-1)

Read carefully before proceeding.

1. Label the plates and the broth tubes, 1 through 6.
2. Aseptically add 1 ml phage suspension to tube 1. Mix by carefully aspirating up and down three times with the pipette. Using a different pipette, transfer 1 ml to the second tube, mix well, and then put 1 ml into the third tube. Why is the pipette changed? ________________
   Continue until the fifth tube. After mixing this tube, discard 1 ml. What dilution exists in each tube? (See Figure 36-1.)

	Tube 1	Tube 2	Tube 3
Dilutions	____	____	____
	Tube 4	Tube 5	Tube 6
	____	____	____

   What is the purpose of tube 6? ____________
3. Add 0.1 ml *E. coli* to the soft agar tubes and place them back in the waterbath. Keep the waterbath at 43° to 45°C at all times. Why? ____________
4. With the remaining pipette, start with broth tube 6 and aseptically transfer 0.1 ml from tube 6 to the soft agar tube. Mix by swirling, and quickly pour the inoculated soft agar evenly over the surface of Petri plate 6. Then, using the same pipette, transfer 0.1 ml broth No.5 to a soft agar tube, mix, and pour over plate 5. Continue until you have completed tube 1.
5. After completing step 4 add 2 loopfuls *E. coli* to each of the broth tubes and mix. Incubate at 35°C. Observe in a few hours.
6. Incubate the plates at 35°C, until plaques develop. The plaques may be visible within hours.
7. Select a plate with between 30 and 300 plaques. Count the number of plaques and multiply by 10; then multiply by the reciprocal of the dilution of that plate to give the number of plaque-forming units (p.f.u.) per milliliter. Why is the number of plaques multiplied by 10? ____________
   Record your results. In the broth tubes, record the highest dilution that was clear as the end point.

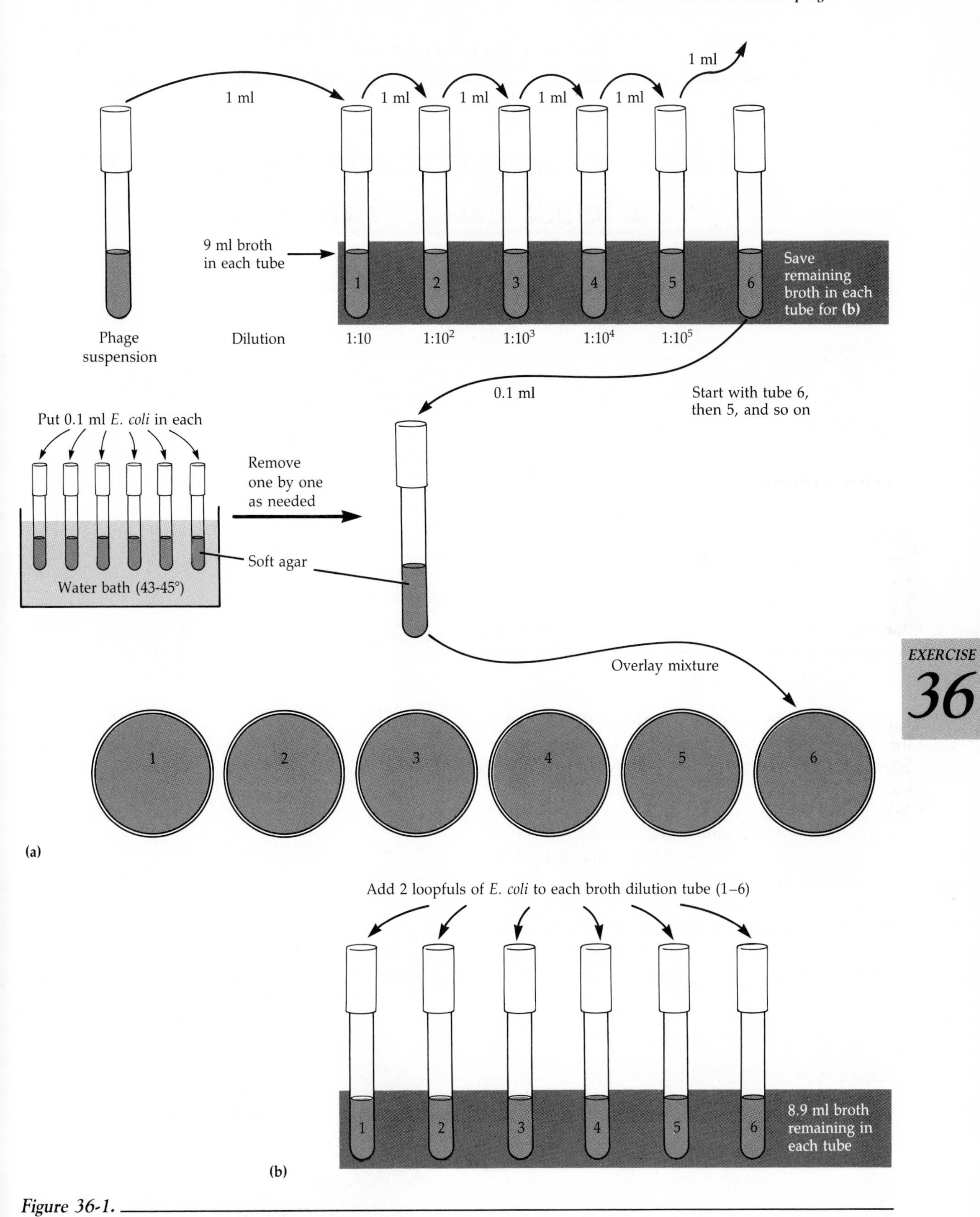

Figure 36-1.
Procedure for titration of bacteriophages. **(a)** Plaque-forming assay. **(b)** Broth-clearing assay.

EXERCISE 36

Turn to the Laboratory Report for Exercise 36.

EXERCISE 37

Identification of Bacteria Using Phage: Phage Typing

Objectives

After completing this exercise you should be able to

1. Describe the purpose of phage typing.
2. Describe phage typing.
3. Identify unknown bacterial cultures that are phage sensitive.

Background

The phage–bacteria interaction is highly specific. Bacteriophages will attack only certain species or strains of bacteria. The virus infection of bacterial cells ultimately leads to the formation of a plaque (Exercise 35).

Due to the specificity of the phage–bacteria interaction, the technique known as **phage typing** can be used to identify strains of bacteria. This tool has been particularly useful in the hospital environment, where frequent outbreaks of *Staphylococcus aureus* occur. Using this method, the precise source of the infecting agent can be determined. For example, *S. aureus,* Type 80/81, is well known for causing serious skin infections in newborns.

Although the technique used in this exercise is different from that used in the clinical laboratory, it will still demonstrate how viruses can be used to type bacterial strains.

Materials

Petri plates containing trypticase soy agar (2)

Sterile cotton swabs (4)

Cultures (one of each)

Broth culture of T_2 phage (1)

Four different unknown bacterial cultures

Techniques Required

Exercises 13 and 35

Procedure*

1. Each Petri plate will be divided into halves and each half will be inoculated with a different unknown culture. Draw a line on the underside of each plate with a marking pen, and label each half with the number of the unknown bacterial culture.
2. Using a sterile cotton swab, swab the entire surface area of half the agar with the strain to be tested (Figure 37-1). This step is critical in the phage-typing procedure. Why? ____________
3. Place the inoculated plates upright in the incubator (35°C) for 5 minutes with the lids slightly ajar to allow the surfaces to dry.
4. Replace lids and remove the plates from the incubator. In the center of each half (on the bottom) draw a dime-sized circle (Figure 37-2).
5. With your inoculating loop, place a loopful of the T_2 bacteriophage culture in the center of each circle (Figure 37-2).
6. Place the plates, lids ajar, in the incubator for 5 minutes to allow the drop to dry, then invert and incubate for 18 to 24 hours.

*Adapted from E. B. Correia and S. L. Burr. 1981. *Introduction to Microbiology Laboratory Procedures.* Yucaipa, Calif.: Crafton Hills College.

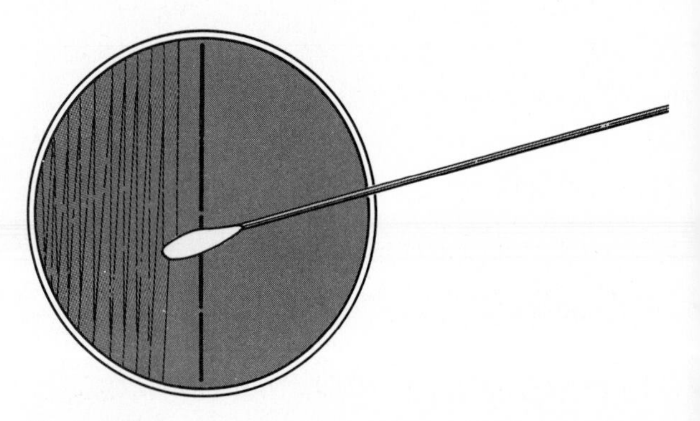

Figure 37-1.
Completely swab one-half of trypticase soy agar plate with the bacterial strain to be tested.

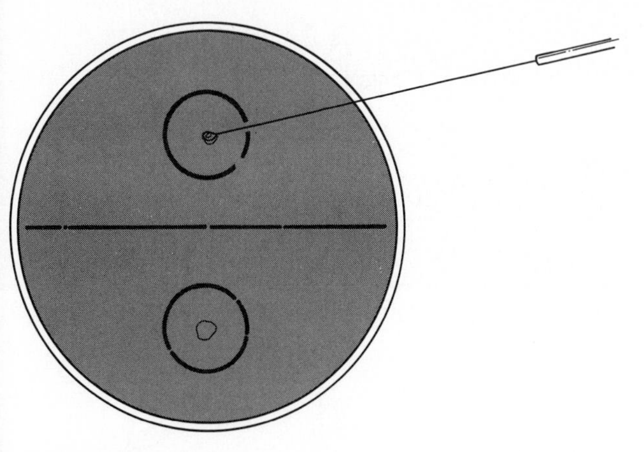

Figure 37-2.
Place a loopful of T_2 bacteriophage in the center of each circle.

7. Examine the plates and record the results as to whether there is a presence or absence of cleared areas.

Turn to the Laboratory Report for Exercise 37.

EXERCISE 38

Plant Viruses

Smoking was definitely dangerous to your health in the 17th century Russia. Czar Michael Federovitch executed anyone on whom tobacco was found. But Czar Alexei Mikhailovitch was easier on smokers; he merely tortured them until they told who their suppliers were.

Objectives

After completing this exercise you should be able to

1. Isolate a plant virus from a natural environment.
2. Describe the cultivation of a plant virus.

Background

Plant viruses are very important economically in that they are the second leading cause of plant disease. (Fungi are the major cause of plant disease.) In order for most plant viruses to infect a plant, the virus must enter through an abrasion in the leaf or stem. Insects called **vectors** carry viruses from one plant to another. Most viruses cause either a **localized infection,** in which the leaf will have necrotic lesions (brown plaques), or a **systemic infection,** in which the infection runs throughout the entire plant. One of the most studied viruses is tobacco mosaic virus (TMV), which was the first virus to be purified and crystallized.

In this exercise, you will attempt to isolate TMV from various tobacco products.

Materials

Tobacco plant

Tomato plant

Tobacco products of various types (bring some yourself if you can)

Mortar and pestle

Fine sand or carborundum

Wash bottle of water

Paper labels or labeling tape

Cotton

Sterile water

Techniques Required

None

Procedure

1. Select a tobacco or tomato plant and label the pot with your name and lab section. Do both tobacco and tomato plants if possible.
2. Place labels around the petioles (Figure 38-1) of several leaves, with names on them corresponding to the various tobacco products available. Label one petiole "C" for control.
3. *Wash your hands carefully before and after each inoculation.*
4. Spray the control leaf with a small amount of water and dust the leaf with carborundum or sand. Gently rub the leaf with a cotton ball dampened in sterile water. Wash the leaf with a wash bottle of water. What is the purpose of this leaf? ________

   ____
5. Select a tobacco product, place a "pinch" of it into a mortar, and add sterile water. Grind into a slurry with a pestle. Spray one leaf with water, dust with carborundum or sand, and gently rub with a cotton ball soaked with the tobacco slurry. Wash the leaf with water. Record the name of the tobacco source.
6. Repeat step 5 for each tobacco product available, cleaning the mortar and pestle with soap and water between each product.

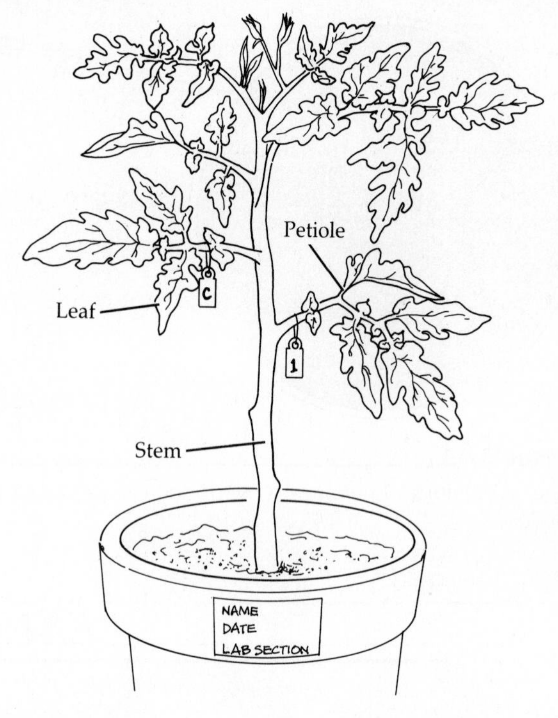

Figure 38-1.
Diagram of a tomato plant, showing its anatomy and placement of labels.

7. Replace the plant in the rack.
8. Check the plant at each laboratory period for up to 3 to 4 weeks for brown plaques (localized infection) or wilting (systemic infection).

Turn to the Laboratory Report for Exercise 38.

PART 8

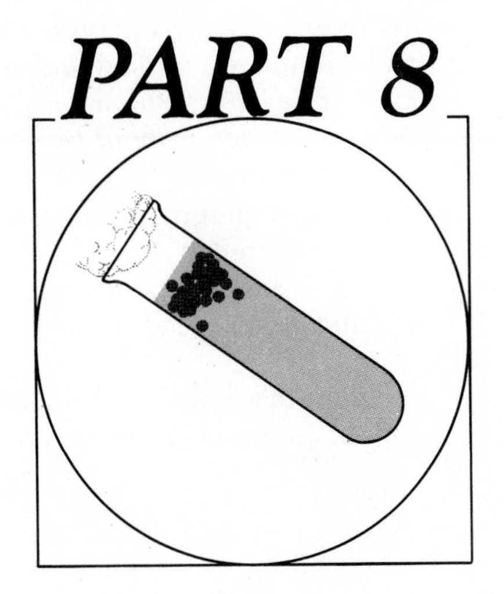

Growth of Microorganisms

EXERCISES

In previous exercises, bacterial growth was determined by observing colonies or turbidity. These visual determinations are made on an all-or-none basis with the naked eye. In Exercise 39, instrumentation will be used to detect the slight changes in bacterial numbers that occur over time—changes that are not visible to the naked eye (see the figure).

In addition to nutritional requirements (Part 3), bacteria have physical and chemical environmental requirements that must be met in order for growth to occur. A knowledge of the conditions necessary for microbial growth can facilitate culturing microorganisms and controlling unwanted organisms.

In 1861, Jean Baptiste Dumas presented a paper to the Academy of Sciences on behalf of Pasteur. In this paper he stated,

> *The existence of infusoria having the characteristics of ferments is already a fact which seems to deserve attention, but a characteristic which is even more interesting is that these infusoria-animalcules live and multiply indefinitely in absence of the smallest amount of air or free oxygen. . . . This is, I believe the first known example of animal ferments and also of animals living without free oxygen gas.*

When we speak of *air* as a growth requirement, we are referring to the oxygen in air. Furthermore, when we refer to oxygen, we usually mean the **molecular oxygen (O_2)** that acts as an electron acceptor in aerobic respiration (Exercise 22). The significance of the presence or absence of oxygen is investigated in Exercise 40.

Physical environmental conditions that influence microbial growth are temperature (Exercise 41), osmotic pressure (Exercise 42), and pH (Exercise 42).

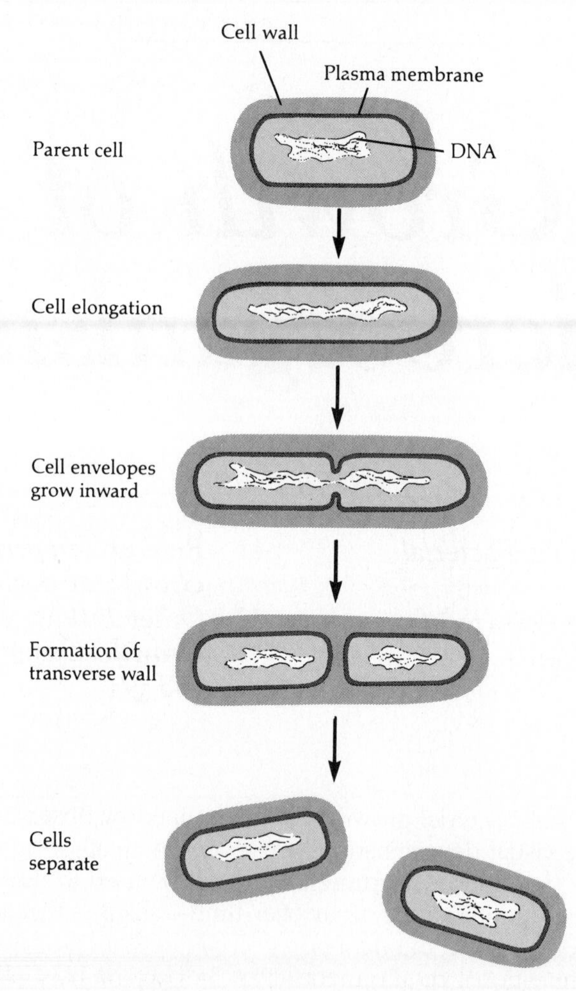

Bacteria normally grow by transverse fission. This diagram illustrates the steps of transverse fission.

EXERCISE 39

Determination of a Bacterial Growth Curve

Objectives

After completing this exercise you should be able to

1. Identify the four typical phases of a bacterial growth curve.
2. Use a spectrophotometer.
3. Measure bacterial growth turbidometrically.
4. Interpret growth data plotted on a graph.

Background

The phases of growth of a bacterial population can be determined by measuring **turbidity** of a broth culture. Turbidity is not a direct measure of bacterial numbers, but increasing turbidity does indicate bacterial growth. Since millions of cells per milliliter must be present to discern turbidity with the eye, a spectrophotometer is used to detect the presence of smaller numbers of bacteria.

In a **spectrophotometer,** a beam of light is transmitted through a bacterial suspension to a photoelectric cell (Figure 39-1). As bacterial numbers increase, the broth becomes more turbid, causing the light to scatter and allowing less light to reach the photoelectric cell. The change in light is registered on the instrument as **percentage of transmission** (the amount of light getting through the suspension) and **absorbance** or **optical density (O.D.)** (a value derived from the percentage of transmission). Optical density is a logarithmic value and is used to plot bacterial growth on a graph.

Operation of the Spectronic 20, a Spectrophotometer (Figure 39-2)

1. Turn on the power and allow the instrument to warm up for 5 minutes.

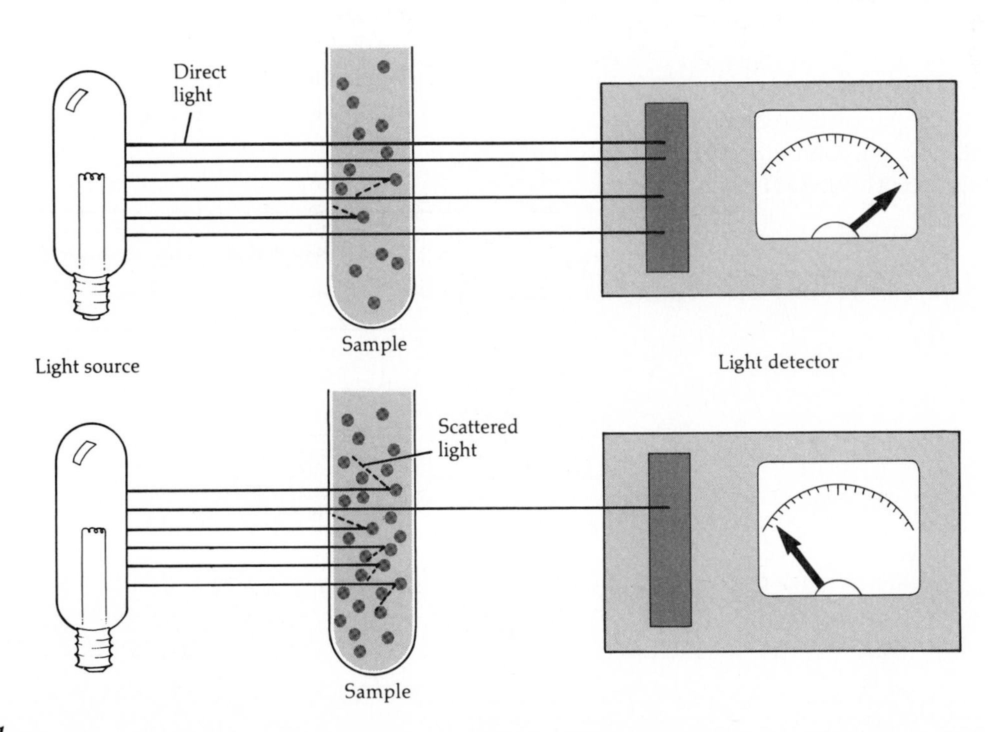

Figure 39-1.
Estimation of bacterial numbers by turbidity. The amount of light picked up by the light detector is proportional to the number of bacteria. The less light transmitted, the more bacteria in the sample.

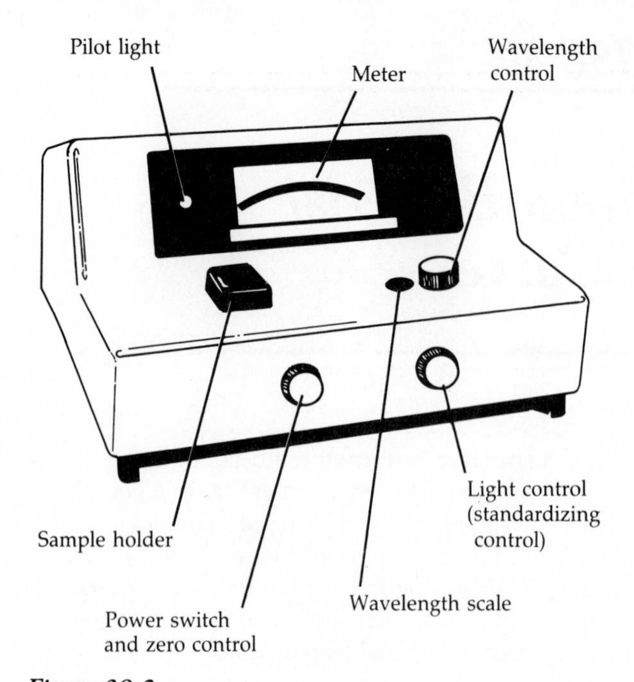

Figure 39-2.
Bausch and Lomb Spectronic 20.

2. Set the wavelength (580 nm) for maximum absorption of *E. coli* and minimal absorption of the culture medium.
3. *Zero* the instrument by turning the zero control until the needle measures 0% transmission. To read the scale, have the needle at eye level and look at it so that the needle is directly over its reflection in the mirror.
4. Place an uninoculated side arm of culture media *(control)* in the sample holder. To *standardize* the instrument, turn the light control until the needle registers 100% transmission.
5. To take a sample measurement, place an inoculated side arm of culture media in the sample holder and record the O.D. from the scale.

Materials

Side arm flask containing glucose-minimal salts broth (1)

Control side arm flask containing glucose-minimal salts broth (1 per lab section)

Sterile 1-ml pipette (1)

Spectrophotometer

Culture

Escherichia coli in glucose-minimal salts broth

Techniques Required

Exercise 13 and Appendices A and D

Procedure

1. Aseptically inoculate a flask of culture medium with 1 ml *E. coli.*
2. Swirl to mix the contents and measure the O.D. on the spectrophotometer. This is the zero time measurement. Record all measurements in your Laboratory Report.
3. Place the flask in the incubator. The O.D. should be measured every 30 minutes for the next 10 to 12 hours.
4. Swirl the flask to mix the contents before taking measurements, and *wipe fingerprints and dust from the side arm with a soft tissue.*
5. Be sure to return the broth from the side arm to the flask after each measurement.
6. Graph the data you obtained in your Laboratory Report. Read Appendix D before drawing your graph.

Turn to the Laboratory Report for Exercise 39

EXERCISE 40

Oxygen and the Growth of Bacteria

Progress is a line through a list.

Objectives

After completing this exercise you should be able to

1. Provide the incubation conditions for each of the following types of organisms: aerobes, obligate anaerobes, aerotolerant anaerobes, microaerophiles, and facultative anaerobes.
2. Describe three methods of culturing anaerobes.
3. Cultivate anaerobic bacteria.

Background

The presence or absence of molecular oxygen (O_2) can be very important to the growth of bacteria. Some bacteria, called obligate **aerobic bacteria,** require oxygen, while others, called **anaerobic bacteria,** do not use oxygen. **Obligate anaerobes** cannot tolerate the presence of oxygen because they lack catalase, and the resultant accumulation of hydrogen peroxide is lethal (Exercise 22). **Aerotolerant anaerobes** cannot use oxygen but tolerate it fairly well although their growth may be enhanced by microaerophilic conditions. Most of these bacteria use fermentative metabolism (Exercise 17).

Some bacteria, the **microaerophiles,** grow best in an atmosphere with increased carbon dioxide (7% to 10%) and lower concentrations of oxygen. Microaerophiles will grow in a solid nutrient medium at a depth to which small amounts of oxygen have diffused into the medium (Figure 40-1). In order to culture microaerophiles on Petri plates and nonreducing media, a **candle jar** is used. Inoculated plates and tubes are placed in a large jar with a lighted candle. After the lid is placed on the jar, the candle will extinguish when the concentration of oxygen decreases (the concentration of carbon dioxide will be raised).

The majority of bacteria are capable of living with or without oxygen; these bacteria are called **facultative anaerobes** (Figure 40-1).

Four genera of bacteria lacking catalase are *Streptococcus, Leuconostoc, Lactobacillus,* and *Clostridium.* Species of *Clostridium* are obligate anaerobes, but the other three are aerotolerant anaerobes. The three aerotolerant anaerobes lack the cytochrome system to produce hydrogen peroxide and therefore do not need catalase. Determining the presence or absence of catalase can be very helpful in identifying bacteria. When a few drops of 3% hydrogen peroxide are

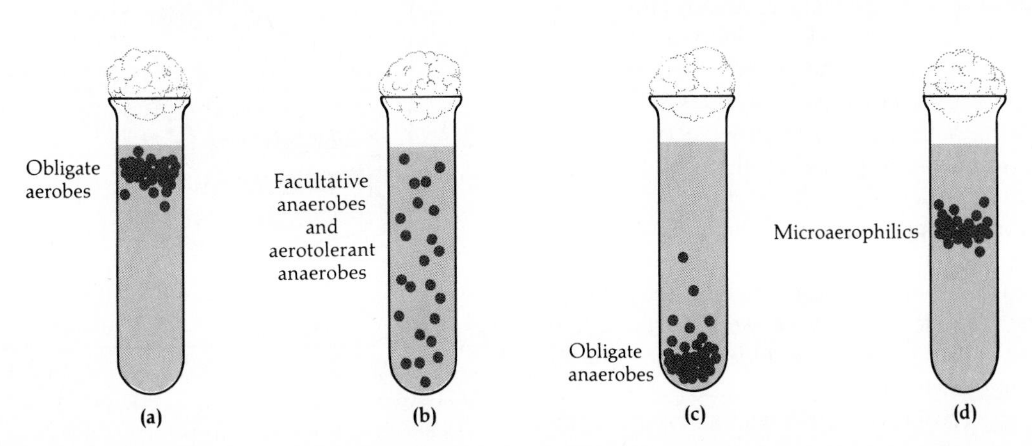

Figure 40-1.
Effect of oxygen concentration on the growth of various types of bacteria in a tube of solid medium. Oxygen diffuses only a limited distance from the atmosphere into the solid medium.

added to a microbial colony and catalase is present, molecular oxygen is released as bubbles.

In the laboratory, anaerobes can be cultured either by excluding free oxygen from the environment or by using reducing media. Many anaerobic culture methods involve both processes. Some of the anaerobic culturing methods are as follows:

1. **Reducing media** contain reagents that chemically combine with free oxygen and reduce the concentration of oxygen in the medium.
   a. **Cooked meat medium** is a reducing medium containing beef heart (50%) and peptone (2%). The medium is boiled before inoculation to expel oxygen. A paraffin plug minimizes further oxygen entry.
   b. In thioglycollate broth, **sodium thioglycollate** ($HSCH_2COONa$) will combine with oxygen. A small amount of agar is added to increase the viscosity, and this reduces the diffusion of air into the medium. Usually a dye is added to indicate where oxygen is present in the medium. Resazurin, which is pink in the presence of excess oxygen and colorless when reduced, or methylene blue (see below) are commonly used indicators.
2. Conventional, nonreducing media can be incubated in an anaerobic environment. Oxygen is excluded from a **Brewer anaerobic jar** by adding hydrogen gas and heating a platinum or palladium catalyst with electricity to catalytically combine the hydrogen with oxygen and form water (Figure 40-2*a*). A safer, more practical Gas-Pak® method has been developed by Brewer and Allgeier. In the Gas-Pak® method, carbon dioxide and hydrogen are given off when water is added to an envelope of sodium bicarbonate and sodium borohydride. An improved palladium catalyst in the cover does not require heat to catalyze the formation of water from hydrogen and oxygen (Figure 40-2*b*). A methylene blue indicator strip is placed in the jar; methylene blue is blue in the presence of oxygen and white when reduced. One of the disadvantages of the Brewer jar is that the jar must be opened to observe or use one plate. An inexpensive modification of the Brewer jar has been developed using disposable plastic bags for each plate. In this technique, the bag is coated with antifogging chemicals, and a wet sodium bicarbonate tablet is added to generate carbon dioxide. Steel wool wetted with dilute copper sulfate (catalyst) is added to achieve an anaerobic environment. How? ______________
   The plate can be observed without opening the bag.
3. **Anaerobic incubators** and **glove boxes** can also be used for incubation. Air is evacuated from

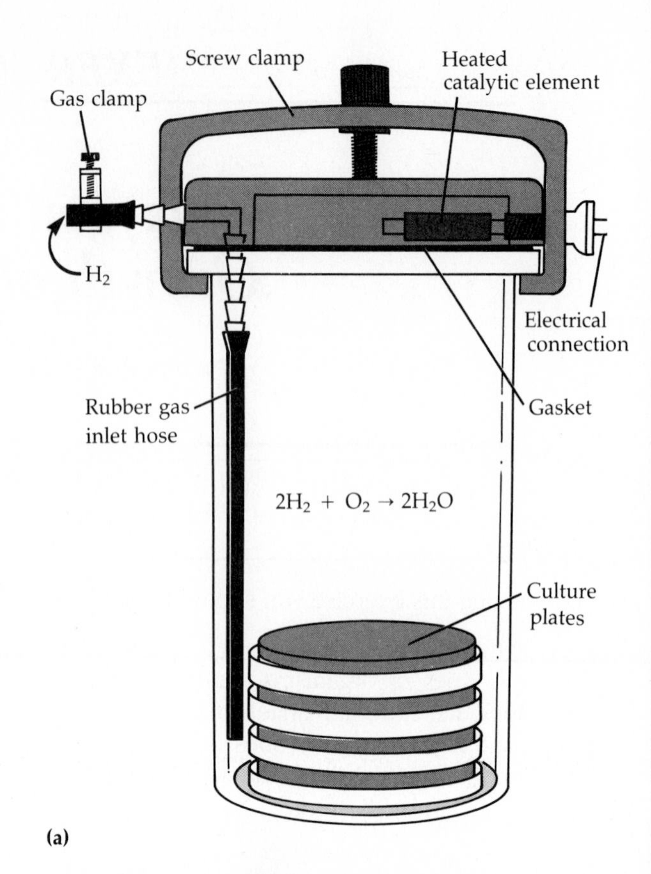

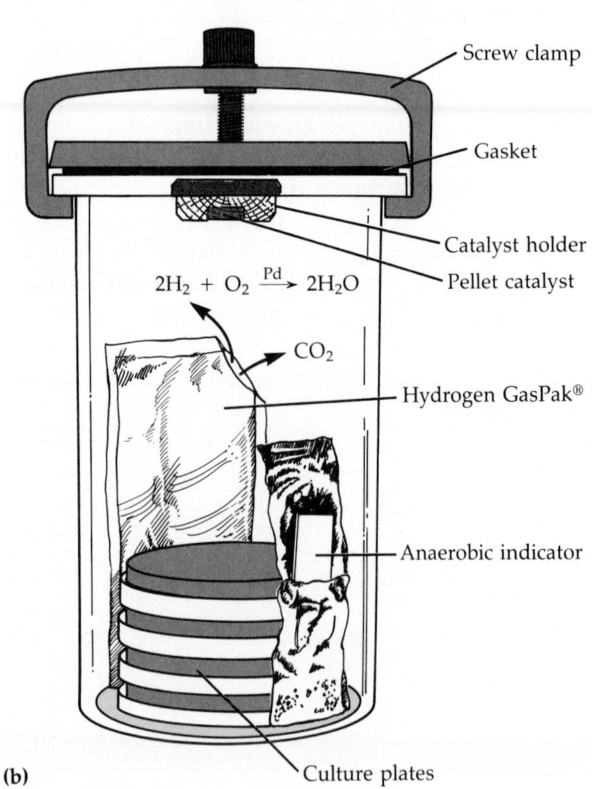

Figure 40-2.
Brewer anaerobic jars. **(a)** Hydrogen gas combines with oxygen to form H_2O and a reduced atmosphere. **(b)** Carbon dioxide and hydrogen are generated by the addition of water to a chemical packet. Hydrogen gas combines with atmospheric oxygen in the presence of a palladium catalyst to form H_2O.

the chamber and can be replaced with a mixture of carbon dioxide and nitrogen.

All methods of anaerobic culturing are only effective if the specimen or culture of anaerobic organisms is collected and transferred in a manner that minimizes exposure to oxygen. In this exercise two methods of anaerobic culture will be attempted and the catalase test will be performed.

Materials

Petri plates containing nutrient agar (2)

Tube containing thioglycollate broth with indicator (1)

Brewer anaerobic jar per lab section (1)

3% hydrogen peroxide (H_2O_2)

Soil or feces slurry that has been previously boiled for 10 minutes and then cooled

Techniques Required

Exercises 13, 14, and 22

Procedure

1. Don't shake the thioglycollate. Why? ________ What should you do to salvage it if you do shake it? ________

   Aseptically inoculate the thioglycollate with a loopful of the cooled soil or feces slurry. Incubate at room temperature until good growth is seen (2 days to a week). Describe the oxygen requirements of the resulting growth. Why was the soil or fecal sample heated? ________
2. Aseptically streak two plates with the soil or feces slurry. Label one "Aerobic" and incubate it inverted at room temperature. Label the other plate "Anaerobic" and place it, inverted, in the Brewer jar.
3. Your instructor will demonstrate how the jar is rendered anaerobic once filled with plates. Ten ml water are added to a Gas-Pak® generator and placed in the jar with a methylene blue indicator. What causes the condensation that forms on the sides of the jar? ________

   The jar will be incubated for about 1 week at room temperature.
4. After incubation, describe the colonies on each plate.
5. Perform the catalase test by adding a few drops of 3% H_2O_2 to the different colonies. A positive catalase test produces a bubbling white froth. A dissecting microscope (Appendix E) can be used if more magnification is required to detect bubbling. The catalase test may also be done by transferring bacteria to a slide and adding the H_2O_2 to it. Bacterial growth on blood agar must be tested this way. Why? ________

Turn to the Laboratory Report for Exercise 40.

EXERCISE 41

Role of Temperature in the Growth of Bacteria

Objectives

After completing this exercise you should be able to

1. Define the following: psychrophile, mesophile, and thermophile.
2. Describe the effect of temperature on the growth of microorganisms.

Background

Bacteria can be classified into three groups, based on their growth at various temperatures (Figure 41-1). **Psychrophiles** can grow between 0° and 20°C, with an optimum temperature around 15°C. **Mesophiles** are capable of growth between 20° and 45°C. The opti-

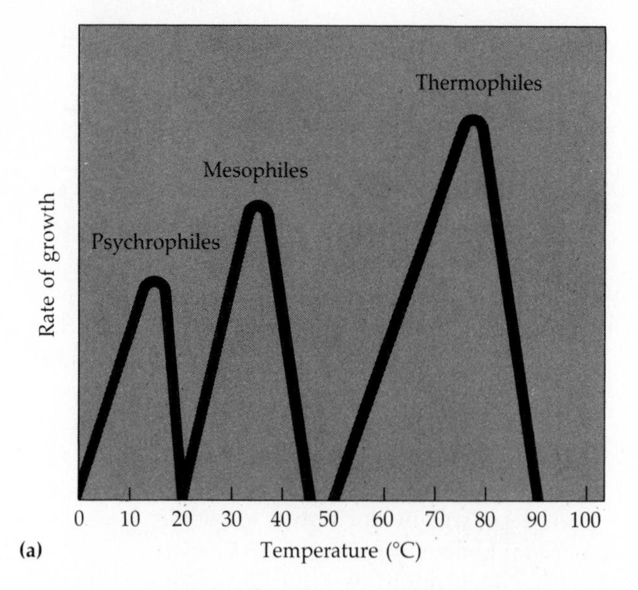

(a)

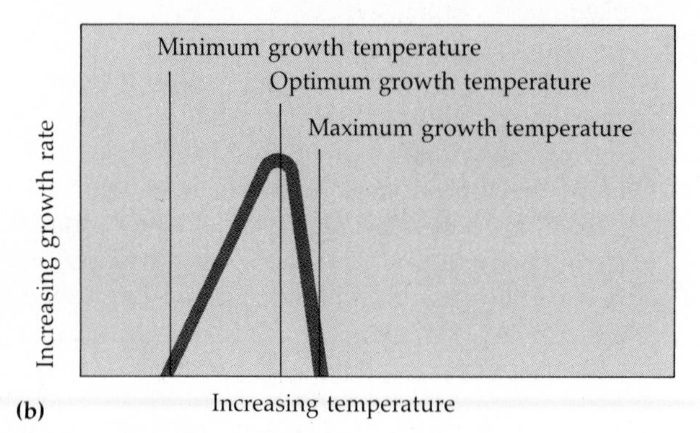

(b)

Figure 41-1.
(a) Typical growth responses to temperature for psychrophiles, mesophiles, and thermophiles. **(b)** Growth response of bacteria within their growing temperature range.

mum temperature for many **thermophiles** is between 50° and 60°C, although some are capable of growth above 90°C. The range of temperature preferred by bacteria is genetically determined, resulting in enzymes with different temperature requirements.

Most bacteria grow within a particular temperature range. The **minimum growth temperature** is the lowest temperature at which a species will grow. A species grows fastest at its **optimum growth temperature.** And the highest temperature at which a species can grow is its **maximum growth temperature.** Some heat is necessary for growth. Heat probably increases the rate of collisions between molecules, thereby increasing enzyme activity. At temperatures near the maximum growth temperature, growth ceases, presumably due to inactivation of enzymes.

Materials

Petri plates containing nutrient agar (4)

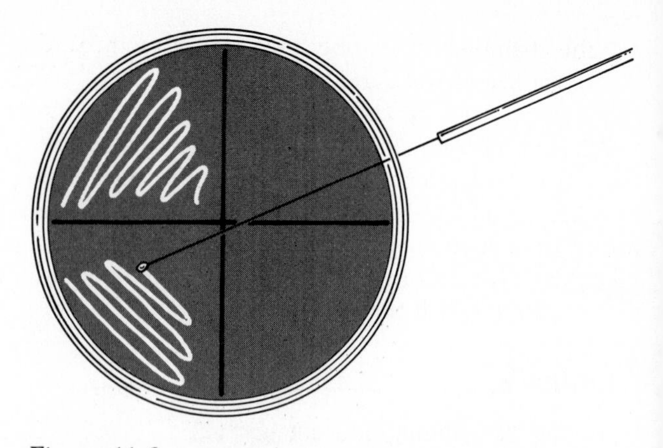

Figure 41-2.
Each quadrant will be inoculated with one bacterium.

Cultures (one of each)

Serratia marcescens

Pseudomonas fluorescens

Micrococcus luteus

Bacillus sterothermophilus

Techniques Required

Exercise 13

Procedure

1. Draw two lines perpendicular to each other to form four quadrants on the bottom of one plate. Each quadrant will be inoculated with one of the organisms and should be labeled accordingly (Figure 41-2). Mark each plate in a similar manner.
2. Using a sterile loop, inoculate *S. marcescens* onto one quadrant on each plate as shown in Figure 41-2. Inoculate the other organisms onto their respective quadrants in the same way.
3. Invert the plates and incubate one plate at each of the following temperatures: 15°C; room temperature; 35°C; and 55°C for 24 to 48 hours. What is room temperature? ________________ °C
4. Record your results. Rate the relative amounts of growth as (−) = no growth; (+) = minimal growth; (2+) = moderate growth; (3+) = heavy growth; (4+) = very heavy (maximum) growth.

Turn to the Laboratory Report for Exercise 41.

EXERCISE 42

Other Influences on Microbial Growth: Osmotic Pressure and pH

Objectives

After completing this exercise you should be able to

1. Define osmotic pressure and explain how it affects a cell.
2. Explain how microbial growth is related to osmotic pressure and pH.
3. Prepare and use a gradient plate.

Background

The osmotic pressure of an environment influences microbial growth. **Osmotic pressure** is the force with which a solvent moves from a solution of lower solute concentration to a solution of higher solute concentration across a semipermeable membrane (Figure 42-1). The addition of solutes, especially salts and sugars, and the resultant increase in osmotic pressure, is used to preserve some foods. Can you name a food that is preserved with salt? ________ With sugar? ________

Bacteria are often able to survive in a **hypotonic** environment in which the concentration of solutes outside the cell is lower than the concentration inside the cell. However, when the concentration of solutes outside the cell is higher than that inside the cell, that is, when it is a **hypertonic** environment, a bacterial cell will undergo plasmolysis. **Plasmolysis** occurs when water leaves the cell and the cytoplasmic membrane draws inward, away from the cell wall. A few bacteria, called **facultative halophiles,** are able to tolerate salt concentrations up to 10%, and **extreme halophiles** require 15% to 20% salt.

The acidity or alkalinity (**pH**) of the environment also influences microbial growth. Optimal bacterial growth usually occurs between pH 6.5 and 7.5. Only a few bacteria grow at an acidic pH below 4.0. Therefore, organic acids are used to preserve foods such as fermented dairy products (Exercise 55).

Many bacteria produce acids that may inhibit their growth. **Buffers,** which produce an equilibrium, are added to culture media to neutralize these acids. The peptones in complex media act as buffers. Phosphate salts are often used as buffers in chemically defined media.

In order to understand environmental influences on growth, bacterial growth will be compared to fungal growth in this exercise. Fungi differ from bacteria morphologically and biochemically. The yeast (Exercise 31) and mold (Exercise 32) used here are microbes belonging to the kingdom Fungi.

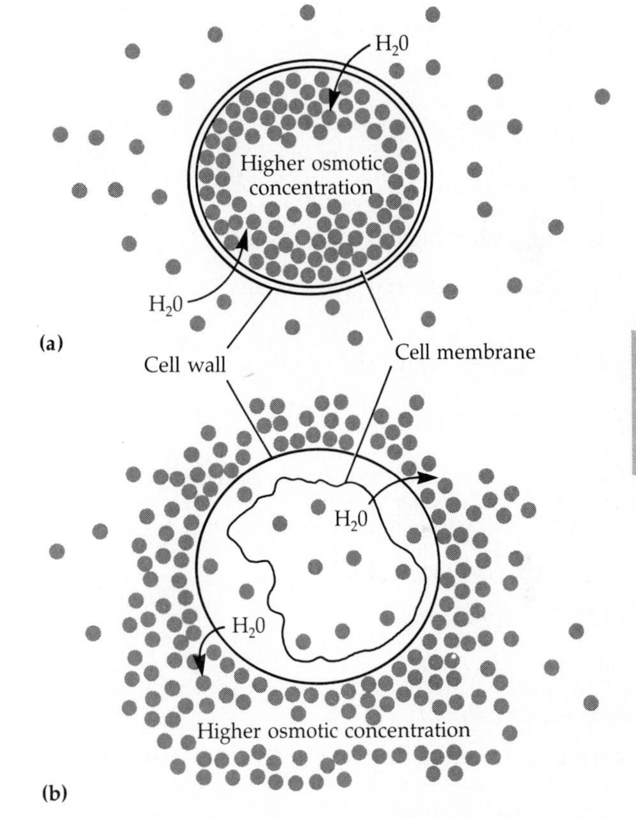

Figure 42-1.
Osmotic pressure. **(a)** Solutes move into the cell in a hypotonic solution. **(b)** Solutes leave the cell in a hypertonic environment.

EXERCISE 42

Materials

Effect of Osmotic Pressure on Growth

(Your instructor will assign salt or sugar.)

Petri plates containing nutrient agar + 5%, 10%, 15% NaCl (1 of each) *or*

Petri plates containing nutrient agar + 10%, 25%, 50% sucrose (1 of each) *or*

Petri plate containing nutrient agar (1)

Cultures (one of each)

Escherichia coli

Staphylococcus aureus

A yeast *(Saccharomyces)*

A mold *(Penicillium)*

Gradient Plate

Sterile Petri plate (1)

Tube containing melted nutrient agar (1)

Tube containing melted nutrient agar + 25% NaCl *or* 50% sucrose

Salt *or* sugar enrichments

Effect of pH on Growth

Tubes containing nutrient broth adjusted to pH 2.5, 5.0, 7.0, 9.5 (1 of each)

Cultures (one of each)

Staphylococcus aureus

Alcaligenes faecalis

Escherichia coli

Serratia marcescens

A mold *(Penicillium)*

A yeast *(Saccharomyces)*

Techniques Required

Exercises 13 and 14

Procedure

Effect of Osmotic Pressure on Growth

1. Obtain a set of nutrient agar + salt *or* nutrient agar + sucrose plates. Draw two crossed lines to form four quadrants on the bottom of one plate. Each quadrant will be inoculated with one of the organisms and should be labeled accordingly (Figure 42-2). Repeat this procedure until all plates are marked. One student group will mark the *control* plate (nutrient agar alone). What is the purpose of the control plate? ______________

____

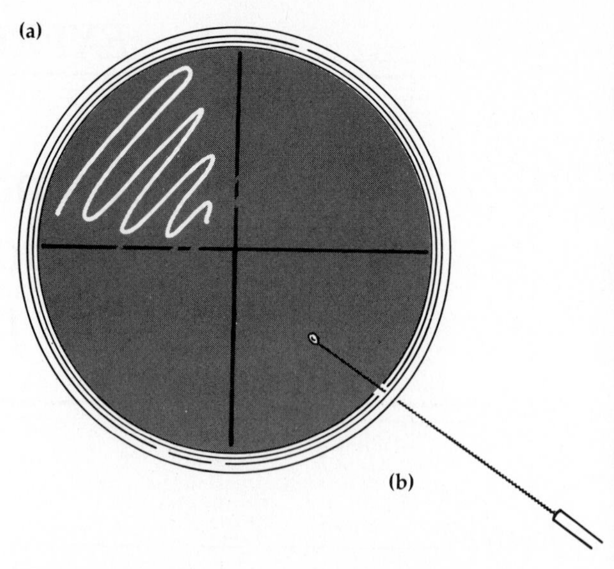

Figure 42-2.
Inoculate each plate with four organisms. **(a)** Streak bacteria and yeast cultures. **(b)** Place a loopful of mold suspension in the center of one quadrant.

2. Using a sterile loop, inoculate *S. aureus* onto one quadrant on each plate as shown in Figure 42-2*a*. Repeat this procedure to inoculate *E. coli* and the yeast.
3. Place a loopful of the mold suspension in the center of the remaining quadrant on each plate (Figure 42-2*b*).
4. Invert the plates and incubate at 35°C for 24 to 48 hours. Record the relative amounts of bacterial growth; the culture showing the "best" growth is given a "4+" and others are evaluated relative to that culture. Incubate the plate at room temperature for 2 to 4 days and record relative amounts of mold growth.

Gradient Plate

1. Pour a layer of nutrient agar into a sterile Petri plate and let it solidify, with the plate resting on a small pencil or a loop handle (Figure 42-3*a*) so that a wedge forms.
2. On the bottom of the plate, draw a line at the end corresponding to the high side of the agar and label it "low." Draw four lines perpendicular to the first line, and, if assigned sucrose, label the lines "0.5," "5," "10," and "15"; if assigned salt, "0.5," "5," "15," and "30" (Figure 42-3*b*).
3. Pour a layer of salt *or* sugar agar over the nutrient agar wedge. Let it solidify. A solute gradient concentration has been prepared. Why was the high side of the agar labeled "low" in step 2? ______________

____

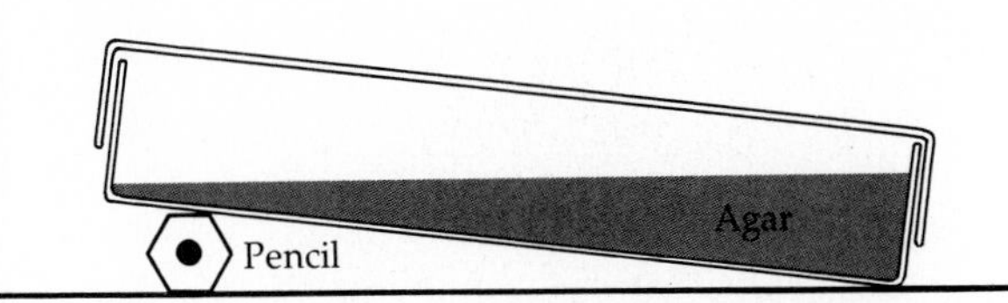

(a) Allow the agar to solidify into a wedge.

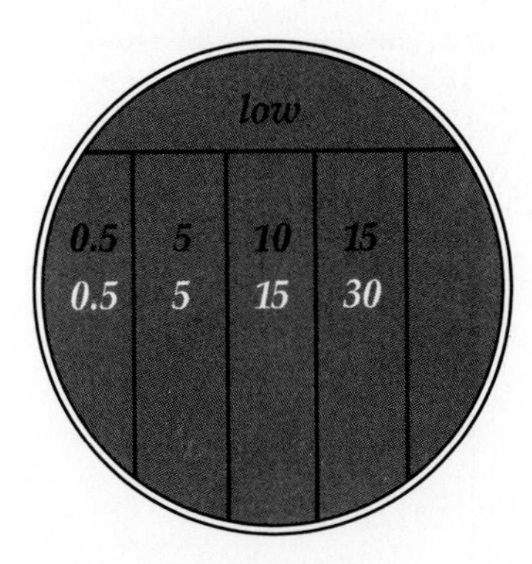

(b) Mark the plate with one line across the high side of the agar (label it ***low***) and four lines perpendicular to it (label them ***0.5, 5, 10,*** and ***15*** if assigned sucrose, and ***0.5, 5, 15,*** and ***30*** if salt).

Figure 42-3.
Gradient plate.

4. The following enrichments (Exercise 15) have been prepared:
   a. Raisins were placed in tubes of nutrient broth + 0.5% sucrose, 5% sucrose, 10% sucrose, and 15% sucrose.
   b. Hamburger was inoculated into tubes of nutrient broth + 0.5% NaCl, 5% NaCl, 15% NaCl, and 30% NaCl.
5. Using the enrichment with the same solute as your gradient plate, streak a loopful of each enrichment onto the agar over the prelabeled lines. Streak from the low to the high end.
6. Incubate the plate inverted at 35°C for 48 hours, and record the growth patterns.

Effect of pH on Growth

1. Inoculate a loopful of your organism into each tube of pH test broth.
2. Incubate the tubes at 35°C for 24 to 48 hours. Record your results and results for the organisms tested by other students.

Turn to the Laboratory Report for Exercise 42.

PART 9

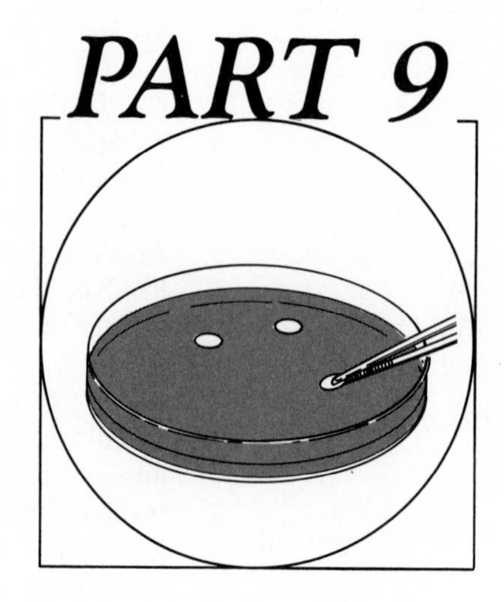

Control of Microbial Growth

EXERCISES

The destruction of microbes by heat was employed by Spallanzani in the 1760s. He heated nutrient broth to kill preexisting life in his attempts to disprove the concept of spontaneous generation. In the 1860s, Pasteur heated broth in specially designed flasks and ended the debate over spontaneous generation (see the figure). In Exercise 43, the effectiveness of heat for killing microbes will be examined.

The surgeon Joseph Lister was greatly influenced by Pasteur's demonstrations of the omnipresence of microorganisms and his proof that microorganisms cause decomposition of organic matter. Lister had observed the disastrous consequences of compound bone fractures (in which the skin is broken) compared to the relative safety of simple bone fractures. He had heard of the treatment of sewage with carbolic acid (phenol) to prevent diseases of cattle that had come into contact with the sewage. In 1867 Lister wrote, *"It appears that all that is requisite is to dress the wound with some material capable of killing these septic germs"* to prevent disease and death due to microbial growth.

Current methods of controlling microbial growth will be investigated in this Part. Special attention is given to the use of disinfectants and antiseptics (Exercise 45), handwashing (Exercise 47), and infection control in the clinical environment (Exercise 48).

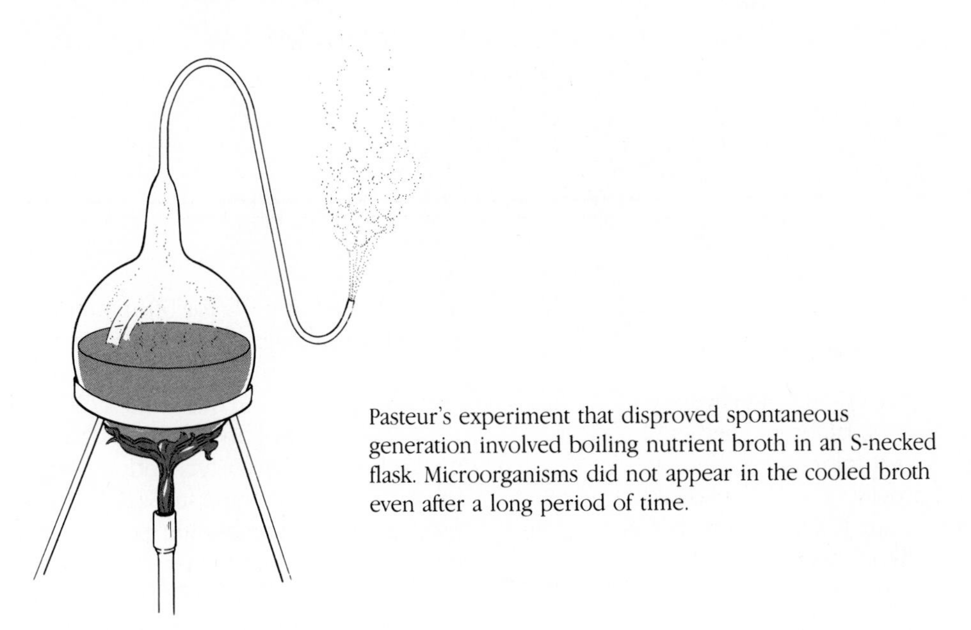

Pasteur's experiment that disproved spontaneous generation involved boiling nutrient broth in an S-necked flask. Microorganisms did not appear in the cooled broth even after a long period of time.

EXERCISE 43

Physical Methods of Control: Heat

The successful man lengthens his stride when he discovers that the sign post has deceived him; the failure looks for a place to sit down.

J. R. ROGERS

Objectives

After completing this exercise you should be able to

1. Compare the bactericidal effectiveness of dry heat and moist heat on different species of bacteria.
2. Evaluate the heat tolerance of microbes.
3. Define and provide a use for each of the following: incineration, hot air oven, pasteurization, boiling, and autoclaving.

Background

The use of extreme temperature to control the growth of microbes is widely employed. Generally, if heat is applied, bacteria are killed; if cold temperatures are utilized, bacterial growth is inhibited.

Heat sensitivity of organisms can be affected by container size, cell density, moisture content, pH, and medium composition (Exercise 11). Bacteria exhibit different tolerances to the application of heat. Heat sensitivity is genetically determined and is partially reflected in the optimal growth ranges (Exercise 41), which are **psychrophilic** (0° to 20°C), **mesophilic** (20° to 45°C), and **thermophilic** (50° to 90°C), and by the presence of heat-resistant endospores (Exercise 9). Overall, bacteria are more heat resistant than most other forms of life.

Heat can be applied as dry or moist heat. **Dry heat,** such as hot air ovens or incineration (for example, flaming loops), denatures enzymes, dehydrates microbes, and kills by oxidation effects. A standard application of dry heat in a hot air oven is 170°C for 2 hours. The heat of hot air is not readily transferred to a cooler body such as a microbial cell. Moisture transfers heat energy to the microbial cell more efficiently than dry air, resulting in the denaturation of enzymes. **Moist heat** methods include pasteurization, boiling, and autoclaving. In **pasteurization** the temperature is maintained at 63°C for 30 minutes or 72°C for 15 seconds to kill designated organisms that are pathogenic or cause spoilage. **Boiling** (100°C) for 10 minutes will kill vegetative bacterial cells; however, endospores are not inactivated. The most effective method of moist heat sterilization is **autoclaving,** the use of steam under pressure (Exercise 11). Increased pressure raises the boiling point of water and produces steam with a higher temperature. (See Table 43-1.) Standard conditions for autoclaving are *15 psi, 121°C* for *15 minutes.* This is sufficient to kill endospores and render materials sterile.

In using heat to control the growth of microbes, two different methods of measuring heat effectiveness are utilized. **Thermal death time (TDT)** is the length of time required to kill all bacteria in a liquid culture at a given temperature. The less common **thermal death point (TDP)** is the temperature required to kill all bacteria in a liquid culture in 10 minutes.

Table 43-1

Relationship Between Pressure and Temperature of Steam

Pressure in Pounds per Square Inch (psi) (In Excess of Atmospheric Pressure)	Temperature in C°
0 psi	100°C
5 psi	110°C
10 psi	116°C
15 psi	121°C
20 psi	126°C
30 psi	135°C

Source: G. J. Tortora, B. R. Funke, and C. L. Case. 1982. *Microbiology.* Menlo Park, Calif.: Benjamin/Cummings.

Materials

Petri plates containing nutrient agar (2)

Thermometer

Empty tube

Beaker

Hot plate or tripod and asbestos pad

Ice

Cultures (as assigned)

Group A:
Old (48–72 hours) *Bacillus subtilis*
Young (24 hours) *Bacillus subtilis*

Group B:
Staphylococcus epidermidis
Escherichia coli

Group C:
Young (24 hours) *Bacillus subtilis*
Escherichia coli

Group D:
Mold *(Penicillium)* spore suspension
Old (48–72 hours) *Bacillus subtilis*

Techniques Required

Exercises 13 and 14

Procedure

Each pair of students should be assigned a pair of cultures and a temperature.

Student Group	Student Group
A—63°C ______	A—72°C ______
B—63°C ______	B—72°C ______
C—63°C ______	C—72°C ______
D—63°C ______	D—72°C ______

Beakers of water can be shared as long as the same temperature is being evaluated.

1. Divide two plates of nutrient agar into five sections each. Label the sections "0," "15 sec," "2 min," "5 min." and "15 min."
2. Set up a water bath in the beaker. The water level should be higher than the level of the broth in the tubes. Do not put the broth tubes into the water bath at this time. Carefully put the thermometer in a test tube of water in the bath.
3. Streak the assigned organisms on the 0-time section of the appropriate plate. Why are we using "old" and "young" *Bacillus* cultures? ______
4. Raise the temperature of the bath to the desired temperature and maintain at that temperature. Ice can be used to adjust the temperature. Why was 63°C selected as one of the temperatures? ______
5. Place the broth tubes of your organism into the bath when the temperature is at the desired point. After 15 sec remove the tubes, resuspend the culture, streak a loopful on the corresponding sections, and return the tubes to the water bath. Repeat at 2, 5, and 15 min. What is the longest time period that any microbe is exposed to heat? ______
6. When done, clean the beaker and return the materials. Incubate the plates inverted at 35°C until the next lab period. Record your results and the results for the other organisms tested: (−) = no growth; (+) = minimum growth: (2+) = moderate growth; (3+) = heavy growth; and (4+) = maximum growth.

Turn to the Laboratory Report for Exercise 43.

EXERCISE 44

Physical Methods of Control: Ultraviolet Radiation

Objectives

After completing this exercise you should be able to

1. Examine the effects of ultraviolet radiation on bacteria.
2. Explain the method of action of ultraviolet radiation and light repair of mutations.

Background

Radiant energy comes to the earth from the sun and other extraterrestrial sources, and some is generated on earth from natural and man-made sources. The **radiant energy spectrum** is shown in Figure 44-1. Radiation differs in wavelength and energy. The shorter wavelengths have more energy. X rays and gamma rays are forms of **ionizing radiation.** Their principal effect is to ionize water into *highly reactive free radicals* (with unpaired electrons) that can break strands of DNA. The effect of radiation is influenced by many variables such as the age of the cells, media composition, and temperature.

Some **nonionizing** wavelengths are essential for biochemical processes. The main absorption for green algae, green plants, and photosynthetic bacteria are shown in Figure 44-1. Animal cells synthesize vitamin D in the presence of light around 300 nm. Nonionizing wavelengths are harmful to most bacteria. The most lethal wavelength is 265 nm, which corresponds to the optimal absorption wavelength of DNA. Ultraviolet light induces *pyrimidine dimers* in the nucleic acid, which results in a deletion mutation. Mutations in critical genes result in the death of the cell unless the damage is repaired. When pyrimidine dimers are exposed to visible light, the enzyme pyrimidine dimerase is activated and splits the dimer. This is called **light repair** or **photoreactivation.**

As a sterilizing agent, ultraviolet radiation is limited by its poor penetrating ability. It is used to sterilize some heat-labile solutions and to decontaminate hospital operating rooms and food-processing areas.

In this exercise we will investigate the effects of ultraviolet radiation and light repair using lamps of the desired wavelenth.

Materials

Petri plates containing nutrient agar (3)

Sterile cotton swabs (3)

Cardboard or prepared templates (13 cm × 13 cm)

Ultraviolet lamp (265 nm)

Scissors

Cultures (one of the following, as assigned)

Serratia marcescens

Bacillus subtilis

Techniques Required

Exercise 13

Procedure

1. Swab the surface of each plate with either *Serratia* or *Bacillus;* to ensure complete covering, swab the surface in two directions. Label the plates "A," "B," and "C."
2. Select a cardboard template or design your own by cutting a pattern in a piece of cardboard with scissors.
3. Place each plate directly under the ultraviolet light about 5 cm from the light with the *cover off,* agar surface up, and the template over the plate bottom according to the following protocol. Why should the cover be removed? ____________

   ____

   Plate A: Expose for 30 sec and put in a dark incubator (22°C).
   Plate B: Expose for 30 sec and incubate in a room with the lights on.
   Plate C: Expose for 90 sec and put in a dark incubator (22°C).
   Do not look at the ultraviolet light. Why not? ____________

   ____

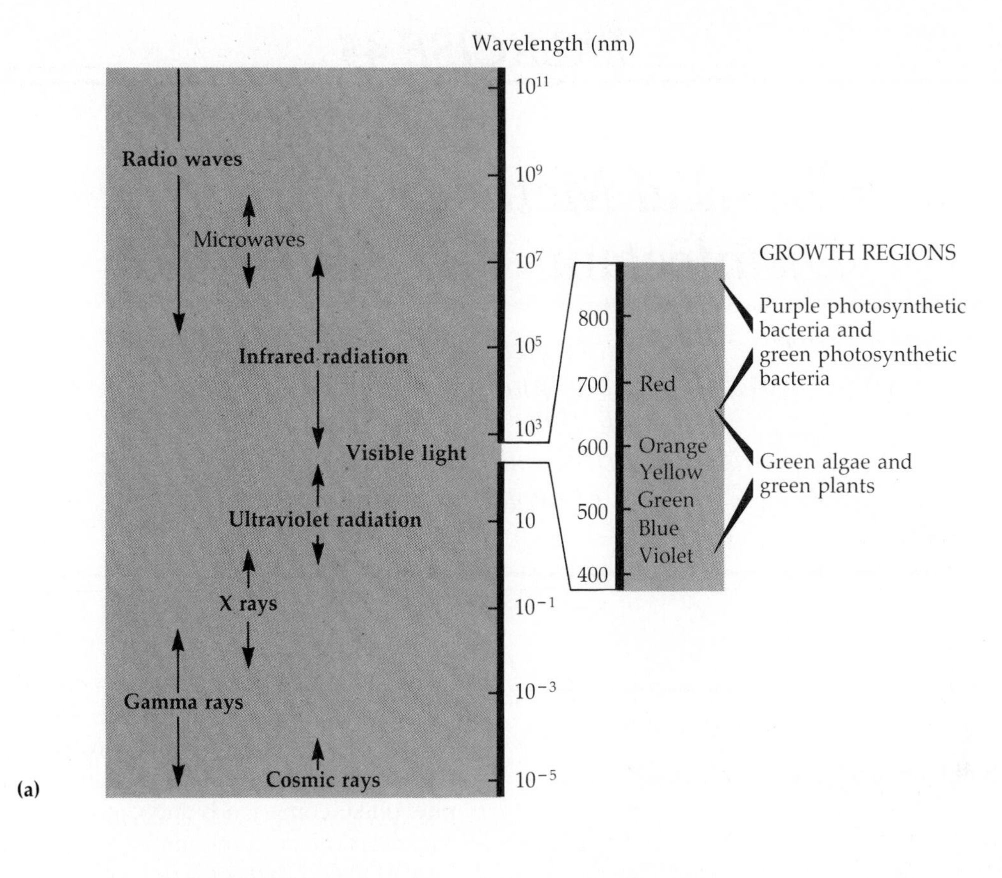

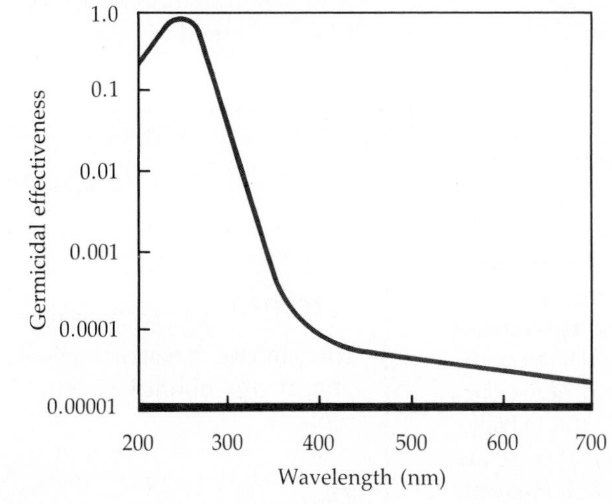

Figure 44-1.

Radiant energy. **(a)** Radiant energy spectrum and absorption of light for photosynthesis (Exercise 53). **(b)** Germicidal effectiveness of radiant energy between 200 and 700 nanometers (nm) (from UV to visible red light).

4. Incubate all three plates inverted at 22°C or at room temperature until the next period.
5. Examine all plates and record your results. Observe the results of students using the other bacteria.

Turn to the Laboratory Report for Exercise 44.

EXERCISE 45

Chemical Methods of Control: Disinfectants and Antiseptics

One 19th-century method of avoiding cholera: Wear a pouch of foul-smelling herbs around your neck. If the odor is bad enough, disease carriers will spare you the trouble of avoiding them.

Objectives

After completing this exercise you should be able to

1. Define the following terms: disinfectant and antiseptic.
2. Evaluate the relative effectiveness of various chemical substances as antimicrobial agents.
3. Calculate a phenol coefficient.

Background

A wide variety of chemicals called **antimicrobial agents** are available for controlling the growth of microbes. Chemotherapeutic agents are used internally and will be evaluated in Exercise 46. **Disinfectants** are chemical agents used on inanimate objects to lower the level of microbes on that surface; **antiseptics** are chemicals used on living tissue to decrease the number of microbes. Disinfectants and antiseptics affect bacteria in many ways. Those that result in bacterial death are called **bactericidal agents.** Those causing temporary inhibition of growth are **bacteriostatic agents.**

No single chemical is the best to use in all situations. Antimicrobial agents must be matched to specific organisms and environmental conditions. Additional variables to consider in selecting an antimicrobial agent include pH, solubility, toxicity, organic material present, and cost. In evaluating the effectiveness of antimicrobial agents, the concentration, length of contact, and whether it is lethal *(-cidal)* or inhibiting *(-static)* are the important criteria. One method employed to measure the effectiveness of a chemical agent is to determine its **phenol coefficient,** where the effectiveness of a phenolic-based chemical is compared to phenol. **Phenol** and its derivatives, called **phenolics,** kill bacteria by damaging the plasma membrane, inactivating enzymes, and denaturing proteins. The phenol coefficient is a comparative test using a standard organism such as *Staphylococcus aureus* as shown in the sample problem (given in the box). The phenol coefficient is limited to bactericidal phenollike compounds and cannot be used to evaluate bacteriostatic compounds.

A modified determination of a phenol coefficient will be performed in this exercise.

Sample Problem: Phenol Coefficient

Purpose

To compare hexachlorophene to phenol and obtain the phenol coefficient of hexachlorophene.

Procedure

1. Serial dilutions of each chemical were set up: 1:10 to 1:500. How many tubes are needed? ____________
2. An equal amount of broth culture of a standard bacterium was added to each tube. Why is *Staphylococcus aureus* used to test hexachlorophene? ____________
3. At 5-min, 10-min, and 15-min intervals, tubes of sterile nutrient broth were inoculated with a loopful of each dilution.
4. The nutrient broth was incubated at 35°C for 24 to 48 hours and examined for growth.
5. Results were recorded as follows: (+) for growth and (−) for no growth. (−) means the agent has killed the bacteria.

Results

Chemical	Dilution	Exposure Time (min) 5	10	15
Phenol	1:10–1:70	–	–	–
	1:80	+	–	–
	1:90	+	+	–
	1:100	+	+	–
	1:110–1:500	+	+	+
Hexachlorophene	1:10–1:150	–	–	–
	1:160	+	–	–
	1:170	+	+	+
	1:180	+	+	+
	1:190–1:500	+	+	+

Interpretation

Circle the dilution of the test substance (that is, hexachlorophene) that killed the bacteria in 10 minutes but not in 5 minutes. Circle the dilution of phenol that killed the bacteria in 10 minutes but not in 5 minutes.

Calculate the phenol coefficient as follows:

$$\frac{\text{Reciprocal of the hexachlorophene dilution circle}}{\text{Reciprocal of the phenol dilution circled}} = \text{Phenol coefficient}$$

Conclusions

What is the phenol coefficient of hexachlorophene? ______

In this sample, was hexachlorophene shown to be more or less effective than phenol? ______

How can you tell? ______

Materials

Glucose fermentation tubes (9)

Sterile water

Sterile tubes (3)

5-ml sterile pipettes (2)

1-ml sterile pipettes (2)

Phenol (0.5%)

Test substance: chemical agents such as bathroom cleaner, hydrogen peroxide, floor cleaner, mouthwash, lens cleaner, acne cream. Bring your own.

Culture

Staphylococcus aureus

Techniques Required

Exercises 13 and 18

Procedure

1. Add 5 ml phenol to one sterile tube. Label the tube. Prepare a dilution of the test substance in sterile water. Add 5 ml of the diluted test substance to the other tube. The test substance should be diluted to the strength at which it is normally used. If a paste, it must be suspended in sterile water. Total volume should be 5 ml in each tube.
2. Label four glucose fermentation tubes "Phenol" and four "Test." Then label one tube of each set "2.5"; label another pair "5"; continue with "10" and "20." (See Figure 45-1.) Inoculate one glucose fermentation tube with a loopful of *S. aureus* and label the tube "0."

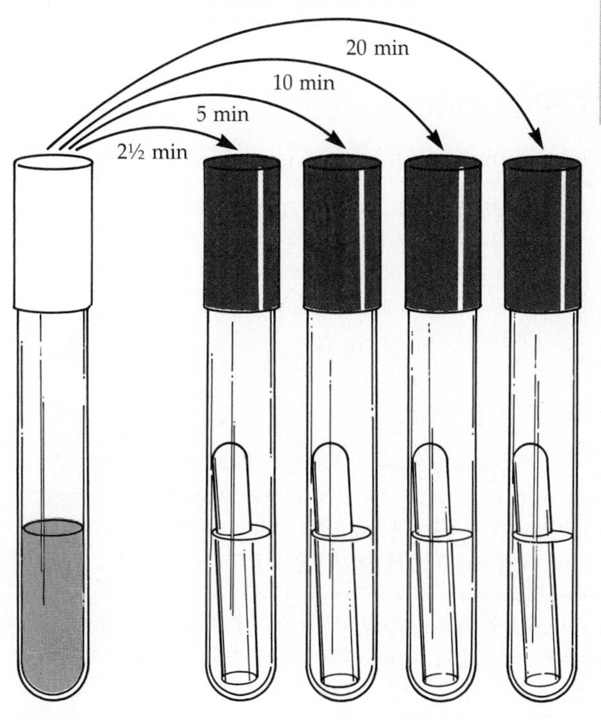

Figure 45-1.

A loopful from the tube containing phenol and *Staphylococcus* is transferred to a glucose fermentation tube at the time intervals shown. The procedure is repeated with a loopful from the tube containing the test substance and *Staphylococcus* into a different set of glucose fermentation tubes.

3. Aseptically add 0.5 ml of the *S. aureus* culture to each tube prepared in step 1. One loopful from each (chemical/*S. aureus*) tube should be transferred to a corresponding glucose fermentation tube at 2.5 min, 5 min, 10 min, and 20 min.
4. Incubate the glucose fermentation tubes at 35°C until the next lab period. (Discard the chemical/bacteria mixtures.)
5. Observe the glucose fermentation tubes for growth and fermentation.

Turn to the Laboratory Report for Exercise 45.

EXERCISE 46

Chemical Methods of Control: Antimicrobial Drugs

The aim of medicine is to prevent disease and prolong life; the ideal of medicine is to eliminate the need of a physician.

W. J. MAYO

Objectives

After completing this exercise you should be able to

1. Define the following terms: antibiotic, chemotherapeutic agent, and MIC.
2. Perform an antibiotic sensitivity test.
3. Provide the rationale for the agar diffusion technique.

Background

The observation that some microbes inhibited the growth of others was made as early as 1874. Pasteur and others observed that infecting an animal with *Pseudomonas aeruginosa* protected the animal against *Bacillus anthracis.* Later investigators coined the word **antibiosis** (against life) for this inhibition and called the inhibiting substance an **antibiotic.** In 1928, Fleming observed antibiosis around a mold *(Penicillium)* growth on a culture of staphylococci. He found that culture filtrates of *Penicillium* inhibited the growth of many gram-positive cocci and *Neisseria* spp. In 1940, Selman A. Waksman isolated the antibiotic streptomycin, produced by an actinomycete. This antibiotic was effective against many bacteria that were not affected by penicillin. Actinomycetes remain an important source of antibiotics. Today, research investigators look for antibiotic-producing actinomycetes and fungi in soil, and have synthesized many antimicrobial substances in the laboratory. Antimicrobial chemicals absorbed or used internally, whether natural *(antibiotics)* or synthetic, are called **chemotherapeutic agents.**

A physician or dentist needs to select the correct chemotherapeutic agent intelligently and administer the appropriate dose in order to treat an infectious disease; then the practitioner must follow that treatment in order to be aware of resistant forms of the organism that might occur. The clinical laboratory isolates the pathogen (disease-causing organism) from a clinical sample and determines its sensitivity to chemotherapeutic agents.

In the **agar diffusion method,** bacteria to be tested are added to melted Mueller Hinton agar and poured onto solidified agar in a Petri plate. Paper disks impregnated with various chemotherapeutic agents are placed on the surface of the agar. (Mueller Hinton agar allows the chemotherapeutic agent to diffuse freely.) During incubation, the chemotherapeutic agent *diffuses* from the disk, from an area of high concentration to an area of lower concentration. An effective agent will inhibit bacterial growth and measurements can be made of the size of the **zones of inhibition** around the disks. The concentration of chemotherapeutic agent at the edge of the zone of inhibition represents its **minimum inhibitory concentration (MIC).** The MIC is determined by comparing the zone of inhibition with MIC values in a standard table (Table 46-1). The zone size is affected by factors such as diffusion rate of the antibiotic and growth rate of the organism. To minimize the variance between laboratories, the standardized **Kirby–Bauer test** for performing agar diffusion tests is employed in many clinical laboratories.

In this exercise a modified agar-disk diffusion method will be evaluated.

Table 46-1

Interpretation of Inhibition Zones of Test Cultures

Disk symbol	Antibiotic	Disk content	Diameter of zones of inhibition mm Resistant	Intermediate	Susceptible
AM	Ampicillin[a] when testing gram-negative microorganisms and enterococci	10 mcg	11 or less	12-13	14 or more
AM	Ampicillin[a] when testing staphylococci and Penicillin G-susceptible microorganisms	10 mcg	20 or less	21-28	29 or more
AM	Ampicillin[a] when testing *Hemophilus* species	10 mcg	19 or less		20 or more
B	Bacitracin	10 units	8 or less	9-12	13 or more
CB	Carbenicillin when testing *Proteus* species and *Escherichia coli*	50 mcg	17 or less	18-22	23 or more
CB	Carbenicillin when testing *Pseudomonas aeruginosa*	50 mcg	12 or less	13-14	15 or more
CR	Cephalothin[b] when reporting susceptibility to cephalothin, cephaloridine, and cephalexin	30 mcg	14 or less	15-17	18 or more
CR	Cephalothin[b] when reporting susceptibility to cephaloglycin	30 mcg	14 or less		15 or more
C	Chloramphenicol (Chloromycetic®)	30 mcg	12 or less	13-17	18 or more
CC	Clindamycin[c] when reporting susceptibility to clindamycin	2 mcg	14 or less	15-16	17 or more
CC	Clindamycin[c] when reporting susceptibility to lincomycin	2 mcg	16 or less	17-20	21 or more
CL	Colistin[d] (Coly-mycin®)	10 mcg	8 or less	9-10	11 or more
E	Erythromycin	15 mcg	13 or less	14-17	18 or more
GM	Gentamicin	10 mcg	12 or less		13 or more
K	Kanamycin	30 mcg	13 or less	14-17	18 or more
ME	Methicillin[e]	5 mcg	9 or less	10-13	14 or more
N	Neomycin	30 mcg	12 or less	13-16	17 or more

Continued

Table 46-1 (continued)
Interpretation of Inhibition Zones of Test Cultures

Disk symbol	Antibiotic	Diameter of zones of inhibition mm			
		Disk content	Resistant	Intermediate	Susceptible
NB	Novobiocin[f]	30 mcg	17 or less	18-21	22 or more
OL	Oleandomycin[g]	15 mcg	11 or less	12-16	17 or more
P	Penicillin G. when testing staphylococci[h]	10 units	20 or less		21 or more
P	Penicillin G. when testing other microorganisms[h,i]	10 units	11 or less	12-21	22 or more
PB	Polymyxin B[d]	300 units	8 or less	9-11	12 or more
R	Rifampin when testing *Neisseria meningitidis* susceptibility only	5 mcg	24 or less		25 or more
S	Streptomycin	10 mcg	11 or less	12-14	15 or more
T	Tetracycline[j]	30 mcg	14 or less	15-18	19 or more
VA	Vancomycin	30 mcg	9 or less	10-11	12 or more

Source: Difco Laboratories, Detroit, Michigan
[a] The ampicillin disk is used for testing susceptibility of both ampicillin and betacillin.
[b] Staphylococci exhibiting resistance to the penicillinase-resistant penicillin class disks should be reported as resistant to cephalosporin class antibiotics. The 30 mcg cephalothin disk cannot be relied upon to detect resistance of methicillin-resistant staphylococci to cephalosporin class antibiotics.
[c] The clindamycin disk is used for testing susceptibility to both clindamycin and lincomycin.
[d] Colistin and polymyxin B diffuse poorly in agar and the accuracy of the diffusion method is thus less than with other antibiotics. Resistance is always significant, but when treatment of systemic infections due to susceptible strains is considered, it is wise to confirm the results of a diffusion test with a dilution method.
[e] The methicillin disk is used for testing susceptibility of all penicillinase-resistant penicillins; that is, methicillin, cloxacillin, dicloxacillin, oxacillin, and nafcillin.
[f] Not applicable to medium that contains blood.
[g] The oleandomycin disk is used for testing susceptibility to oleandomycin and trioleandomycin.
[h] The penicillin G. disk is used for testing susceptibility to all penicillinase-susceptible penicillins except ampicillin and carbenicillin: that is, penicillin G., phenoxymethyl penicillin, and phenethicillin.
[i] This category includes some organisms such as enterococci and gram-negative bacilli that may cause systemic infections treatable with high doses of penicillin G. Such organisms should only be reported susceptible to penicillin G. and not to phenoxymethyl penicillin or phenethicillin.
[j] The tetracycline disk is used for testing susceptibility to all tetracyclines; that is chlorotetracycline, demeclocycline, doxycycline, methacycline, oxytetracycline, rolitetracycline, minocycline, and tetracycline.

Materials

Petri plates containing Mueller Hinton Agar (3)

Sterile cotton swabs (3)

Antibiotic dispenser and disks

Forceps

Alcohol

Garlic powder, elemental copper, or a penny

Ruler

Cultures (one of each)

Staphylococcus aureus

Escherichia coli

Pseudomonas aeruginosa

Techniques Required

Exercise 13

Procedure

1. Label one plate for each culture. Aseptically swab

each culture onto the appropriate plate. Swab in three directions to ensure complete plate coverage. Why is complete coverage essential? ______

______

Let stand at least 5 minutes.

2. Follow procedure (a) or (b), as given:
   a. Place the chemotherapeutic impregnated disks by pushing the dispenser over the agar. Sterilize your loop and touch each disk to give better contact with the agar. Record the agents and the disk code in your laboratory report. Circle the corresponding chemicals in Table 46-1.
   b. Sterilize forceps by dipping in alcohol and burning off the alcohol. Obtain a disk impregnated with a chemotherapeutic agent and place the disk on the surface of the agar (Figure 46-1*a*). Gently tap the disk to give better contact with the agar. Repeat, placing five to six disks the same distance apart on the Petri plate. See the location of the disks in Figure 46-1*b*. Record the agents and the disk code in your Laboratory Report. Circle the corresponding chemical in Table 46-1.
3. If you have space on your plate, add a small amount of garlic, elemental copper, or a penny to an unused area.
4. Incubate inverted until the next period. Measure the zones of inhibition in millimeters, using a ruler on the underside of the plate (Figure 46-1*b*). Record the zone size and, based on the values in Table 46-1, indicate whether the organism is sensitive, intermediate, or resistant.

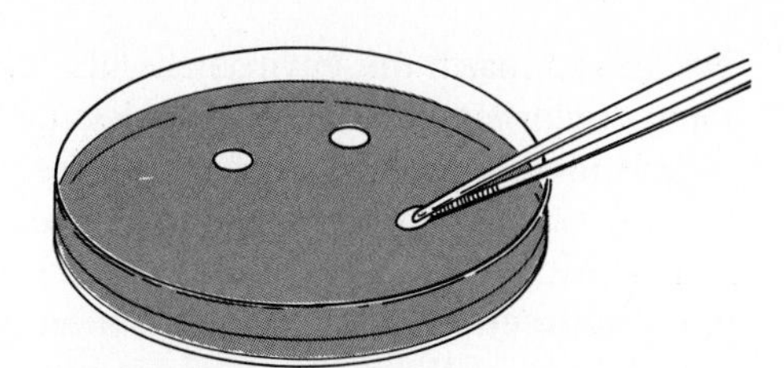

(a) Place disks impregnated with chemotherapeutic agents on an inoculated culture medium with sterile forceps to get the pattern shown in **(b)**.

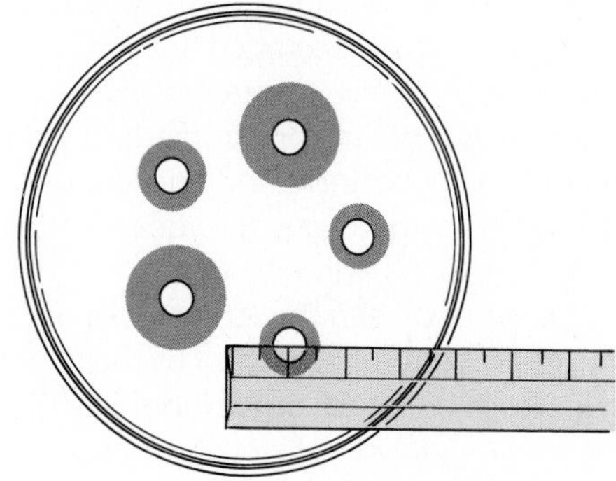

(b) After incubation, measure diameter of zone of inhibition.

Figure 46-1. ______
Agar diffusion method.

Turn to the Laboratory Report for Exercise 46.

EXERCISE 47

EXERCISE 47

Effectiveness of Hand Scrubbing

People are sick because they are poor, they become poorer because they are sick and they become sicker because they are poorer.

AUTHOR UNKNOWN

Objectives

After completing this exercise you should be able to

1. Evaluate the effectiveness of hand washing and a surgical scrub.
2. Explain the importance of aseptic technique in the hospital environment.

Background

The skin is sterile during fetal development. When a baby enters the birth canal, the skin becomes colonized by bacteria and is colonized by many bacteria for the rest of its life. As an individual ages and changes environments, the microbial population changes to match the environmental conditions. The microorganisms that are more or less permanent are called **normal flora.** Microbes that are present only for days or weeks are referred to as **transient flora.**

Discovery of the importance of hand and skin surface **disinfection** in disease prevention is credited to Semmelweis at Vienna General Hospital in 1846. He noted that the lack of aseptic methods was directly related to the incidence of puerperal fever and other diseases. Medical students would go directly from the autopsy room to the patient's bedside and assist in child delivery without washing their hands. Less puerperal sepsis occurred in patients attended by nurses who did not touch cadavers. Semmelweis established a policy for the medical students of hand washing with a chloride of lime solution that resulted in a drop in the death rate due to puerperal sepsis from 12% to 1.2% in one year.

A layer of oil and the structure of the skin prevents the removal of all bacteria by hand washing. Soap helps remove the oil and scrubbing will maximize the removal of bacteria. Hospital procedures require personnel to wash their hands before attending a patient, and a complete surgical scrub—removing the transient and many of the resident microflora—is done before surgery. Transient flora are usually removed after 7 to 8 minutes of scrubbing with an antiseptic and soap. The surgeon's skin is never sterilized. Only burning or scraping it off would achieve that.

In this experiment, the effectiveness of washing skin with soap and water will be examined. Only organisms capable of growing aerobically on nutrient agar will be examined. Since organisms with different nutritional and environmental requirements will not grow, this experimental procedure will involve only a minimum number of the skin microflora.

Materials

Petri plates containing nutrient agar (2)

Techniques Required

Exercise 12

Procedure

1. Divide two nutrient agar plates into four quadrants. Label the sections of each plate 1 through 4. Label one plate "Water," the other "Soap."
2. Do the "water" plate first. Touch section 1 with your fingers, wash well *without* soap, shake off excess water, and, while still wet, touch section 2. Do not dry your fingers with a towel. Wash again, and while wet touch section 3. Wash a final time and touch section 4.
3. Use your other hand on the plate labeled "soap." Repeat the procedure in step 2 except wash each time with soap, rinse, shake off excess water, then touch the designated sector.
4. Incubate the plates inverted at 35°C until the next period.
5. Record results.

Turn to the Laboratory Report for Exercise 47.

EXERCISE 48

Infection Control: Equipment Sampling

Objectives:

After completing this exercise you should be able to

1. Define the following terms: nosocomial and fomite.
2. Provide a rationale for equipment sampling.
3. List the functions of a hospital infection control unit.

Background

It is estimated that one-third of all infections in hospitalized patients are **nosocomial.** That is, the infections were not present when the patients entered the hospital. A patient's lowered resistance increases susceptibility to infections by microorganisms introduced by hospital staff, by visitors, and on fomites. A **fomite** is a nonliving object that can spread infection.

Aseptic technique and disinfection or sterilization of equipment are essential to minimize the incidence of nosocomial infection. **Hospital infection control** personnel are responsible for collecting and analyzing data on hospital infections in order to direct attention to areas of greatest need. Infection control personnel also routinely oversee equipment and procedures.

One type of equipment sampling will be used in this exercise. Samples will be inoculated onto the following selective and differential media (Exercise 15).

EMB (eosin-methylene blue) **agar** is selective because the EMB dyes inhibit the growth of gram-positive organisms, allowing the growth of gram-negative bacteria. EMB is differential in that lactose-fermenting bacteria give colored colonies and nonlactose fermenters produce colorless colonies. This medium is used to culture members of the Enterobacteriaceae found in the human intestines. Lactose fermenters are considered normal intestinal flora, but many nonlactose fermenters are pathogens (Exercise 61).

Mannitol salt agar selects for bacteria that tolerate high (7.5%) concentrations of sodium chloride. Phenol red is used to indicate fermentation of mannitol. A yellow halo surrounds colonies of mannitol-fermenting bacteria. This medium is primarily used to select for skin microflora and to differentiate between mannitol fermenters such as *Staphylococcus aureus* and nonfermenters such as *Micrococcus* spp. or *Staphylococcus epidermidis*.

Materials

Petri plate containing trypticase soy agar (1)

Petri plate containing EMB *or* mannitol salt agar (1)

Sterile cotton swabs (2)

Sterile water

Gram staining reagents

Techniques Required

Exercises 7, 13, and 15

Procedure

1. Select a piece of equipment to sample. Your instructor will show you appropriate nursing, respiratory therapy, or laboratory equipment.
2. Aseptically remove one swab and moisten it in sterile water. Squeeze excess moisture from the swab by pressing it against the inside of the tube (Figure 48-1*a*).
3. Swab the surface to be tested.
4. Swab one-half the surface of the trypticase soy agar plate, rotating the swab on the plate.

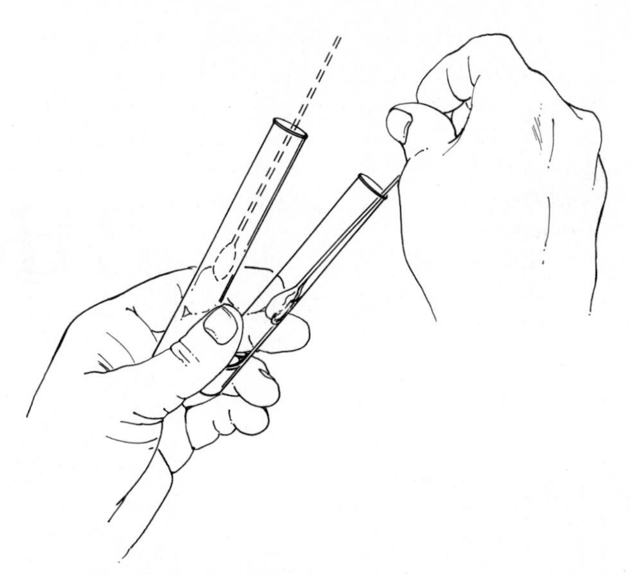

(a) Aseptically moisten a sterile cotton swab in water.

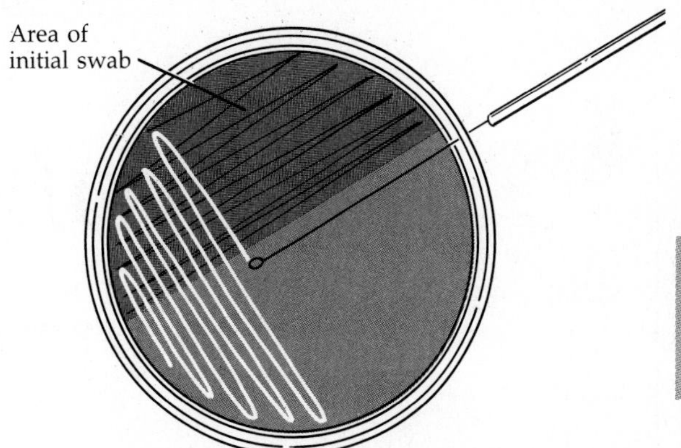

(b) Using a sterile loop, spread the inoculum over the remaining half of the plate.

Figure 48-1.
Taking a sample from a fomite.

5. Repeat the procedure on another section of the equipment and inoculate the EMB or mannitol salt plate.
6. Using a sterile loop, spread the inoculum over the remaining half of each plate (Figure 48-1*b*), streaking at right angles to the initial swab.
7. Incubate the plates inverted at 35°C for 24 to 48 hours.
8. Examine the plates and record the numbers of different bacteria.
9. Prepare and examine Gram stains from isolated colonies.

Turn to the Laboratory Report for Exercise 48.

PART 10

Microbes in the Environment

EXERCISES

It should be evident from previous exercises that microorganisms are omnipresent in our environment. Although the presence of some microorganisms is undesirable, the activities of most microbes are beneficial. In Exercise 55, selected microorganisms will be used to produce desired flavors in foods and to preserve foods. In Exercises 49 and 50, we will use the presence of nonpathogenic bacteria to indicate water pollution. The presence and number of bacteria in foods may indicate that foods were contaminated during processing; these bacteria may result in food spoilage (Exercise 54).

The numbers of bacteria in the soil will be estimated in Exercise 51, and in Exercises 52 and 53 their metabolism will be investigated. The activities of these soil bacteria are essential to the maintenance of life on earth. Nitrogen fixation (Exercise 52) was first described by Sergei Winogradsky, who entered into the study of biogeochemical cycles because he was

> *impressed by the incomparable glitter of Pasteur's discoveries. I started (in 1893) to investigate the great problem of fixation of atmospheric nitrogen. I succeeded without too much difficulty in isolating an anaerobic rod called* Clostridium . . . *that would perform this function.*

Winogradsky named the organism *Clostridium pasteurianum* in honor of Pasteur.

The experiment performed in Exercise 53 is a modification of the work done by Winogradsky in the 1890s. Winogradsky isolated and identified autotrophic bacteria. **Autotrophs** are organisms that use carbon dioxide as their principal carbon source and inorganic molecules or light as their energy source. He proved that autotrophs use carbon dioxide as their main carbon source and that the **chemoautotrophs** use inorganic chemicals as their energy source, while **photoautrophs** use light (see the table).

Nutritional Classification of Organisms

Nutritional Type	Energy Source	Carbon Source	Examples
Photoautotroph	Light	Carbon dioxide (CO_2)	Photosynthetic bacteria (green sulfur and purple sulfur bacteria), cyanobacteria, algae, plants
Photoheterotroph	Light	Organic compounds	Purple nonsulfur and green nonsulfur bacteria
Chemoautotroph	Inorganic compounds	Carbon dioxide (CO_2)	Hydrogen, sulfur, iron, and nitrifying bacteria
Chemoheterotroph	Organic componds	Organic compounds	Most bacteria, fungi, protozoans, and all animals

Source: G. J. Tortora, B. R. Funke, and C. L. Case. 1982. *Microbiology.* Menlo Park, Calif.: Benjamin/Cummings, p. 119.

EXERCISE 49

Microbes in Water: Multiple-Tube Technique

Objectives

After completing this exercise you should be able to

1. Define coliform.
2. Provide the rationale for determining the presence of coliforms.
3. List and explain each step in the multiple-tube technique.

Background

Tests to determine the bacteriologic quality of water have been developed to prevent transmission of waterborne diseases of fecal origin. However, it is not practical to look for pathogens in water supplies. When present, pathogens are in small numbers and might be missed by sampling. Moreover, when pathogens are detected, it is usually too late to prevent occurrence of the disease. Rather, the presence of **indicator organisms** is used to detect fecal contamination of water. The most frequently used indicator organisms are the coliform bacteria. **Coliforms** are aerobic or facultatively anaerobic, gram-negative, nonendospore-forming, rod-shaped bacteria that ferment lactose with acid and gas formation within 48 hours at 35°C. Coliforms are not restricted to the human gastrointestinal tract but may be found in other animals and in the soil. Tests to determine the presence of fecal coliforms (of human origin) have been developed. Coliforms are not usually pathogenic, although they can be opportunistic pathogens (Exercise 62).

Established public health standards specify the maximum number of coliforms in each 100 ml of water, depending on the intended use of the water (for example, drinking, water-contact sports, or treated wastewater for irrigation or for discharge into a bay or river).

Coliforms can be detected and enumerated in the **multiple-tube technique.** In this method, coliforms are detected in three stages (Figure 49-1). In the **presumptive test,** dilutions from a water sample are added to lactose fermentation tubes. The lactose broth is selective for gram-negative bacteria because of the addition of lauryl sulfate or brilliant green and bile. Fermentation of lactose to acid and gas is a positive reaction.

Samples from the positive presumptive tube at the highest dilution are streaked onto EMB agar (Exercises 48 and 61) in the **confirmed test.** Colored colonies on EMB agar is a positive confirmed test. In the **completed test,** isolated lactose-positive colonies from EMB agar are inoculated into lactose broth and onto a nutrient agar slant. If acid and gas are produced in the lactose broth, and the isolated bacterium is a gram-negative non–endospore-forming rod, it is a positive completed test.

The number of coliforms is determined by a statistical estimation called the **most probable number (MPN)** method. In the presumptive test, 15 tubes of lactose broth are inoculated with samples of the water being tested: 5 tubes receive 10 ml of water each; 5 tubes, samples of 1 ml each; and 5 tubes, samples of 0.1 ml each. A count of the number of tubes showing acid and gas is then taken and the figure compared to statistical tables shown in Table 49-1. The number is the *most probable number* of coliforms per 100 ml of water.

We will use a modification of the multiple-tube technique in this exercise. The presumptive test will be performed with only 3 tubes instead of 15.

Materials

9-ml single strength lactose fermentation tubes (2)

10-ml double strength lactose fermentation tube (1)

Petri plate containing EMB agar (1)

Nutrient agar slant (1)

Sterile 10-ml pipette (1)

Sterile 1-ml pipette (1)

Gram staining reagents

Water sample, 15 ml (Bring your own from a pond or stream.)

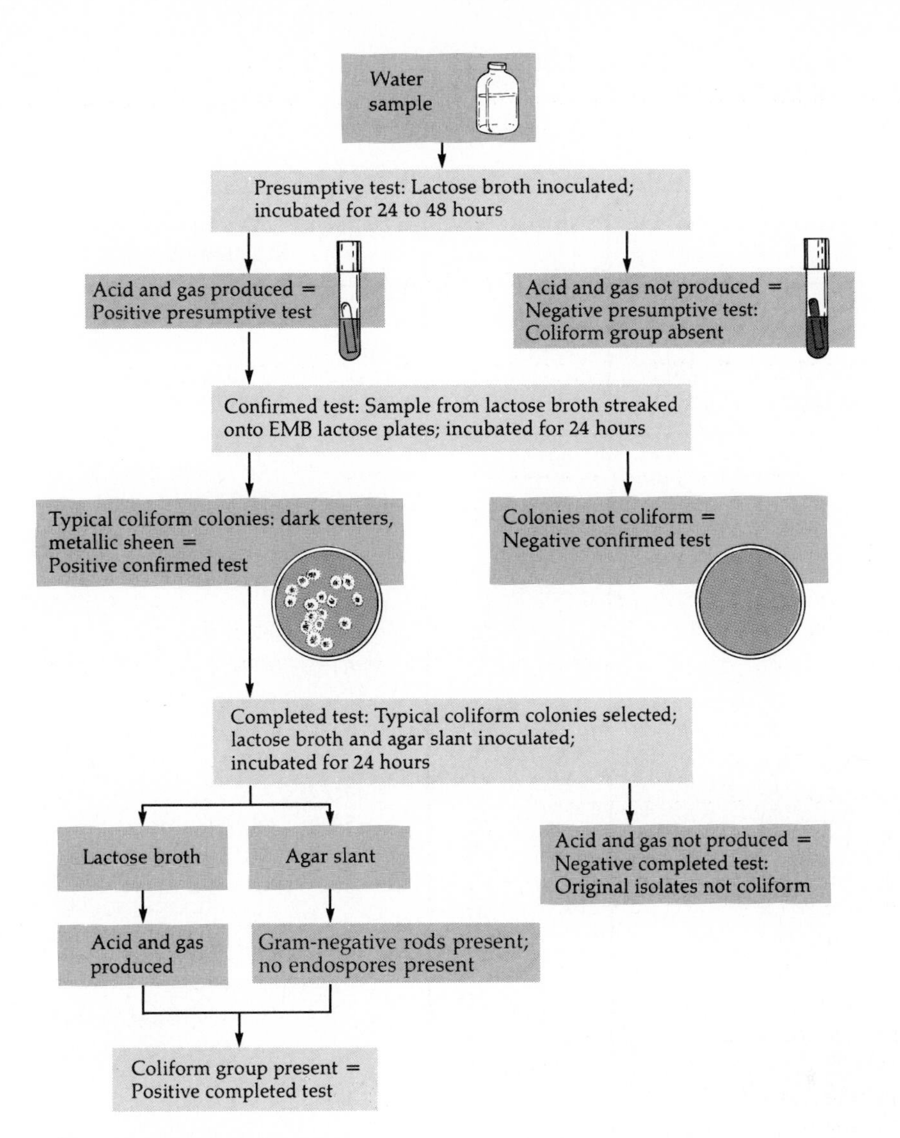

Figure 49-1.
Analysis of drinking water for coliforms by the multiple-tube fermentation technique.

Demonstration

MPN test

Techniques Required

Exercises 7, 13, 18, 48, and Appendix A

Procedure

1. Inoculate the single strength lactose broth with 1 ml of the water sample.
2. Inoculate the double strength lactose broth with 10 ml of the water sample. Why is double strength lactose broth used for this step? ________________

   ____
3. Incubate both tubes for 24 to 48 hours at 35°C.
4. Record the results of your presumptive test (Figure 49-1). Which tube has the highest dilution of the water sample? ________________

   ____

   If a tube has gas, streak the EMB agar with the positive broth. Incubate the plate inverted for 24 to 48 hours at 35°C.
5. Record the results of your confirmed test (Figure 49-1). How can you tell whether coliform colonies are present? ________________

   ____

   If coliform colonies are present, inoculate a single strength lactose broth tube and nutrient agar slant from an isolated colony. Incubate the broth and slant for 24 to 48 hours at 35°C and check for the presence of coliforms.
6. Gram stain the organisms found on the slant.
7. Examine the demonstration MPN test and determine the number of coliforms per 100 ml of the original sample.

Table 49-1

Most Probable Numbers (MPN) Index for Various Combinations of Positive and Negative Results When Five 10-ml Portions, Five 1-ml Portions, and Five 0.1-ml Portions Are Used

No. of Tubes Giving Positive Reaction out of			MPN Index per 100 ml	No. of Tubes Giving Positive Reaction out of			MPN Index per 100 ml
5 of 10 ml Each	5 of 1 ml Each	5 of 0.1 ml Each		5 of 10 ml Each	5 of 1 ml Each	5 of 0.1 ml Each	
0	0	0	<2	4	2	1	26
0	0	1	2	4	3	0	27
0	1	0	2	4	3	1	33
0	2	0	4	4	4	0	34
1	0	0	2	5	0	0	23
1	0	1	4	5	0	1	31
1	1	0	4	5	0	2	43
1	1	1	6	5	1	0	33
1	2	0	6	5	1	1	46
2	0	0	5	5	1	2	63
2	0	1	7	5	2	0	49
2	1	0	7	5	2	1	70
2	1	1	9	5	2	2	94
2	2	0	9	5	3	0	79
2	3	0	12	5	3	1	110
3	0	0	8	5	3	2	140
3	0	1	11	5	3	3	180
3	1	0	11	5	4	0	130
3	1	1	14	5	4	1	170
3	2	0	14	5	4	2	220
3	2	1	17	5	4	3	280
3	3	0	17	5	4	4	350
4	0	0	13	5	5	0	240
4	0	1	17	5	5	1	350
4	1	0	17	5	5	2	540
4	1	1	21	5	5	3	920
4	1	2	26	5	5	4	1600
4	2	0	22	5	5	5	≥2400

Source: Standard Methods for the Examination of Water and Wastewater, 13th ed., American Public Health Association, New York, 1971. From G. J. Tortora, B. R. Funke, and C. L. Case. 1982. *Microbiology.* Menlo Park, Calif.: Benjamin/Cummings, p. 724.

Turn to the Laboratory Report for Exercise 49.

EXERCISE 50

Microbes in Water: Membrane Filter Technique

Objectives

After completing this exercise you should be able to

1. Explain the principle of the membrane filter technique.
2. Provide the purpose of the membrane filter technique.
3. Perform a coliform count using the membrane filter technique.

Background

You have seen (in Exercise 49) that fecal contamination of water is determined by the number of coliforms present in a water sample. The multiple-tube technique was introduced in Exercise 49. Coliforms can also be detected by the membrane filter technique.

In the **membrane filter technique,** water is drawn through a thin filter. (See Appendix F.) Filters with a variety of pore sizes are available. Pores of 0.45 μm are used for filtering out most bacteria. Bacteria are retained on the filter, which is then placed on a pad of suitable nutrient medium. Nutrients that diffuse through the filter can be metabolized by bacteria trapped on the filter. Each bacterium that is trapped on the filter will develop into a colony. Bacterial colonies growing on the medium can be counted. When a selective or differential medium (Exercise 15) is used, desired colonies will have a distinctive appearance. Endo medium (Table 50-1) is frequently used as a selective and differential medium with the membrane filter technique. Desoxycholate and lauryl sulfate select for coliforms. Coliforms produce acid from lactose, causing colonies to have a metallic sheen.

Materials

47-mm Petri plate containing differential medium (EMB or Endo)

Sterile membrane filter apparatus

Sterile 0.45 μm filter

Table 50-1
Bacto-*m* Endo Broth MF®

Yeast extract	0.15%
Casitone	0.5%
Thiopeptone	0.5%
Tryptose	1.0%
Lactose	1.25%
Sodium desoxycholate	0.01%
Dipotassium phosphate	0.4375%
Monopotassium phosphate	0.1375%
Sodium chloride	0.5%
Sodium lauryl sulfate	0.005%
Sodium sulfite	0.21%
Basic fuchsin	0.105%

Source: Difco Laboratories. Prepared according to the Millipore Corporation.

Forceps

Alcohol

Sterile pipette or graduated cylinder, as needed

Sterile rinse water

Water sample (Bring your own from a pond or stream.)

Techniques Required

Exercise 13 and Appendices A and F

Procedure

1. Set up filtration equipment (see Appendix F). Remove wrappers as each piece is fitted into place. Why shouldn't all the wrappers be removed at once? ________
   a. Attach the filter trap to the vacuum source. What is the purpose of the filter trap? ________

   ____

b. Place the filter holder base (with stopper) on the filtering flask. Attach the flask to the filter trap.

c. Disinfect forceps by burning off the alcohol. Using the forceps, place a filter on the filter holder. Why must the filter be centered exactly on the filter holder? ______________________

____

d. Set the funnel on the filter holder and fasten in place.

2. Filtering

a. Shake the water sample and pour or pipette a measured volume into the funnel. Your instructor will help you determine the volume. *(For samples of 10 ml or less, pour 20 ml sterile water into the funnel first.)*

b. Turn on vacuum and allow the sample to pass into the filtering flask. Leave the vacuum on.

c. Pour sterile rinse water into the funnel *(use the same volume as the sample).* Allow the rinse water to go through the filter. Turn the vacuum off.

3. Inoculation

a. Carefully remove the filter from the filter holder using sterile forceps. Why does the filter have to be "peeled" off? ______________________

____

b. Carefully place the filter on the culture medium. Do not bend the filter; place one edge down first, then carefully set the remainder down (Figure 50-1). Place the filter on the plate as it was in the filter holder.

4. Invert the plate and incubate for 24 hours at 35°C.

5. Examine the plates for the presence of coliforms. On Endo or EMB, coliforms will form pink to red colonies, and some may have a green metallic sheen. Count the number of coliform colonies:

Number of coliforms per 100 ml of water =

$$100 \times \frac{\text{Number of coliform colonies}}{\text{Volume of sample filtered}}$$

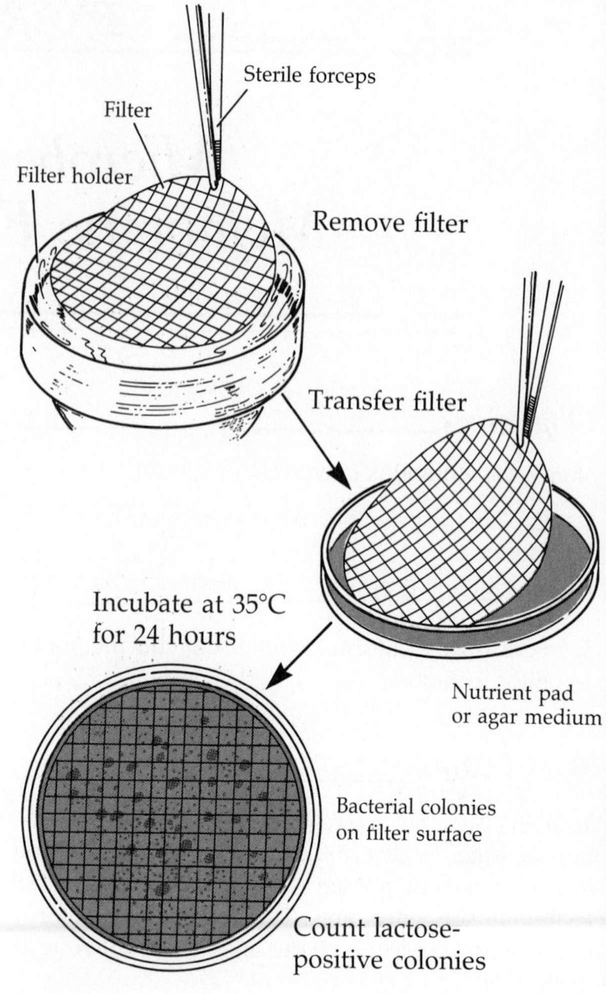

Figure 50-1. ______________________
Inoculation. Using sterile forceps remove the filter from the filter holder. Place the filter on the culture medium, gradually laying it down from one edge to the other.

Turn to the Laboratory Report for Exercise 50.

EXERCISE 51

Microbes in Soil: Quantification

Objectives

After completing this exercise you should be able to

1. Estimate the number of microorganisms in soil using the plate count method.
2. Describe the general activity of microorganisms in the soil.
3. Distinguish between bacterial, actinomycete, and fungal colonies.

Background

The soil is one of the main reservoirs of microbial life. Typical garden soil has millions of bacteria in each gram. The most numerous microbes in soil are bacteria (Table 51-1). Although actinomycetes are bacteria, they are listed separately because conidiospores make their dry, powdery colonies easily recognizable (Figure 51-1). Soil bacteria include aerobes and anaerobes with a wide range of nutritional requirements, from photoautotrophs to chemoheterotrophs. As usable nutrients and suitable environmental conditions (such as light, aeration, temperature) become available, the microbial populations and their metabolic activity rapidly increase until the nutrients are depleted or physical conditions change, and then they return to lower levels.

Human pathogens, with the exception of endospore-forming bacteria, are uncommon in the soil. Soil microorganisms are responsible for recycling elements so they can be used over and over again. The cycles of elements are called **biogeochemical cycles.** In the **carbon cycle** (Figure 51-2), carbon atoms are transferred from organism to organism. Carbon dioxide is incorporated into organic compounds by photoautotrophs, and these organic compounds are digested by chemoheterotrophs, which release carbon dioxide in respiration to be reused by photoautotrophs. Other biogeochemical cycles, such as the

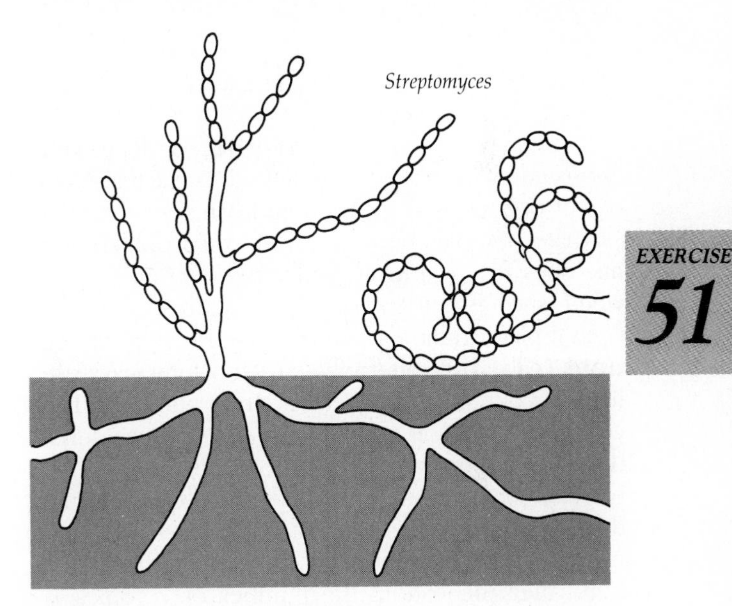

Figure 51-1.
Actinomycetes. *Streptomyces* form long filaments that penetrate the agar. Aerial filaments produce conidiospores.

EXERCISE 51

Table 51-1
Distribution of Microorganisms in Numbers Per Gram of Typical Garden Soil at Various Depths

Depth (cm)	Bacteria	Actinomycetes	Fungi	Algae
3–8	9,750,000	2,080,000	119,000	25,000
20–25	2,179,000	245,000	50,000	5,000
35–40	570,000	49,000	14,000	500
65–75	11,000	5,000	6,000	100
135–145	1,400	—	3,000	—

Source: Adopted from M. Alexander. 1977 *Introduction to Soil Microbiology*, 2nd Ed. New York: Wiley. From G. J. Tortora, B. R. Funke, and C. L. Case. 1982. *Microbiology*. Menlo Park, Calif.: Benjamin/Cummings, p. 666.

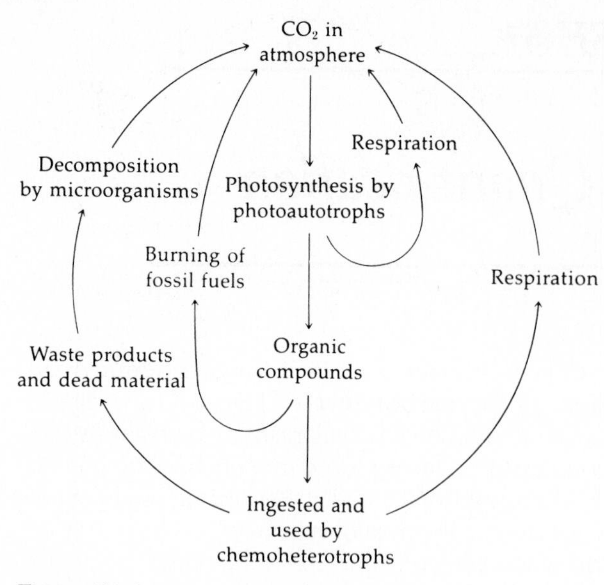

Figure 51-2.
The carbon cycle.

nitrogen cycle, will be studied in Exercise 52 and the sulfur cycle in Exercise 53.

The numbers of bacteria and fungi in soil are usually estimated by the plate count method. The actual number of organisms is probably much higher than the estimate, however, because a plate count only detects microbes that will grow under the conditions provided (such as nutrients and temperature).

In a plate count, the number of **colony-forming units (c.f.u.)** are determined. Each colony may arise from a group of cells rather than from one individual cell. The initial soil sample is diluted through serial dilutions (Appendix B) in order to obtain a small number of colonies on each plate. A known volume of the diluted sample is placed in a sterile Petri plate and melted, cooled nutrient agar is poured over the inoculum. After incubation, the number of colonies is counted. Plates with between 30 and 300 colonies are suitable for counting. A plate with fewer than 30 colonies is inaccurate because a single contaminant could influence the results. A plate with greater than 300 colonies is extremely difficult to count. The microbial population in the original soil sample can then be calculated using the following equation:

$$\text{Organisms per gram of soil} = \frac{\text{Number of colonies}}{\text{Amount plated} \times \text{dilution*}}$$

*"Dilution" refers to the tube prepared by serial dilutions (Appendix B). For example, if 250 colonies were present on the $1:10^7$ plate, the calculation would be

$$\begin{aligned}\text{Colony-forming units/gram} &= \frac{250 \text{ colonies}}{0.1 \text{ ml} \times 10^{-6}} \\ &= 250 \times 10^6 \times 10^1 \\ &= 2{,}500{,}000{,}000 \\ &= 2.5 \times 10^9\end{aligned}$$

Materials

Balance

Sterile 99-ml dilution blanks (3)

Sterile 1-ml pipettes (3)

Sterile Petri plates (4)

Sterile melted nutrient agar or Sabouraud agar, 45 ml

Soil sample (Bring your own. Collect from 5 to 30 cm below the surface.)

Techniques Required

Exercises 14 and 32, and Appendices A, B, and C

Procedure (Figure 51-3)

1. Using the same soil source, one group of students should use nutrient agar and one group should use Sabouraud agar. Label each plate with one of the following dilutions: $1:10^4$, $1:10^5$, $1:10^6$, $1:10^7$.
2. Label the dilution blanks "$1:10^2$," "$1:10^4$," "$1:10^6$."
3. Categorize your soil sample as sandy soil, clayey soil, or organic-rich soil. Place 1 gram of soil in the "$1:10^2$" bottle. Why is this a $1:10^2$ dilution? ______ Let the bottle sit for 5 min. Why? ______
Shake the bottle 20 times, with your elbow resting on the table as shown in Figure 51-3*a*.
4. Aseptically pipette 1 ml from the $1:10^2$ dilution into the "$1:10^4$" bottle. Mix thoroughly as before.
5. Using a different pipette, prepare a $1:10^6$ dilution. Why is it necessary to change pipettes? ______
6. With the remaining pipette, transfer 0.1 ml of the $1:10^6$ dilution into the "$1:10^7$" plate and 1.0 ml into the "$1:10^6$" plate. *Note:* 0.1 ml of the $1:10^6$ dilution results in a final dilution of $1:10^7$.
7. Using the same pipette, transfer 0.1 ml and 1.0 ml into the remaining plates. Why can one pipette be used for all the inoculations? ______
8. Obtain a flask of melted agar from the water bath. Check the temperature of the water and the flask. Be sure the agar is cool enough for pouring. Why shouldn't the agar be too hot? ______
9. Carefully pour agar into one plate. The agar should just barely cover the bottom of the plate. Replace the cover and gently rotate the plate to

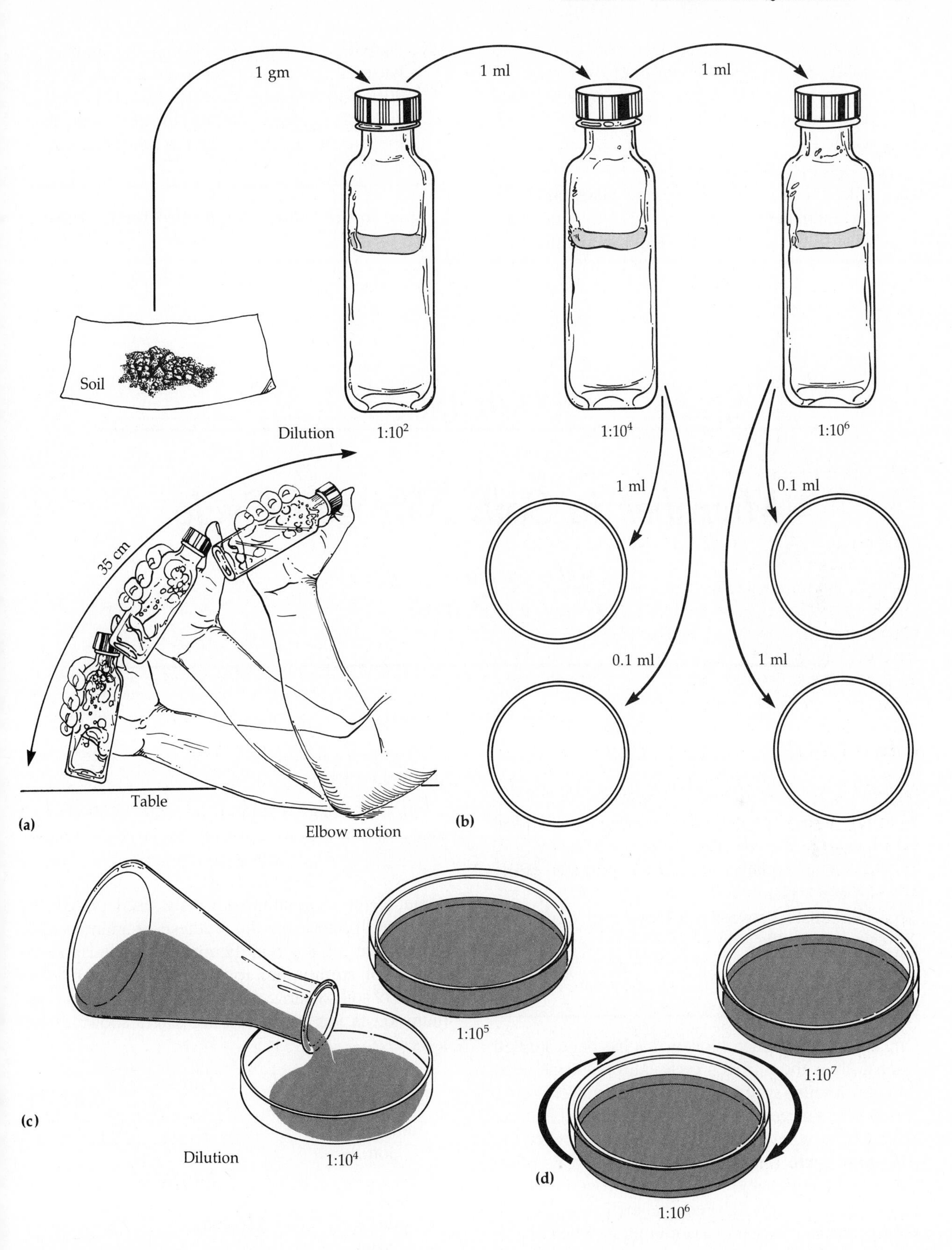

Figure 51-3.

Plate count procedure. **(a)** Shake the dilution bottle 20 times through a 35-cm arc. **(b)** Inoculate plates with measured volumes. **(c)** Pour agar into the plates. **(d)** Rotate each plate to mix the agar and inoculum.

mix the inoculum in the agar. Take care not to splash the agar onto the cover or over the sides of the Petri plate. What would happen if you didn't mix the contents of the plate? ________

____

____

10. When the agar is solid, invert the plates and incubate them between 20°C and 25°C for 4 to 7 days. What is the purpose of using this temperature range? ________

11. Examine the plates and count the number of bacterial colonies on the plate having between 30 and 300 colonies. Count the number of fungal and actinomycete colonies. Obtain results from the pair of students using the other medium.

Turn to the Laboratory Report for Exercise 51.

EXERCISE 52

Microbes in Soil: Nitrogen Cycle

Soil is the placenta of life.

P. FARB (1959)

Objectives

After completing this exercise you should be able to

1. Diagram the nitrogen cycle showing the chemical changes that occur at each step.
2. Provide an explanation for the importance of the nitrogen cycle.
3. Differentiate between symbiotic and nonsymbiotic nitrogen fixation.

Background

One aspect of soil microbiology that has been studied extensively is the nitrogen cycle. All organisms need nitrogen for the synthesis of proteins, nucleic acids, and other nitrogen-containing compounds. The recycling of nitrogen by different organisms is called the **nitrogen cycle** (Figure 52-1). Microbes play a fundamental, irreplaceable role in the nitrogen cycle by participating in many different metabolic reactions involving nitrogen-containing compounds. When plants, animals, and microorganisms die, microbes decompose them by proteolysis and ammonification.

Proteolysis is the hydrolysis of proteins to form amino acids (Exercise 20). **Ammonification** liberates ammonia by deamination of amino acids (Exercise 21) or catabolism of urea to ammonia (Exercise 20). In most soil, ammonia dissolves in water to form ammonium ions:

$$\underset{\text{Ammonia}}{NH_3} + \underset{\text{Water}}{H_2O} \rightarrow \underset{\text{Ammonium hydroxide}}{NH_4OH} \rightarrow \underset{\text{Ammonium ion}}{NH_4^+} + \underset{\text{Hydroxyl ion}}{OH^-}$$

Some of the ammonium ions are used directly by plants and bacteria for the synthesis of amino acids.

The next sequence in the nitrogen cycle is the oxidation of ammonium ions in **nitrification.** Two genera of soil bacteria are capable of oxidizing ammonium ion in the two successive stages shown as follows:

$$\underset{\text{Ammonium ion}}{2NH_4^+} + \underset{\text{Oxygen}}{3O_2} \xrightarrow{\textit{Nitrosomonas}} \underset{\text{Nitrite ion}}{NO_2^-}$$

$$\underset{\text{Nitrite ion}}{2NO_2^-} + \underset{\text{Oxygen}}{O_2} \xrightarrow{\textit{Nitrobacter}} \underset{\text{Nitrate ion}}{2NO_3^-}$$

These reactions are used to generate energy (ATP) for the cells. Nitrifying bacteria are chemoautotrophs, and many are inhibited by organic matter.

Nitrates are an important source of nitrogen for plants. **Denitrifying bacteria** reduce nitrates and re-

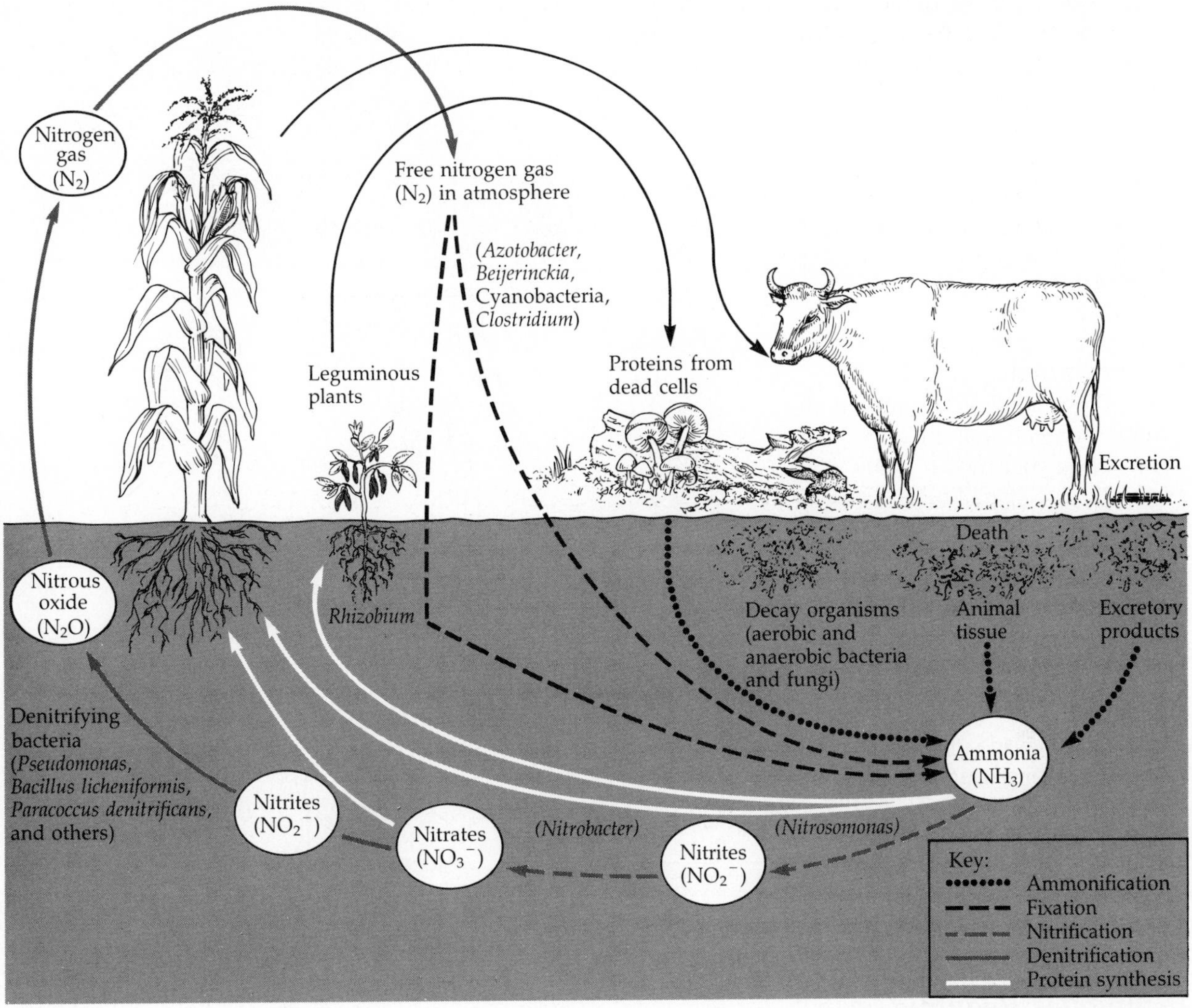

Figure 52-1.
Nitrogen cycle.

move them from the nitrogen cycle. **Denitrification** is the reduction of nitrates to nitrites and nitrogen gas (Exercise 22). This conversion may be represented as follows:

$$\underset{\text{Nitrate ion}}{NO_3^-} \rightarrow \underset{\text{Nitrite ion}}{NO_2^-} \rightarrow \underset{\text{Nitrous oxide}}{N_2O} \rightarrow \underset{\text{Nitrogen gas}}{N_2}$$

Denitrification is also called **anaerobic respiration.** Many genera of bacteria including *Pseudomonas* and *Bacillus* are capable of denitrification under anaerobic conditions (Exercise 22).

Atmospheric nitrogen can be returned to the soil by the conversion of nitrogen gas into ammonia, a process called **nitrogen fixation.** Cells possessing the nitrogenase enzyme can fix nitrogen under anaerobic conditions as follows:

$$\underset{\text{Nitrogen gas}}{N_2} + \underset{\text{Hydrogen ions}}{6H^+} + \underset{\text{Electrons}}{6e^-} \rightarrow \underset{\text{Ammonia}}{2NH_3}$$

Some free-living procaryotic organisms such as *Azotobacter,* clostridia, and cyanobacteria can fix nitrogen. Many of the nitrogen-fixing bacteria live in close association with the roots of grasses in the **rhizosphere,** where root hairs contact the soil.

Symbiotic bacteria serve a more important role in nitrogen fixation. One such symbiotic relationship is a mutualistic relationship between *Rhizobium* and the roots of legumes (such as soybeans, beans peas, alfalfa, and clover). In addition to the cultivated agricultural crops, there are thousands of other legumes, many of which are able to grow in the poor soils found in tropical rain forests or arid deserts. *Rhizobium* species are specific for the host legume that they infect. When a root hair and rhizobia make contact in the soil, a root nodule forms on the plant. The nodule provides the anaerobic environment necessary for nitrogen fixation.

Symbiotic nitrogen fixation also occurs in the roots of nonleguminous plants. The actinomycete *Frankia* forms root nodules in alders.

Any break in the nitrogen cycle would be critical to the survival of all life. The steps of the nitrogen cycle will be investigated in this exercise.

Materials

Ammonification

Tube containing peptone broth (1)
Nessler's reagent
Spot plate
Soil (Bring your own.)

Denitrification

Tubes containing nitrate-salts broth (2)
Nitrate reagents A and B
Soil (Bring your own.)

Culture

Pseudomonas aeruginosa

Nitrogen fixation

Petri plate containing mannitol–yeast extract agar (1)
Methylene blue
Sterile razor blade (1)
Legumes (Bring your own [Figure 52-2].)

Demonstrations

Rhizobium inoculum used by farmers
Slides of stained root nodules

Techniques Required

Exercises 5, 13, and 22

Procedure

Ammonification

1. Obtain an inoculum of soil on a wetted loop and inoculate the peptone broth. What is peptone? ____
2. Incubate the peptone broth at room temperature and test for ammonia at 2 days and at 7 days.
3. Test for ammonia by placing a drop of Nessler's reagent in a spot plate, add a loopful of peptone broth, and mix (Figure 52-3). A yellow to brown color indicates the presence of ammonia. Use an uninoculated tube of peptone broth as a control. Why is a control necessary? ____

Denitrification

1. Inoculate one nitrate-salts broth with soil using a wetted loop to pick up the soil. Inoculate the other tube with *P. aeruginosa.*
2. Incubate both tubes at room temperature for 1 week.

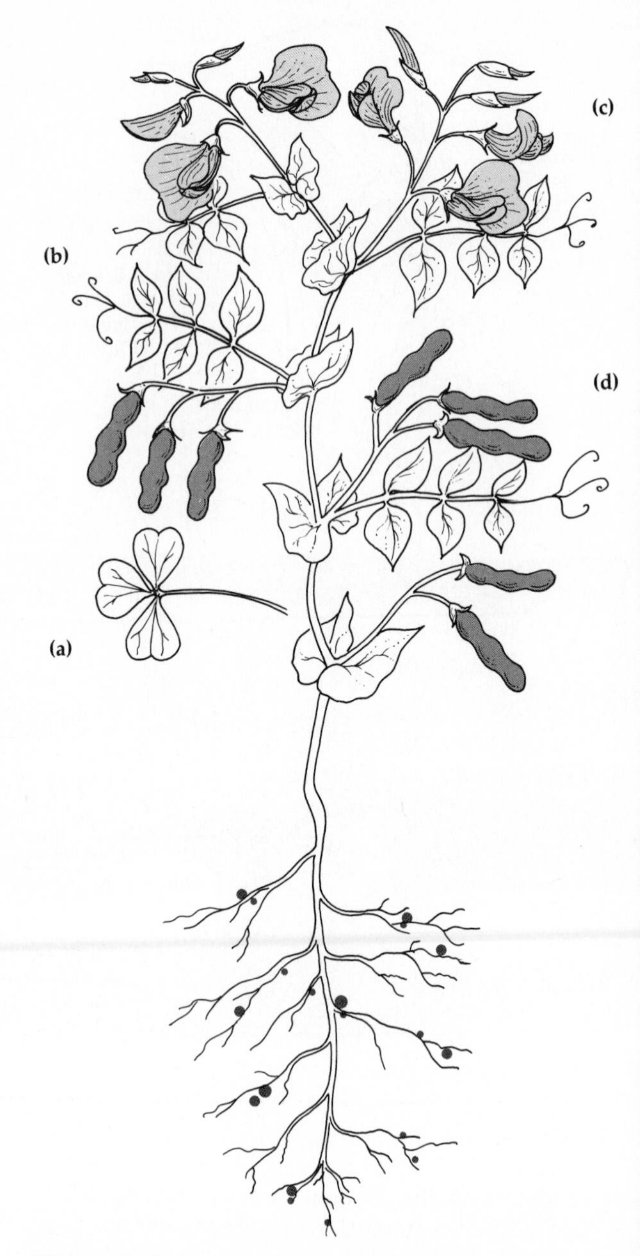

Figure 52-2.
Some characteristic features of members of the Legume family. The leaves are opposite and may be **(a)** palmately or **(b)** pinnately compound. Flowers **(c)** have five asymmetrical petals. Fruits **(d)** are legumes (pealike).

3. Test for nitrate reduction. Remember how? Describe. ____

Hint: Refer to Exercise 22. Record your results.

Nitrogen Fixation

1. Cut off a nodule from a legume and wash it under tap water. Observe.
2. Cut the nodule in half with a sterile razor blade. Crush the nodule between two slides and make a smear by rotating the slides together.

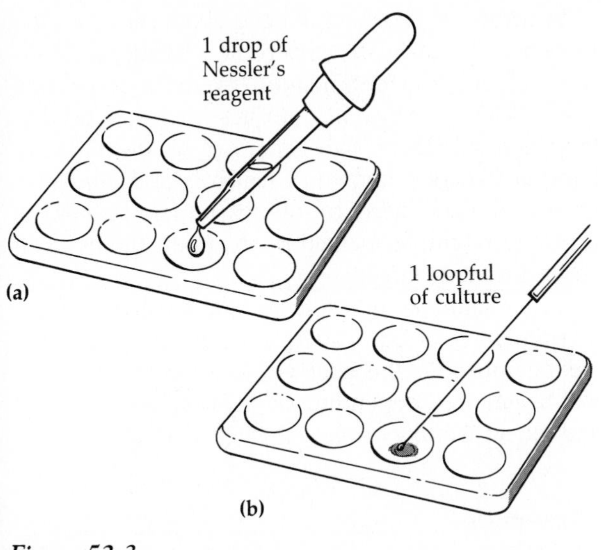

Figure 52-3.
Place a drop of the reagent in a well of a spot plate. Add a loopful of the sample to be tested, and mix. Observe for a color change.

3. Streak a loopful of the crushed nodule on the mannitol–yeast extract agar. Incubate the plate at room temperature for 7 days.
4. Stain the slide for 30 seconds with methylene blue and observe under oil immersion.
5. Observe the demonstrations. What is in the farmer's inoculum? ______
6. Observe the growth on the plate and make a simple stain of the growth. Compare this stain to the stain prepared from the nodule.

Turn to the Laboratory Report for Exercise 52.

EXERCISE 53

Microbes in Soil: Photosynthetic Bacteria

Objective

After completing this exercise you should be able to

1. Define the following terms: photosynthesis, light reaction, and dark reaction.
2. Differentiate between oxygen-producing photosynthesis and bacterial photosynthesis.
3. Diagram the sulfur cycle as it occurs in a Winogradsky column.

Background

Photoautotrophs use light as a source of energy, and carbon dioxide as their chief source of carbon. The process by which photoautotrophs transform carbon dixoide into carbohydrates for use in catabolic processes is called **photosynthesis.** Photosynthesis can be divided into two parts: the light reaction and the dark reaction. In the **light reaction,** light energy is converted into chemical energy (ATP) using light-trapping pigments. Chlorophyll molecules trap light energy and provide electrons that are used to *generate* ATP. Carbon dioxide is reduced to a carbohydrate in the **dark reaction.** Carbon dioxide reduction or **carbon dioxide fixation** requires an electron donor and energy.

The two types of photosynthesis are classified according to the way ATP is generated and the source of electrons. Cyanobacteria, algal protists, and green plants use *chlorophyll a* to generate ATPs. The resulting oxygen is produced by hydrolysis (splitting water) of the electron donor, *water*. This photosynthetic reaction is summarized as follows:

$$\underset{\text{Carbon dioxide}}{6CO_2} + \underset{\text{Water}}{6H_2O} \xrightarrow{\text{Chlorophyll } a} \underset{\text{Sugar}}{C_6H_{12}O_6} + \underset{\text{Oxygen}}{6O_2}$$

In addition to cyanobacteria, there are several other photosynthetic procaryotes. These are classified in *Bergey's Manual* Part I, "Phototrophic Bacteria." Most photosynthetic bacteria use **bacteriochlorophylls** to generate electrons for ATP synthesis and use *sulfur, sulfur-containing compounds, hydrogen gas,* or *organic molecules* as electron donors. The generalized equation for bacterial photosynthesis is:

$$\underset{\text{Carbon dioxide}}{CO_2} + \underset{\text{Hydrogen sulfide}}{H_2S} \xrightarrow{\text{Bacteriochlorophyll}} \underset{\text{Sugar}}{C_6H_{12}O_6} + \underset{\text{Sulfide ion}}{S^{2-}}$$

Some photosynthetic bacteria store sulfur granules in or on their cells as a result of the production of sulfide ions. The stored sulfur can be used as an electron donor in photosynthesis, resulting in the production of sulfates.

Bacterial photosynthesis differs from green plant photosynthesis in that **bacterial photosynthesis:**

1. Occurs in an anaerobic environment
2. Does not produce oxygen

Green photosynthetic bacteria are colored by bacteriochlorophylls although they may appear brown due to the presence of red accessory photosynthetic pigments called **carotenoids. Purple photosynthetic bacteria** appear purple or red because of large amounts of carotenoids. Purple bacteria also have bacteriochlorophylls.

Purple and green bacteria are involved in another biogeochemical cycle (Exercise 51), the **sulfur cycle** (Figure 53-1). In this exercise we will enrich for bacteria involved in an *anaerobic* sulfur cycle.

In nature, hydrogen sulfide is produced from the reduction of sulfates in anaerobic respiration (Exercise 22) and the degradation of sulfur-containing amino acids (Exercise 21). Sulfates can be reduced to hydrogen sulfide by five genera of *sulfate-reducing* bacteria (the best known of which is *Desulfovibrio*). Carbon dioxide used by photosynthetic bacteria is provided by the fermentation of carbohydrates in an anaerobic environment.

An enrichment culture technique involving a habitat-simulating device called a Winogradsky column will be utilized in this exercise. A variety of organisms will be cultured depending on their exposure to light and the availability of oxygen (Figure 53-2).

Materials

Mud mixture (mud, $CaCO_3$, hay or paper, and $CaSO_4$)

Large test tube or graduated cylinder

Glass rod

Plain mud

Winogradsky buffer (NH_4Cl, Na_2S, KH_2PO_4, K_2HPO_4)

Aluminum foil

Light source

Techniques Required

Exercise 2

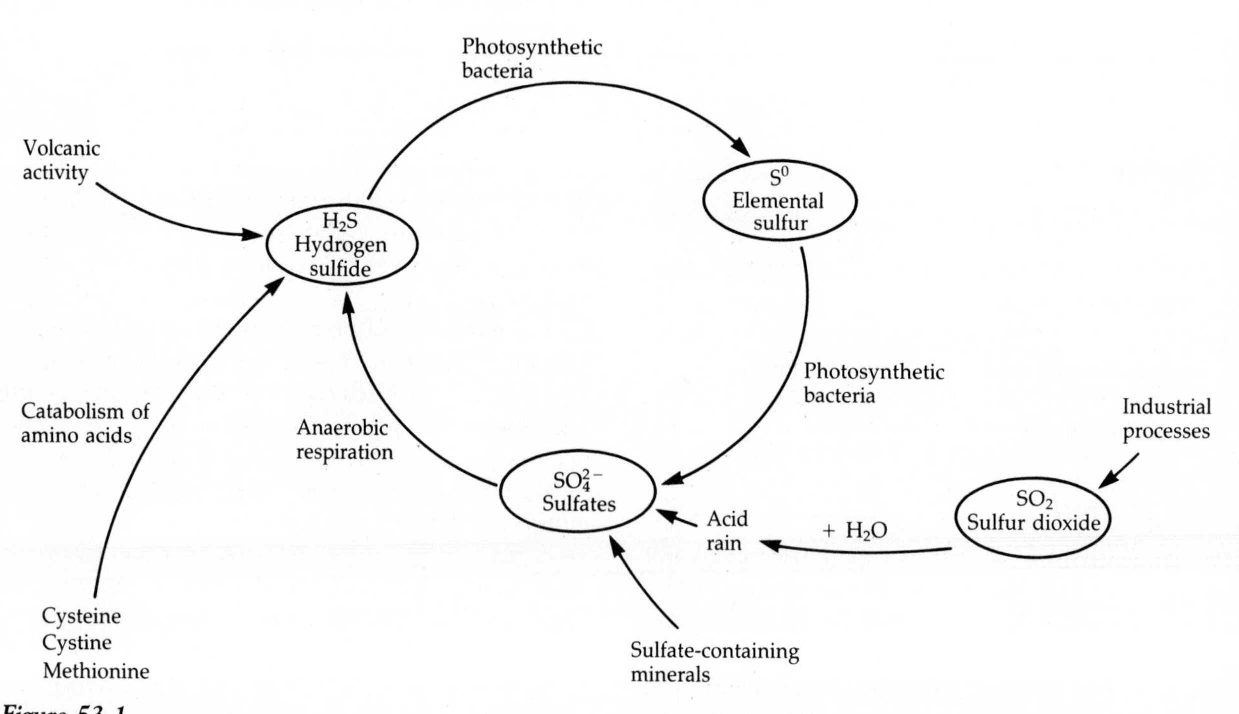

Figure 53-1.
Anaerobic sulfur cycle.

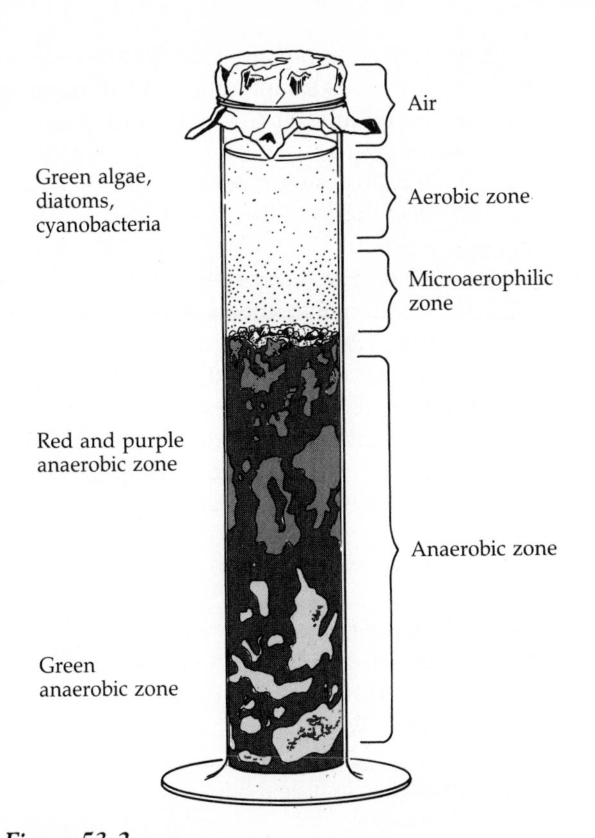

Figure 53-2.
Winogradsky column.

Procedure

1. Pack a large test tube two-thirds full with the mud mixture. Pack it to eliminate air bubbles. Why? ________
2. Carefully pack a narrow layer of plain mud on top of the first layer.
3. Gently pour the buffer down the side of the tube taking care not to disturb the mud surface. Fill the tube as full as possible.
4. Cover the top of the tube with foil and place in front of a light source. The instructor may assign different light sources (such as incandescent, fluorescent, red, or green).
5. Observe the tubes at weekly intervals for 4 weeks. Record the appearance of colored areas. Aerobic mud will be brownish and anaerobic mud will be black. Why? ________
6. After 4 weeks, prepare wet mounts from the purple or green patches. Record the microscopic appearance of the bacteria and whether sulfur granules are present.

Turn to the Laboratory Report for Exercise 53.

EXERCISE 54

Microbes in Food: Contamination

Objectives

After completing this exercise you should be able to

1. Determine the number of bacteria in a food sample using a standard plate count.
2. Provide reasons for monitoring the bacteriologic quality of foods.
3. Explain why the standard plate count is used in food quality control.

Background

Illness and food spoilage can result from microbial growth in foods. The sanitary control of food quality is concerned with testing foods for the presence of pathogens. During processing (grinding, washing, and packaging), food is contaminated with soil microbes and flora from animals, food handlers, and machinery.

Foods are the primary vehicle responsible for the transmission of diseases of the digestive system. For this reason, they are examined for the presence of coliforms. Recall that the presence of coliforms indicates fecal contamination (Exercise 49).

Standard plate counts are routinely performed on food and milk by food processing companies and public health agencies. The **standard plate count** is used to determine the total number of viable bacteria in a food sample. The presence of large numbers of bacteria is undesirable in most foods because it increases the likelihood that pathogens will be present, and it increases the potential for food spoilage.

A limitation of the standard plate count is that only bacteria capable of growing in the culture medium and environmental conditions provided will be counted. A medium that supports the growth of most heterotrophic bacteria is used.

Materials

Melted standard plate count or nutrient agar, cooled to 45°C

Sterile 1-ml pipettes (Part A, 2; Part B, 3)

Sterile Petri plates (Part A, 4; Part B, 4)

Sterile 99-ml dilution blanks (Part A, 1; Part B, 2)

Food sample, diluted 1:10 (1)

Techniques Required

Exercise 14 and Appendices A and B. Read the Background section of Exercise 51.

Procedure

First Period

A. Bacteriologic Examination of Milk

1. Obtain a sample of either raw or pasteurized milk that has been diluted 1:10.
2. Using a sterile 1-ml pipette, aseptically transfer 1 ml of the 1:10 milk sample into a 99-ml dilution blank, label the tube "$1{:}10^3$" and discard the pipette (Figure 54-1*a*).
3. Mark the bottoms of four sterile Petri plates with the dilutions: "1:10," "$1{:}10^2$," "$1{:}10^3$" and "$1{:}10^4$" (Figure 54-1*b*).
4. Using a 1-ml pipette, aseptically transfer 0.1 ml of the $1{:}10^3$ dilution into the bottom of the plate marked "$1{:}10^4$." *Note:* 0.1 ml of the $1{:}10^3$ dilution results in a $1{:}10^4$ dilution of the original sample. Using the same pipette, transfer 1.0 ml of the $1{:}10^3$ into the plate so marked. Pipette 0.1 ml and 1.0 ml from the 1:10 dilution into the "$1{:}10^2$" and "1:10" plates, respectively, with the same pipette (Figure 54-2*a*). Why is it important to proceed from the highest to the lowest dilution? ____________
____
5. Check the temperature of the water bath containing the nutrient agar. Test the water and agar container with your hand to make certain the agar is cooled to 45°C. Why? ____________
____
6. Pour melted nutrient agar into one of the plates (to about one-third full) (Figure 54-2*b*). Cover the plate and swirl it gently (Figure 54-2*c*) to distribute the milk sample evenly through the agar. Continue until all the plates are poured.

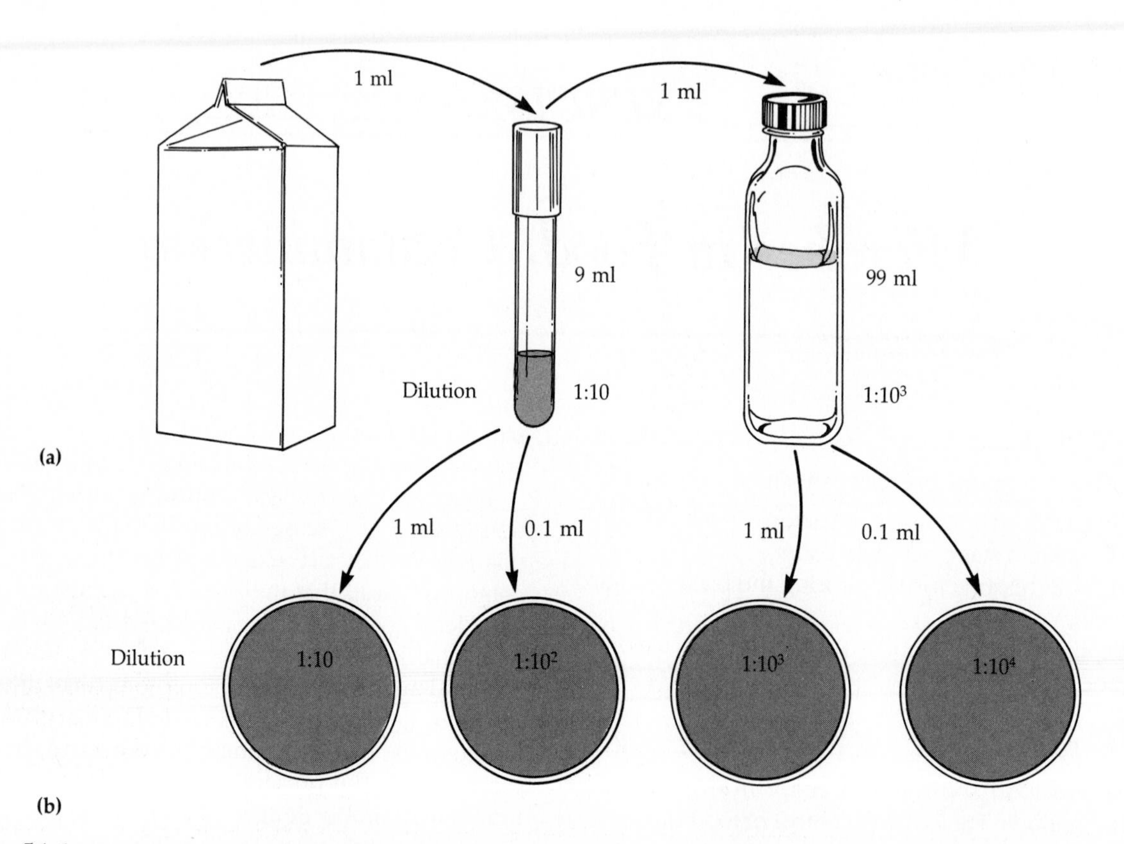

Figure 54-1.
Standard plate count of milk. **(a)** Make serial dilutions of a milk sample. **(b)** Mark four Petri plates with the dilutions.

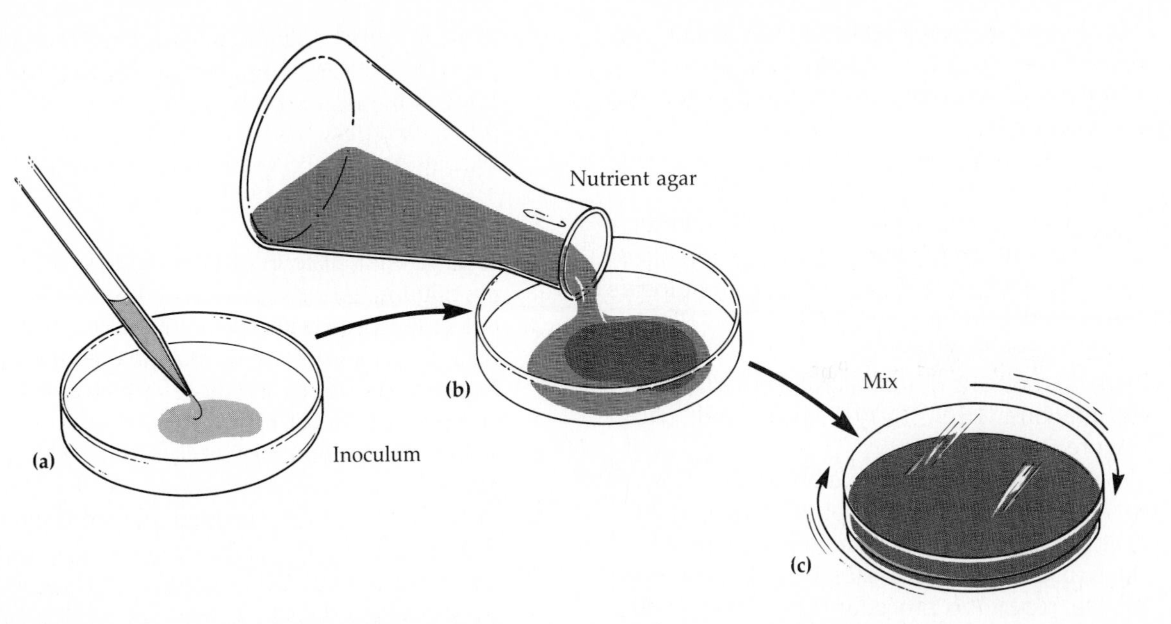

Figure 54-2.
Pour plate. **(a)** Inoculum is pipetted into a Petri plate. **(b)** Liquefied nutrient agar is added, and **(c)** the agar is mixed with the inoculum by gentle swirling of the plate.

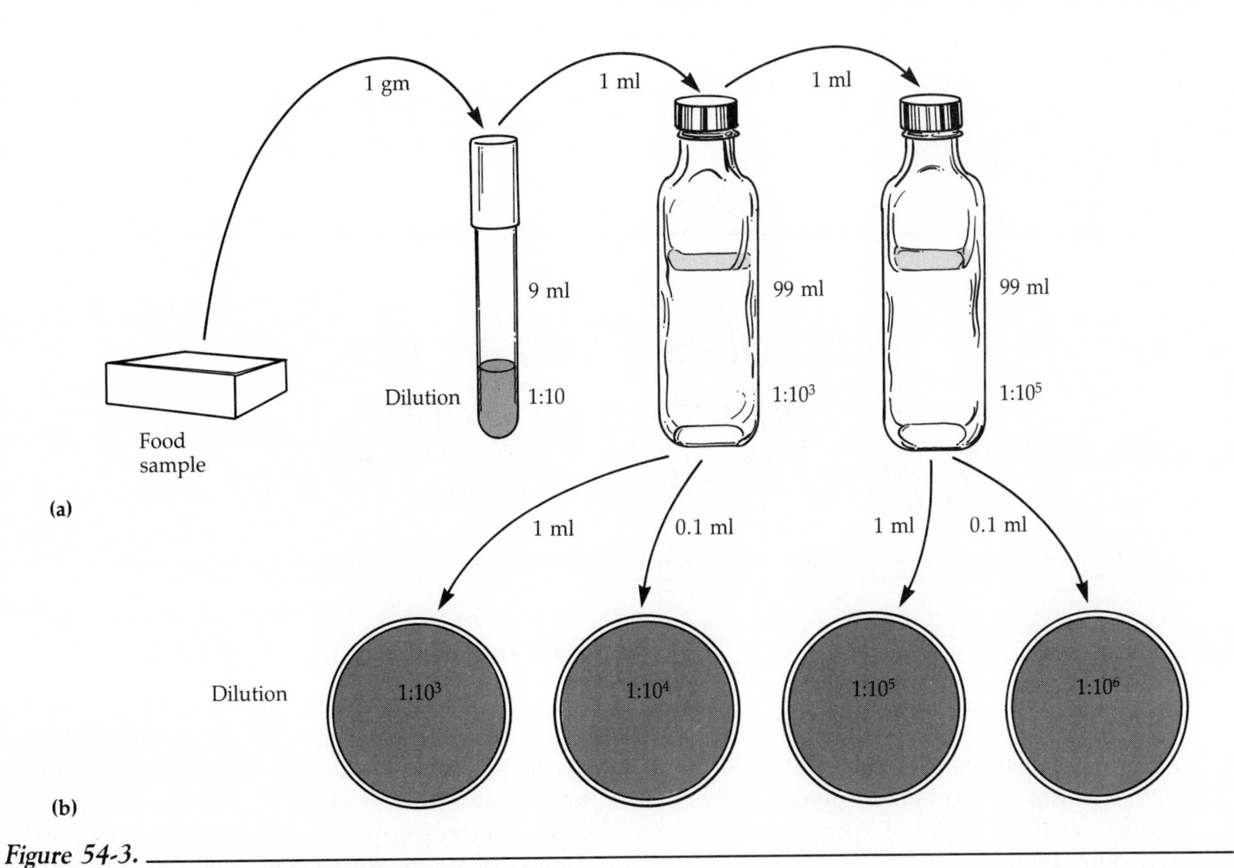

Figure 54-3.
Standard plate count on food. **(a)** Prepare serial dilutions of a food sample. **(b)** Mark four Petri plates with the dilutions to be plated.

7. When each plate has solidified, invert it, and incubate all plates at 35°C for 24 to 48 hours.

B. Bacteriologic Examination of Hamburger and Frozen Vegetables

1. Obtain a sample of raw hamburger or frozen vegetables diluted 1:10.
2. Using a sterile 1-ml pipette, aseptically transfer 1 ml of the 1:10 sample into a 99-ml dilution blank, label the bottle "1:10^3," and discard the pipette. Make a 1:10^5 dilution using another 99-ml dilution blank (Figure 54-3*a*).
3. Mark the bottoms of four sterile Petri plates with the dilutions: "1:10^3," "1:10^4," "1:10^5," and "1:10^6," (Figure 54-3*b*).
4. Using a 1-ml pipette, aseptically transfer 0.1 ml of the 1:10^5 dilution into the plate marked "1:10^6." *Note:* 0.1 ml of a 1:10^5 dilution results in a 1:10^6 dilution of the original sample. Using the same pipette, repeat this procedure with the 1:10^3 dilution until all the plates have been inoculated (Figure 54-2*a*). Why can the same pipette be used for each transfer? ______________________________

____

____
5. Check the temperature of the water bath containing the nutrient agar. Test the water and agar container with your hand to make certain the agar is cooled to 45°C. Why? ______________________________

____
6. Pour melted nutrient agar into one of the plates (to about one-third full) (Figure 54-2*b*). Cover the plate and swirl it gently (Figure 54-2*c*) to distribute the sample through the agar evenly. Continue until all the plates are poured.
7. When each plate has solidified, invert it, and incubate all plates at 35°C for 24 to 48 hours.

Second Period

1. Arrange each plate in order from lowest to highest dilution.
2. Select the plate with 30 to 300 colonies. Record data for plates with fewer than 30 colonies as *too few to count (TFTC)* and those with more than 300 colonies, *too many to count (TMTC)*.
3. Count the number of colonies on the plate selected.
4. Multiply the number of colonies by the dilution of that plate to determine the number of bacteria in the original food. For example, if 129 colonies were counted on a 1:10^3 dilution:

$$\frac{\text{129 colonies}}{\text{1 ml} \times 10^{-3}} = 129{,}000$$
$$= 1.29 \times 10^5 \text{ bacteria/ml or gram of food}$$

Turn to the Laboratory Report for Exercise 54.

EXERCISE 55

Microbes Used in the Production of Foods (Yogurt, Root Beer, and Wine)

Obectives

After completing this exercise you should be able to

1. List four examples of the beneficial activities of microorganisms in food production.
2. Explain how the activities of microorganisms are used to preserve food.
3. Define fermentation.
4. Produce an enjoyable product.

Background

Microbial fermentations are used to produce a wide variety of foods. In industrial usage, **fermentation** is any large-scale microbial process occurring with or without air. To a biochemist, **fermentation** is the group of metabolic processes that release energy from a sugar or other organic molecule, do not require oxygen or an electron transport system, and use an organic molecule as a final electron acceptor. In this exercise we will examine three types of microbial fermentations used in the production of food. One of these fermentations uses oxygen, one does not use oxygen even when it is present, and in the third, oxygen must be eliminated entirely.

In dairy fermentations such as yogurt production, microorganisms use lactose and produce lactic acid without using oxygen. In a nondairy fermentation such as wine production, yeast use sucrose to produce ethyl alcohol and carbon dioxide under anaerobic conditions. If oxygen is available, the yeast will grow aerobically, liberating carbon dioxide and water as metabolic end products.

Historically, milk has been selectively fermented, with the resulting acidity preventing spoilage by acid-intolerant microbes. These "sour" milks have varied from country to country depending on the source of milk, conditions of culture, and bacteria "starter" used. Milk from donkeys to zebras has been used, with the Russian *kumiss* (horse milk), containing 2% alcohol, and Swedish *surjolk* (reindeer milk) being unusual examples. The bacteria yield lactic acid and yeast produce ethyl alcohol. Currently, two fermented cow milk products, buttermilk and yogurt, are widely used.

Buttermilk is the fluid left after cream is churned into butter. Today, buttermilk is actually prepared by souring true buttermilk or by adding bacteria to skim milk and then flavoring it with butterflake. *Streptococcus lactis* ferments the milk producing lactic acid (sour) and neutral fermentation products (diacetyls) are produced by *Leuconostoc*. Yogurt originated in the Balkan countries, with goat milk being the primary source of milk. Yogurt is milk that has been concentrated by boiling and fermented at elevated temperatures. *Streptococcus* produces lactic acid and *Lactobacillus* produces the flavors and aroma of yogurt.

Wine was discovered long ago. The ancient Greeks cultivated wine grapes and made wine long before fermentation was explained by Pasteur. Ancestors of the Greeks probably ate fallen, fermented fruits and found the taste (or the effect) enjoyable. Wine can be made from a *must* (juice and crushed pulp) of almost any fruit, vegetable, or flower, although it is usually made from the juice of grapes *(Vitis vinifera)*. The tastes of wines are due to the strain (or variety) of grape, the amount of sugar they contain, and the strain of yeast used. The alcohol concentration of wine is regulated in part by the amount of sugar in the must and the tolerance of the yeast. The yeast will continue producing alcohol until the concentration inhibits their growth: most yeast strains tolerate up to 14% alcohol.

Since the time of George Washington, Americans have had many "coffee breaks" drinking homemade root beer. A classic recipe from the *American Encyclopedia of Formulas* is given here for those who can find the ingredients.

ROOT BEER

5 gal boiling water
1½ gal molasses
¼ lb bruised sassafras bark*
¼ lb wintergreen bark
¼ lb sarsaparilla root
½ pt fresh yeast
Water

Combine 5 gallons boiling water and molasses. Allow it to stand for 3 hours. Add sassafras bark, wintergreen bark, sarsaparilla root, and yeast. Add enough water to make 15 to 17 gallons. Allow to ferment at room temperature for 12 hours, then draw off liquid and bottle.

Techniques Required

Exercises 5, 7, 13, and 14, and Appendices A and C

YOGURT

Materials

Homogenized milk

Nonfat dry milk

Large beaker

Stirring rod

Thermometer

Hot plate or ring stand and asbestos pad

5-ml pipette

Styrofoam cups with lids

Plastic spoons

Petri plate containing trypticase soy agar

pH paper

Gram stain reagents

Optional: jams, jelly, honey, and so on

Culture

Commercial yogurt or *Streptococcus thermophilus* and *Lactobacillus bulgaricus*

*Sassafrol® is the synthetic substitute for sassafras that is presently used in commercial root beer.

EXERCISE 55

Procedure

1. Add 100 ml milk per person (in your group) to a wet beaker (wash out beaker first with water to decrease sticking of the milk).
2. Heat milk on a hot plate or over a burner on an asbestos pad placed on a ring stand to about 80°C for 40 min. Stir occasionally. Do not let it boil. Why is the milk heated? ______
3. Cool to about 65°C and add 3 grams nonfat dry milk per person. Stir to dissolve. Why is dry milk added? ______
4. Rapidly cool to about 45°C. Pour milk equally into the cups.
5. Inoculate each cup with 1–2 teaspoons of commercial yogurt or 2.5 ml *S. thermophilus* and 2.5 ml *L. bulgaricus.* Cover and label.
6. Incubate at 45°C for 4 to 18 hours or until firm (custardlike).
7. Cool yogurt to about 5°C, then taste with a clean spoon. Add jam or some other flavor if you desire. Eat! Save a small amount for steps 8 through 10.
8. Determine the pH of the yogurt.
9. Make a smear, and after heat fixing, Gram stain it. Record your results.
10. Streak for isolation on nutrient agar. Incubate the plate, inverted at 45°C.
11. After distinct colonies are visible, record your observations. Prepare Gram stains from each different colony.

WINE

Materials

250 g fresh or frozen berries

230 g sucrose

Water

Optional:
- 0.56 g yeast nutrient
- 0.56 g acid mix
- 0.3 g pectinase

Triple beam balance

Enamel kettle (about 3 liters)

Hot plate

1-liter beaker

Wooden spoon

Strainer or cheesecloth

Clean 1-liter bottles (3)

Balloons

100-ml cylinder

Hydrometer

2-m rubber tubing

Simple stain reagents

Culture

Yeast *(Saccharomyces cerivisiae* var. *ellipsoides)*

Procedure

1. Rinse berries in tap water and crush with the *heel of your hand* in enamel kettle. Add water to bring the volume to about 1 liter using the beaker to measure.
2. Put all ingredients except yeast into the kettle and heat to dissolve.
3. When the sucrose is dissolved, pour some of the mixture, called **must,** through a strainer or cheesecloth into a 100-ml cylinder.
4. To determine the specific gravity, gently place the hydrometer into the cylinder (Figure 55-1). **Specific gravity** is a measure of the relative weight (density) of a substance compared to the weight (density) of water. The specific gravity of

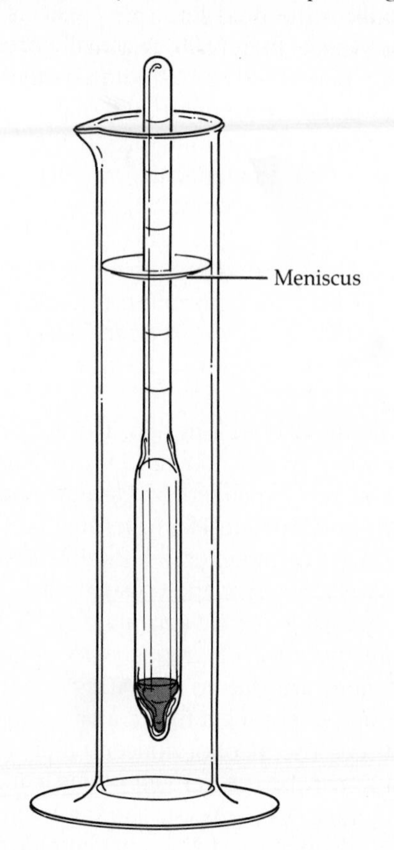

Figure 55-1.
Use of a hydrometer. Fill a cylinder with the solution to be tested. Gently place the hydrometer in the cylinder. The hydrometer should float (if not, a larger volume is required). Read the specific gravity off the hydrometer at the meniscus.

water is 1.000. As solutes are added to water, the specific gravity increases. What is the specific gravity of the must? ______
If you do not have a wine hydrometer, refer to Table 55-1 for the relationship between specific gravity and sugar concentration. The amount of alcohol that could be produced from the sugar **(percent potential alcohol)** is also given. What is the percent potential alcohol of your must? ______

5. The starting specific gravity should be 1.090 to 1.095. Return the sample to the kettle. Add sucrose, making sure it is thoroughly dissolved, until the correct specific gravity is reached. What is the percent potential alcohol at a specific gravity of 1.090? ______
6. Add 2.5 grams yeast when the must has cooled to 22° to 25°C.
7. Pour the must into a 1-liter bottle, filling to about three-fourths full. Why isn't the bottle filled to the top? ______
8. Label the bottle and attach a balloon to the mouth. What is the purpose of the balloon? ______
9. Incubate at room temperature for 5 to 6 days or until the specific gravity is 1.040.
10. Strain out the fruit pulp as the wine is poured into a clean bottle. Fill the bottle to the neck.

Table 55-1

Specific Gravity–Potential Alcohol. The table shows the alcohol yield you may expect in relation to the specific gravity of the must.

Specific Gravity	% Sugar	% Potential Alcohol
1.000	0	0
1.010	2.5	0.9
1.020	5.5	2.3
1.030	8.0	3.7
1.040	11.0	5.1
1.050	12.5	6.5
1.060	16.0	7.8
1.080	20.0	10.6
1.090	23.0	12.0
1.100	25.0	13.4
1.110	27.5	14.9
1.120	29.0	16.3
1.130	32.0	17.7

Why is it important to completely fill the bottle? ______

11. After 3 weeks, siphon the wine into a new bottle leaving the yeast, called **lees,** in the first bottle. Determine the specific gravity of the wine. How much alcohol has been made? ______

    Initial % potential alcohol
    − Ending % potential alcohol
    % alcohol produced

12. Prepare a simple stain of the lees. Record your observations.

*ROOT BEER**

Materials

Hires Root Beer Extract®

Sucrose

Bottles, washed and sterilized

Wine corks or caps and crowner

Enamel kettle or dishpan (about 20 liters)

Triple beam balance

Liter container (for measuring)

250-ml beaker

Graduated cylinder

Petri plate containing nutrient agar

Simple stain reagents

Optional: raisins

Culture

Dried or compressed cake yeast *(Saccharomyces cerevisiae)*

Procedure

Decrease all amounts by one-half.

1. Pour 1 bottle of Hires Root Beer Extract® over 1800 grams sugar in kettle. Mix well.
2. Dissolve mixture in 19 liters lukewarm water.
3. Mix 2.5 grams dried yeast or one-half cake compressed yeast in 500 ml lukewarm water. Let stand for 5 minutes. Add 5 grams or 1 cake yeast when the temperature is 21°C.
4. Add yeast to sugar–extract mixture, mix, and pour into bottles immediately. A raisin may be added if

*Courtesy of The Proctor & Gamble Company.

you like a great deal of carbonation. Fill to within 1 cm of the top. Excess air will spoil the result. Why? ________________________________

____

5. Cork securely or seal with a capper.
6. Place bottles at 21° to 27°C for 5 to 7 days. Place the bottles on their sides. Why? ____________

____

7. Refrigerate before drinking. Why isn't alcohol usually produced (final alcohol = 0.03%)? ______

____

8. Streak a nutrient agar plate with the sediment and incubate inverted at room temperature for 24 to 48 hours.
9. Prepare a simple stain of an isolated colony.

Turn to the Laboratory Report for Exercise 55.

PART 11

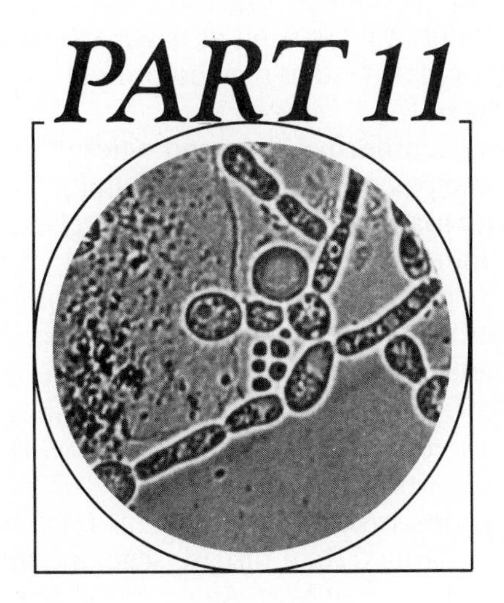

Medical Microbiology

EXERCISES

In 1883, after having determined the number of microorganisms in the environment (Part 10), Robert Koch asked, *"What significance do these findings have? Can it be stated that air, water, and soil contain a certain number of microorganisms and yet are without significance to health?"* Koch felt that the **etiologic** (causative) agents of infectious diseases could be found in the study of microorganisms. He proved that anthrax and tuberculosis were caused by specific bacteria, and his work provided the framework for the study of the etiology of any infectious disease. Today we refer to Koch's protocol for identifying etiologic agents as **Koch's postulates** (Exercise 57).

Many microorganisms grow abundantly both inside and on the surface of the normal adult body. The microorganisms that establish more or less permanent residence without producing diseases are known as **normal flora** (see the figure). Microorganisms that may be present for a few days or months are called **transient flora.**

The relationship between normal flora and a healthy person may be classified as one of two types of **symbiosis** (meaning "living together"): **commensalism** or **mutualism.** In a *commensal* relationship, one organism is benefited and the other is unharmed. For example, many of the bacteria living in the large intestine are supplied with continual food and a constant temperature, while the host is neither benefited nor harmed. In a *mutualistic* relationship, both organisms are benefited. Some bacteria that live on and in our bodies receive a constant supply of nutrients, while they, in exchange, supply something of benefit to us. For example, *E. coli* synthesizes vitamin K and certain B vitamins for its human host.

A third kind of symbiosis is **parasitism.** In this relationship, one organism gains while the other is harmed. In a disease state, the microorganism is a parasite and is harming its human host. An organism like this is called a **pathogen.** Some members of the normal flora become parasites under certain conditions. These organisms are called **opportunistic pathogens.**

In order to cause a disease, a pathogen must gain access to the human body. The route of access is called the **portal of entry.** Some microbes cause disease only when they gain access by the "correct" portal of entry. The study of how and when diseases occur is called **epidemiology** and will be the topic of Exercise 56.

In Exercises 58 through 62, we will examine the variety of bacteria associated with the human body. In a clinical laboratory, samples from diseased tissue are cultured and pathogens must be distinguished from normal and transient flora. Bacteria in a simulated clinical sample will be identified in Exercise 63.

In Exercises 64 through 67, microorganisms other than bacteria that are associated with the human body will be studied.

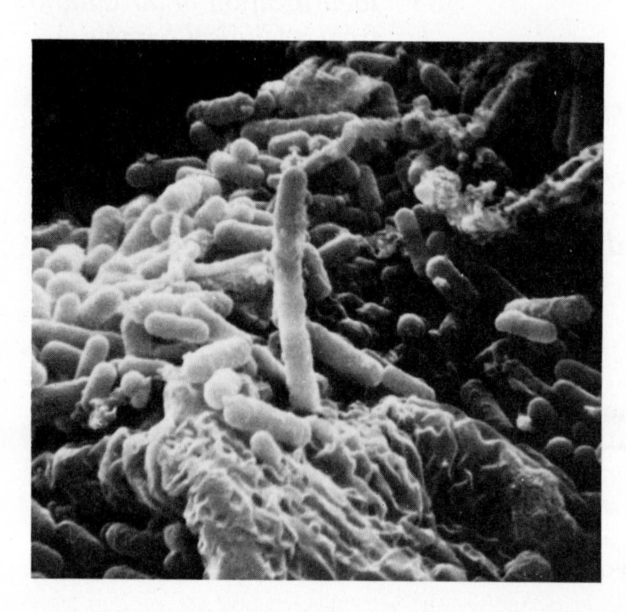

This scanning electron micrograph shows mucous membranes of the large intestine with bacteria in the crypts (× 420).

EXERCISE 56

Epidemiology

Catch As Catch Can

Everyone's desirous
Of hiding from the virus
And caution starts the very day we're born.
We're filled with fear of kisses
From infected sirs or misses
(We know full well the germs
that this will spawn!)
A sorry state, then, this is
When we're cloaked in cowardices,
But it's better to be careful than to mourn.
Yes, of catchy things we're leery,
But I think it's pretty eerie—
There is nothing more contagious
than a yawn!

LOLA SCHANCER

Objectives

After completing this exercise, you should be able to

1. Define the following terms: epidemiology, epidemic, reservoir, and carrier.
2. Describe three methods of transmission.
3. Determine the source of a simulated epidemic.

Background

In every infectious disease, the disease-producing microorganism, the **pathogen,** must come in contact with the **host,** the organism that harbors the pathogen. **Communicable diseases** can be spread from one host to another. Some microorganisms cause disease only if the body is weakened or if a predisposing event such as a wound allows them to enter the body. Such diseases are called **noncommunicable diseases;** that is, they cannot be transmitted from one host to another. The science that deals with when and where diseases occur and how they are transmitted in the human population is called **epidemiology. Sporadic diseases** are those that occur occasionally in a population; an example is polio. **Endemic diseases** are constantly present in the population. When many people in a given area acquire the disease in a relatively short period of time, it is referred to as an **epidemic disease.** Influenza often achieves epidemic status.

Diseases can be transmitted by **direct contact** between hosts. **Droplet infection,** when microorganisms are carried on liquid drops from a cough or sneeze, is a method of direct contact. Diseases can also be transmitted by contact with contaminated inanimate objects, or **fomites** (Exercise 48). Drinking glasses, bedding, and towels are examples of fomites that can be contaminated with pathogens from feces, sputum, or pus.

Some diseases are transmitted from one host to another by vectors. **Vectors** are insects and other arthropods that carry pathogens. In **mechanical transmission,** insects carry a pathogen on their feet and may transfer the pathogen to a person's food. For example, houseflies may transmit typhoid fever from the feces of an infected person to food. Transmission of a disease by an arthropod's bite is called **biological transmission.** An arthropod ingests a pathogen while biting an infected host and then transfers the pathogen to a healthy person in its feces or saliva.

The continual source of an infection is called the **reservoir.** Humans who harbor pathogens but who do not exhibit any signs of disease are called **carriers.**

An **epidemiologist** compiles data on the incidence of a disease and its method of transmission and tries to locate the source of infection in order to decrease the incidence.

In this experiment, an epidemic will be simulated. You will be the epidemiologist, who, by deductive reasoning and with luck, determines the source of the epidemic.

Materials

Petri plate containing nutrient agar (1)

Small plastic bag (sandwich size)

1 Tootsie Roll® per student. (One Tootsie Roll® has *Serratia marcescens* on it.)

Techniques Required

Exercise 12

Procedure

1. Divide the Petri plate into five sectors labeled 1 to 5.
2. Open the paper wrapper of your Tootsie Roll® carefully, without touching the candy. Put the plastic bag on your left hand. Holding the Tootsie Roll® by the wrapper with your right hand, rub it on the palm of your left hand (that is, on the plastic bag). Discard the Tootsie Roll® and wrapper in a container of disinfectant.
3. Shake hands (using your left hand) with five of your classmates when the instructor gives the signal—shake hands so your fingers touch their palm and vice versa. After each shake touch your fingers to the corresponding sectors of the nutrient agar. Label with the person's name and Tootsie Roll® number.
4. Discard the plastic bag in disinfectant. Incubate the plate inverted at room temperature.
5. At the next lab period, deductively try to determine who had the contaminated fomite.

Turn to the Laboratory Report for Exercise 56.

EXERCISE 57

Koch's Postulates

Objectives

After completing this exercise you should be able to

1. Define the following terms: etiologic agent, pathogenicity, and virulence.
2. List and explain Koch's postulates.

Background

The **etiologic agent** is the cause of a disease. Microorganisms are the etiologic agents of a wide variety of infectious diseases in all forms of life. Microbes that cause diseases are called pathogens, and the process of disease initiation and progress is called **pathogenesis.** The interaction between the microbe and host is complex. Whether or not a disease occurs depends on our vulnerability or **susceptibility** to the pathogen and on the virulence of the pathogen. **Virulence** is the degree of pathogenicity. Factors influencing virulence include the number of pathogens, the portal of entry into the host, and toxin production.

The actual cause of many diseases is hard to determine. Although many organisms can be isolated from a diseased tissue, their presence does not prove that any or all of them caused the disease. A microbe may be a secondary invader or part of the normal flora or transient flora of that area. While working with anthrax and tuberculosis, Robert Koch established four criteria, now called **Koch's postulates,** to help identify a particular organism as the causative agent for a particular disease. Koch's postulates are the following:

1. The same organism must be present in every case of the disease.
2. The organism must be isolated from the diseased tissue and grown in pure culture in the laboratory.

3. The organism from the pure culture must cause the disease when inoculated into healthy, susceptible laboratory animals.
4. The organism must again be isolated—this time from the inoculated animals—and must be shown to be the same pathogen as the original organism.

These criteria are used by most investigators, but they cannot be applied to all infectious diseases. For example, viruses cannot be cultured on artificial media and they are not readily observable in a host. Moreover, many viruses cause diseases only in humans and not in laboratory animals.

In this exercise, you will demonstrate Koch's postulates with one of three different bacteria:

Vibrio anguillarium isolated from Chinook salmon

Erwinia carotovora isolated from soft rot of carrots

Agrobacterium tumefaciens isolated from crown gall (tumors) of a variety of plants

VIBRIO–Goldfish*

Materials

Petri plate containing nutrient agar (1)

Petri plate containing salt-starch agar (1)

Scalpel and scissors

2 syringes, one containing broth (without bacteria) and the other an 18-hour culture of *Vibrio anguillarium*

Goldfish (2)

Extra fish tank (for inoculated fish)

Demonstration

Dissected control fish

Techniques Required

Exercises 7, 14, and 17

Procedure

1. Divide each Petri plate into three sectors and label A, B, and C. Streak a drop of broth from the syringe onto sector A and a drop of *Vibrio* from the other syringe onto sector B of each plate. Incubate the plates inverted at room temperature.
2. Make a smear of *Vibrio* and heat fix; store in your drawer.
3. Remove one fish with a wet paper towel. Do not touch the fish directly. Why? ________________

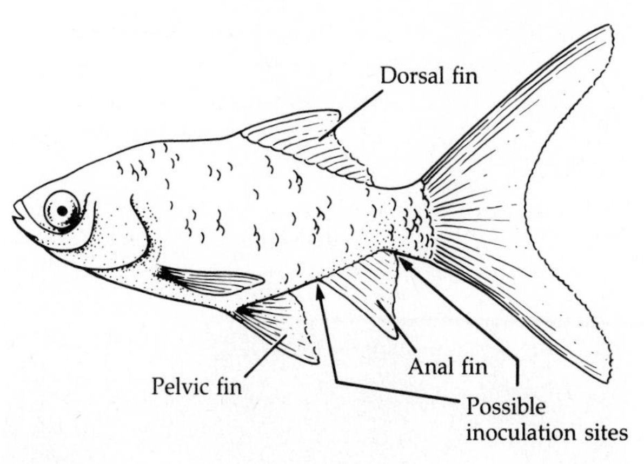

Figure 57-1.
Inoculate goldfish in front of (anterior to) the anal fin or just behind (posterior to) the anal fin.

Holding the fish on its back, inject it with *Vibrio* in front or behind the anal fin (Figure 57-1). Hold the syringe at about a 45° angle and inject about 0.05 ml. Put the fish into the tank for inoculated fish.

4. Inject the remaining fish in the same manner with 0.05 ml broth. What is the purpose of this step?

____

Treat your fish with care. Return it to the tank for uninoculated fish.

5. Your inoculated fish should die within 24 to 48 hours. Once dead, place in a wet towel and keep in the refrigerator until you have time to necropsy the fish. *If your fish does not die, you have a new pet.* Provide some reasons for its survival.
6. Observe the viscera of the dissected control fish.
7. Work with your fish on a disinfectant-saturated towel. Disinfect your instruments by putting in alcohol and burning off the alcohol immediately before and after use. Slit the belly open from anus to gills, then make horizontal cuts to expose the viscera (Figure 57-2). Examine and record your findings. Transfer a loopful of the peritoneum exudate (fluid or pus) to sector C on your plates and incubate at room temperature for 48 hours. Make a smear of exudate on the same slide you used in step 2. Discard the fish as instructed.
8. Gram stain the smears and examine the growth on the plates. If growth occurred on nutrient agar, a contaminant was present. Growth on the plates can also be Gram stained. Flood the salt-starch plate with Gram's iodine. Record your results.

ERWINIA—Carrot Soft Rot*

Materials

Petri plate containing nutrient agar

*Modified from W. Umbreit and E. Ordal. 1972. "Infection of Goldfish with *Vibrio anguillarium.*" *ASM News* 38: 93–98. With the permission of the American Society for Microbiology.

*Modified from R. S. Hogue. 1971. "Demonstration of Koch's Postulates." *American Biology Teacher 33: 174–175.*

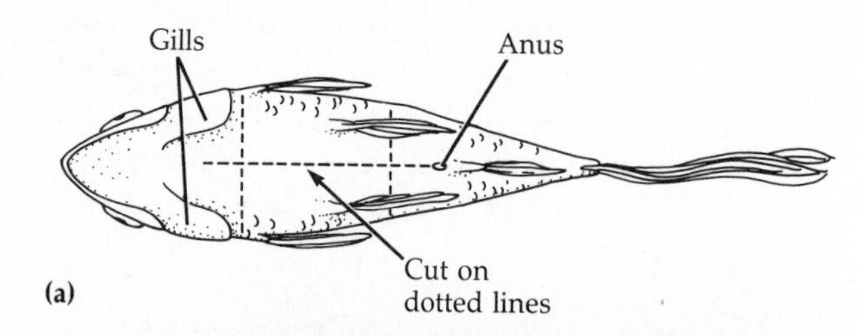

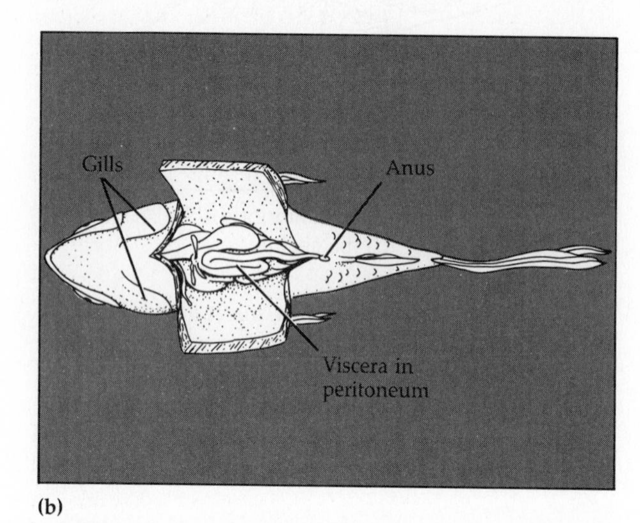

(b)

Figure 57-2.
Goldfish necropsy. **(a)** Dissect as shown. **(b)** Anatomy.

Carrot

Scalpel or razor blade, potato peeler

Forceps

Alcohol

Disinfectant

Sterile water

Sterile Petri plate with filter paper

Culture

Erwinia carotovora

Techniques Required

Exercises 7 and 14

Procedure

1. Wash the carrot well, and *peel* it to eliminate the outer surface; then dry, wash with disinfectant, and rinse with sterile saline. Dip your scalpel in alcohol, burn off the alcohol, and cut the carrot into four slices that are 5 to 8 mm thick.
2. Put the four slices on filter paper in the bottom of a Petri plate.
3. Inoculate the center of three slices with a loopful of the *Erwinia* culture. Why not all four?

   ____

   Saturate the filter paper with sterile water. Incubate the plate right side up at room temperature until soft rot appears. More sterile water may be added if the disease process is slow.
4. Divide the nutrient agar plate in half. Inoculate one-half of the nutrient agar with the *Erwinia* broth, incubate inverted for 48 hours at room temperature, and then refrigerate. Make a smear of the *Erwinia*. Heat fix the smear and store in your drawer.
5. Streak an inoculum from the diseased carrot on the remaining half of the Petri plate. Incubate inverted at room temperature for 48 hours.
6. Make a smear from the diseased carrot. Gram stain both smears and record your observations.
7. Observe the nutrient agar plate and record your results. Prepare a Gram stain from the nutrient agar cultures if time permits.

AGROBACTERIUM–Crown Gall

Materials

Petri plate containing nutrient agar (1)

Carrot and/or tomato plants (2)

Sterile Petri plate with filter paper

Sterile water

White tape or labels

Mortar and pestle

Syringe containing nutrient broth and syringe containing *Agrobacterium tumefaciens*

Demonstration

Slide of an *Agrobacterium*-induced gall

Techniques Required

Exercises 7 and 14

Procedure

1. Divide the nutrient agar plate in half. Inoculate one-half with *Agrobacterium.* Incubate inverted for 48 hours at room temperature and then refrigerate for use in step 3. Make a smear of the *Agrobacterium.*
2. Inoculate either or both of the test organisms as assigned by your instructor.
   a. Tomato: For the first plant, inject three or four nodes (Figure 57-3) with about 0.05 ml *Agrobacterium.* Inject three or four nodes of the

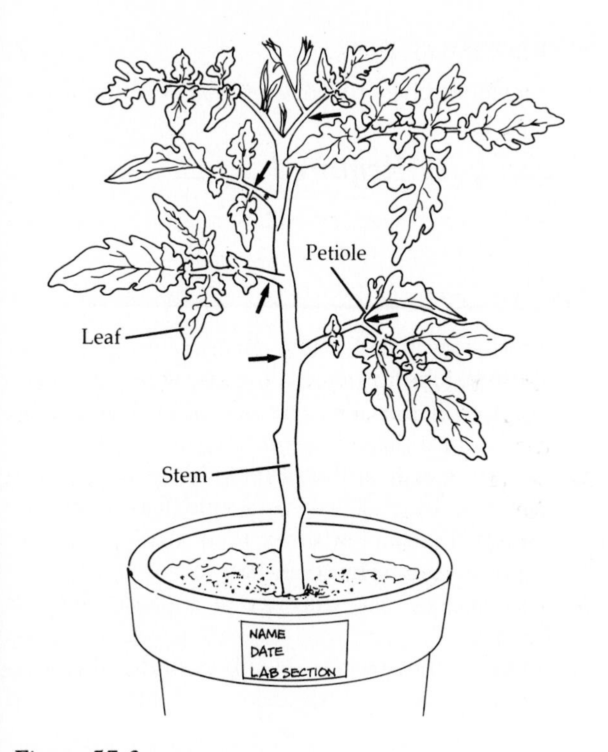

Figure 57-3.
Inoculate tomato plants at any of the locations marked with an arrow. Place labels close to the injection sites.

second plant with nutrient broth. Place a label around the petiole next to each site. Label and identify the inoculation sites on Figure 57-3. Observe weekly. When good-sized galls appear, describe their appearance, and then cut them off.

b. Carrot: Prepare slices as described in step 1 of the *Erwinia* procedure. Place four slices on the filter paper and wet with sterile water. Inoculate three of the slices with a loopful of *Agrobacterium*. Incubate in your drawer, and examine and add water periodically. When galls appear, cut them off.

3. Cut up the galls and grind them into a slurry in a mortar with sterile water. Inoculate the remaining one-half of the nutrient agar plate with the slurry. Incubate the plate as before. Make a smear from the slurry.

4. Gram stain both smears, and observe. Observe growth on the plate and Gram stain.

Turn to the Laboratory Report for Exercise 57.

EXERCISE 58

Bacteria of the Skin

Objectives

After completing this exercise you should be able to

1. Isolate and identify bacteria from the human skin.
2. Provide an example of normal skin flora.
3. List characteristics of the staphylococci.
4. Explain why many bacteria are unable to grow on human skin.

Background

The skin is generally an inhospitable environment for most microorganisms. The dry layers of keratin-containing cells that make up the epidermis (outermost layer of the skin) are not easily colonized by most microbes. Sebum secreted by oil glands inhibits bacterial growth, and salts in perspiration create a hypertonic environment. Perspiration and sebum are nutritive for certain microorganisms, however, establishing them as part of the normal flora of the skin.

Normal flora of the skin vary according to skin region. The flora of the face reflect that of the throat just behind the mouth, and the flora of the anal region are influenced by microorganisms of the lower gastrointestinal tract. More bacteria are found in moist areas such as the axilla and the side of the nose than on the dry surface of arms or legs. Transient flora are present on hands and arms in contact with the environment.

Propionibacterium live in hair follicles on sebum from oil glands. The propionic acid they produce maintains the pH of the skin between 3 and 5, which suppresses the growth of other bacteria. Most bacteria on the skin are gram-positive. Catalase-positive, salt-tolerant members of gram-positive cocci (*Bergey's*

Manual, Part 14), which includes three families, are the most frequently encountered bacteria on the skin.

Staphylococcus aureus is part of the normal flora of the skin and is also considered a pathogen. Strains that produce **coagulase,** an enzyme that coagulates (clots) the fibrin in blood, are pathogenic. However, since the majority of *S. aureus* strains are coagulase-positive, a test for the presence of coagulase is used to distinguish *S. aureus* from other species of *Staphylococcus.*

Although many different bacterial genera live on human skin, you will attempt to isolate and identify a member of the family Micrococcaceae (catalase-positive) in this exercise.

Materials

Petri plates containing mannitol salt agar (2)

Sterile cotton swab (1)

Sterile saline

3% hydrogen peroxide

Gram staining reagents

Fermentation tubes, coagulase plasma (as needed)

Techniques Required

Exercises 7, 14, 18, 22, and 42

Procedure

1. Wet the swab with saline, push against the wall of the test tube to express excess saline. Swab any surface of your skin. The side of the nose, axilla, elbow, or a pus-filled sore are possible areas.
2. Swab one-half of the plate with the swab. Using a sterile loop streak back and forth into the swabbed area a few times, then streak away from the inoculum (Figure 48-1*b*).
3. Incubate the plate inverted at 35°C for 24 to 48 hours.
4. Examine the colonies. Members of the Micrococcaceae usually form large opaque colonies. Record

Table 58-1
Key to the Micrococcaceae

Gram-positive cocci
↓
Catalase-positive
↓

I. Cells arranged in tetrads	*Micrococcus*
A. Glucose not fermented	
1. Colonies have yellow pigment	*M. luteus*
2. Colonies have red pigment	*M. roseus*
3. Colonies have yellow-red pigment	*Planococcus*
II. Cells arranged in grapelike clusters	*Staphylococcus*
A. Acid produced from mannitol	
1. Coagulase-positive	*S. aureus*
2. Coagulase-negative	
a. Acid produced from xylose	*S. xylosus*
b. No acid from xylose	*S. cohnii*
B. No acid from mannitol	
1. Acid produced from fructose	
a. Acid produced from mannose	*S. epidermidis*
b. No acid from mannose	*S. caseolyticus*
2. No acid from fructose	*S. saprophyticus*

the appearance of the colonies and any mannitol fermentation (yellow halos) (Exercise 48). Test colonies for catalase production (Exercise 22).

5. Subculture a catalase-positive, gram-positive coccus on another mannitol salt plate. Do not attempt to subculture a colony that has not been tested for catalase. Why? ______________________
6. Using the key to the Micrococcaceae (Table 58-1), proceed to identify your isolate.
   a. To test for coagulase, place a loopful of rehydrated coagulase plasma on a clean slide. Add a loopful of water and make a heavy suspension of the bacteria to be tested. Observe for clumping of the bacterial cells (clumping = coagulase-positive; no clumping = coagulase-negative).
   b. Inoculate the appropriate fermentation tubes.

Turn to the Laboratory Report for Exercise 58.

EXERCISE 59

Bacteria of the Respiratory Tract

Objectives

After completing this exercise you should be able to

1. List representative normal flora of the respiratory tract.
2. Differentiate the pathogenic streptococci based on biochemical testing.
3. List a characteristic used to identify *Neisseria, Corynebacterium, Mycobacterium,* and *Bordetella.*

Background

The respiratory tract can be divided into two systems: the upper and lower respiratory systems. The **upper respiratory system** consists of the nose and throat, and the **lower respiratory system** consists of the larynx, trachea, bronchial tubes, and alveoli. The lower respiratory tract is normally sterile because of the efficient functioning of the ciliary escalator. The upper respiratory system is in contact with the air we breathe—air contaminated with microorganisms.

The throat is a moist, warm environment, allowing many bacteria to establish residence. Species of many different genera, such as *Staphylococcus, Streptococcus, Neisseria,* and *Haemophilus,* can be found living as normal flora in the throat. Despite the presence of potentially pathogenic bacteria in the upper respiratory system, the rate of infection is minimized by microbial **antagonism.** That is, certain microorganisms of the normal flora suppress the growth of other microorganisms through competition for nutrients and production of inhibitory substances.

Streptococcal species are the predominant organisms in throat cultures, and some species are the major cause of bacterial sore throats (acute pharyngitis). Streptococci are identified by biochemical characteristics including hemolytic reactions and antigenic characteristics (Lancefield's system). Hemolytic reactions are based on hemolysins that are produced by streptococci while growing on blood-enriched agar. Blood agar is usually made with defibrinated sheep blood (5.0%), sodium chloride (0.5%) to minimize spontaneous hemolysis, and nutrient agar. Three patterns of hemolysis can occur on blood agar.

1. **Beta-hemolysis:** Complete hemolysis, giving a clear zone with a clean edge around the colony.
2. **Alpha-hemolysis:** Incomplete hemolysis, producing methemoglobin and a green, cloudy zone around the colony.
3. **Gamma-hemolysis:** No hemolysis, and no change in the blood agar around the colony.

Alpha- and gamma-hemolytic streptococci are usually normal flora whereas beta-hemolytic streptococci are frequently pathogens.

The streptococci can be antigenically classified into groups by specific antigens in their cell walls. Over 90% of streptococcal infections are caused by Group A beta-hemolytic streptococci.

Other bacteria found in the throat will be examined in the third part of this exercise.

Materials

Throat Culture

Petri plate containing blood agar (1)

Sterile cotton swab (1)

Gram stain reagents

Hydrogen peroxide, 3%

Streptococcus

Petri plate containing blood agar (1)

Sterile cotton swabs (2)

Forceps and alcohol

Optochin disk

Bacitracin disk

10% bile salts

Tube containing 2 ml nutrient broth (1)

Cultures (one of each)

Streptococcus pyogenes

Streptococcus pneumoniae

Selected Characteristics of Other Bacteria of the Respiratory Tract

Petri plate containing tellurite agar (1)

Petri plate containing brain-heart infusion agar (1)

Gram stain reagents

Methylene blue (simple stain)

Acid-fast stain reagents

Oxidase reagent

CO_2 jar

Cultures (one of each)

Corynebacterium diphtheriae (avirulent)

Mycobacterium smegmatis on Lowenstein Jensen medium

Bordetella bronchiseptica on Bordet-Gengou medium

Neisseria sicca

Techniques Required

Exercises 5, 7, 8, 14, 22, 36 and 46

Procedure

Throat Culture

1. Have your lab partner swab your throat with a sterile cotton swab. The area to be swabbed is between the "golden arches" (glossopalatine arches) as shown in Figure 59-1. Do not hit the tongue.

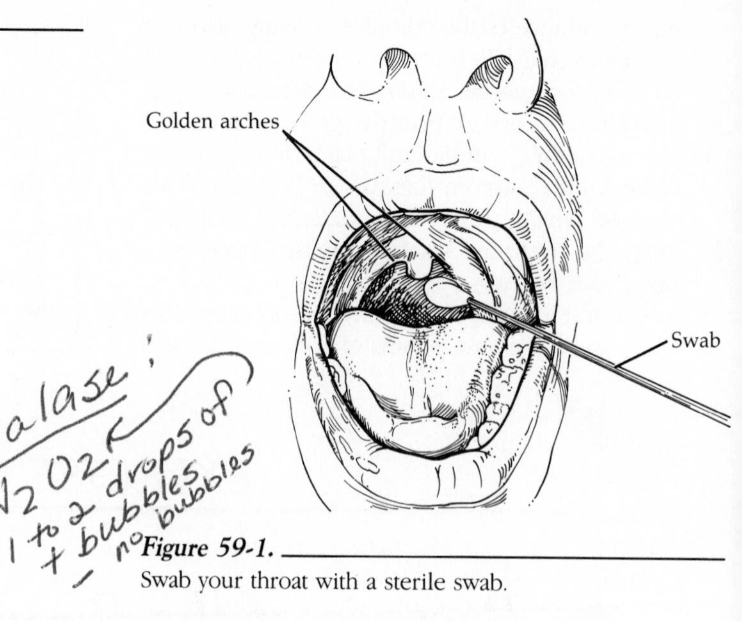

Figure 59-1.
Swab your throat with a sterile swab.

2. After obtaining an inoculum from the throat, swab one-half of a blood agar plate. Streak the remainder of the plate with a sterile loop (Figure 48-1*b*).
3. Incubate the plate inverted at 35°C for 24 hours. Observe the plate for hemolysis. Transfer some colonies to a slide and perform a catalase test. Why can't the catalase test be done on blood agar?

____

____

Streptococcus

1. Inoculate each half of a blood agar plate, one side with *S. pyogenes* and the other half with *S. pneumoniae.* Use a swab for confluent growth. *Be careful, these bacteria are pathogens.*
2. Dip forceps in alcohol and burn off the alcohol. Using the forceps, place and press a Bacitracin disk and an Optochin disk on each half. Space the disks so that zones of inhibition (Exercise 46) may be observed.
3. Incubate the plates inverted at 35°C for 24 hours. Observe for hemolysis and inhibition of growth by Bacitracin and Optochin.
4. Prepare Gram stains from smears of each organism. Do these organisms differ microscopically?

   ____
5. Using a sterile loop, prepare a suspension of each organism in a tube of nutrient broth.
6. Add a few drops of 10% bile salts to each tube. Observe the tubes after 15 minutes for lysis of the cells. Which culture is "bile soluble"? ________

   ____

Selected Characteristics of Other Bacteria of the Respiratory Tract

1. Perform a Gram stain on smears of each organism and an acid-fast stain (Exercise 8) on *Mycobacterium* and one other organism.

2. Record the appearance of *Mycobacterium* on Lowenstein Jensen medium.
3. Prepare a simple stain of *Bordetella.* Do the cells stain darker in one region? ________
4. Inoculate *Corynebacterium* onto tellurite agar, incubate inverted at 35°C for 24 hours. Record any pigment production. Do a simple stain of *Corynebacterium.* Describe the shape of the bacterium.
5. Inoculate brain-heart infusion agar with *Neisseria.* Incubate the plate inverted in a CO_2 jar at 35°C for 24 hours. A CO_2 jar (Figure 59-2) is prepared by placing the plates to be incubated with a candle into a large jar. The candle is lit and the lid screwed on the jar. The candle will extinguish when the CO_2 concentration in the jar reaches 5% to 10%.
6. Perform an oxidase test on *Neisseria* (Exercise 22). To test for the production of cytochrome oxidase do one of the following:
   a. Flood an agar culture with oxidase reagent (1% tetramethyl-*p*-phenylene-diamine). Observe for a change in the color of the *colonies,* not the agar. The presence of oxidase is indicated by pink to black colonies.
   b. Moisten an oxidase disk and place it *on* colonies on a plate. Incubate the plate for 15 minutes at 35°C. The disk and colonies turn black in oxidase-positive cultures.

Figure 59-2.
CO_2 jar. The candle is lighted and the lid is screwed onto the jar. Do not let the flame touch the Petri plates.

Turn to the Laboratory Report for Exercise 59.

EXERCISE 60

Bacteria of the Mouth

Objectives

After completing this exercise you should be able to

1. List characteristics of streptococci found in the mouth.
2. Describe the formation of dental caries.
3. Explain the relationship between sucrose and caries.

Background

The mouth may contain millions of bacteria in each milliliter of saliva. Some of these bacteria are transient flora carried on food. Some species of *Streptococcus* are part of the normal flora of the mouth. An identification scheme for streptococci found in the mouth is shown in Table 60-1. *Streptococcus mutans, S. salivarius,* and *S. sanguis* produce a sticky polysaccharide, **dextran,** from sucrose. A bacterial exoenzyme hydrolyzes sucrose into its component monosaccharides, glucose and fructose, and the energy released in the hydrolysis is used for polymerization of the glucose to form dextran. Fructose released from the sucrose is fermented to produce lactic acid.

The dextran capsule enables the bacteria to adhere to surfaces in the mouth. *S. salivarius* colonizes the

Table 60-1
Key to *Streptococcus* spp. Found in the Mouth

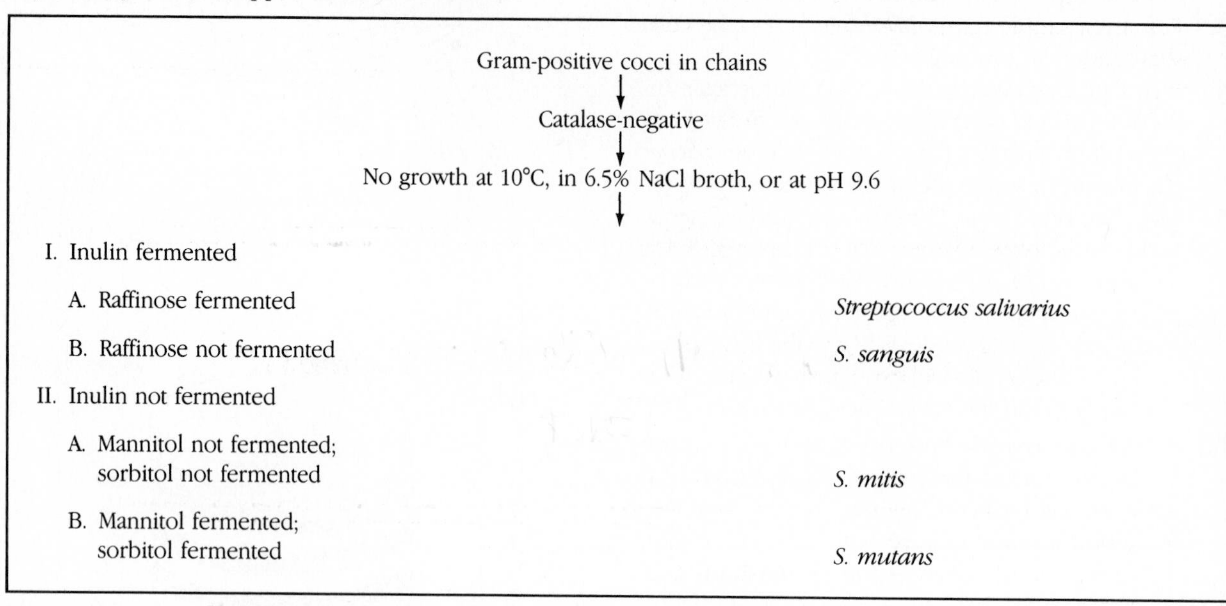

Gram-positive cocci in chains
↓
Catalase-negative
↓
No growth at 10°C, in 6.5% NaCl broth, or at pH 9.6
↓

I. Inulin fermented	
A. Raffinose fermented	*Streptococcus salivarius*
B. Raffinose not fermented	*S. sanguis*
II. Inulin not fermented	
A. Mannitol not fermented; sorbitol not fermented	*S. mitis*
B. Mannitol fermented; sorbitol fermented	*S. mutans*

surface of the tongue, and *S. sanguis* and *S. mutans,* the teeth. Masses of bacterial cells, dextran, and debris adhering to the teeth constitute **dental plaque.** Production of lactic acid by bacteria in the plaque initiates **dental caries** by eroding tooth enamel. Streptococci and other bacteria, such as *Lactobacillus,* are able to grow on the exposed dentin and tooth pulp.

Other carbohydrates such as glucose or starch may be fermented by bacteria, but they are not converted to dextran and hence do not promote plaque formation. "Sugarless" candies contain mannitol or sorbitol, which cannot be converted to dextran although they may be fermented by normal flora such as *Streptococcus mutans.* Acid production increases the size of the dental carie.

The **Snyder test** can be used to measure the rate of acid production from glucose by acid-tolerant normal flora such as *Lactobacillus.* The Snyder test utilizes a medium having a pH of about 4.8 and bromcresol green indicator, which turns from green to yellow with further acid production. Enamel decalcification begins at pH 5.5, so a significant amount of acid production may indicate the presence of cariogenic bacteria. Interpretation of a Snyder test depends on the number of samples that have been tested. A single sample represents what is happening at the time the sample is collected. Additional data such as personal hygiene and diet must also be considered. The test is run over a 3-day period to detect the rate of acid production. Generally, more rapid production of acid correlates to increased susceptibility to dental caries.

Materials

Tube containing melted Snyder test agar (1)

Petri plate containing sucrose gelatin agar (1)

Small piece of paraffin

50-ml beaker (1)

Stirring rod (1)

1-ml pipette (1)

3% hydrogen peroxide

Techniques Required

Exercises 13, 18, and 22 and Appendix A

Procedure

1. Allow a small piece of paraffin to soften under your tongue and then chew it for 3 minutes. Do not swallow your saliva or the paraffin. Instead collect all of the saliva in a 50-ml beaker. Don't be embarrassed; everyone has saliva!
2. Stir the saliva vigorously with a glass rod for 30 seconds to disperse the organisms evenly.
3. Pipette 0.2 ml of saliva into a tube of melted Snyder test agar. Roll the tube between your hands to mix the contents (Figure 14-4). Let solidify into a deep.
4. Incubate the Snyder agar for 72 hours at 35°C. Examine the tube every 24 hours for a change in the indicator and record your observations.
5. Inoculate a sucrose gelatin agar plate by one of the following:
   a. Using a loop, streak a loopful of the saliva you collected on the sucrose gelatin agar.
   b. Swab your gums, teeth, and tongue, and inoculate the sucrose gelatin agar as shown in Figure 48-1*b*.
6. Incubate the plate inverted for 24 to 48 hours at 35°C.

7. Examine the plate for the production of large mucoid capsules. Determine whether catalase (Exercise 22) is produced by the capsule-forming bacteria.

Turn to the Laboratory Report for Exercise 60.

EXERCISE 61

Bacteria of the Gastrointestinal Tract

EXERCISE 61

Objectives

After completing this exercise you should be able to

1. List the bacteria commonly found in the gastrointestinal tract.
2. Define the following terms: coliform, enteric, and enterococci.
3. Interpret results from TSI slants and SF broth.

Background

The stomach and small intestine have relatively few microorganisms due to the hydrochloric acid produced by the stomach and the rapid movement of food through the small intestine. Microbial populations in the large intestine are enormous, exceeding 10^{11} bacteria per gram of feces. Most of these intestinal organisms are commensals and some are in mutualistic relationships with their human hosts. Some intestinal bacteria synthesize useful vitamins, such as folic acid and vitamin K. The normal intestinal flora prevent colonization of pathogenic species by producing antimicrobial substances and competition.

The population of the large intestine is mostly anaerobes of the genera *Bacteriodes, Bifidobacterium, Lactobacillus,* and facultative anaerobes such as *Escherichia, Enterobacter, Citrobacter,* and *Proteus.* Species of *Streptococcus* belonging to Lancefield Group D called the **enterococci** are also present.

Most diseases of the gastrointestinal system result from the ingestion of food or water that contains pathogenic microorganisms. Good sanitation practices, modern methods of sewage treatment, and disinfection of drinking water help break the fecal–oral cycle of disease. A number of tests have been developed to identify bacteria associated with fecal contamination (see Exercises 25, 49, and 50). Gram-negative facultatively anaerobic rods are a large and diverse group of bacteria that include the **enteric family** (Enterobacteriaceae).

Media have been developed to differentiate between lactose-fermenting enterics and nonlactose-fermenting enterics. The lactose-fermenters are called **coliforms** (Exercise 25) and are generally not pathogenic. The nonlactose-fermenting group includes such pathogens as *Salmonella* and *Shigella.* One of the most common media is eosin-methylene blue (EMB) agar (Exercise 48). EMB agar is selective in that the eosin-methylene blue dyes are inhibitory to gram-positive organisms; thus, they allow the media to selectively culture gram-negative organisms. EMB is differential in that lactose-fermenting organisms (coliforms) give colored colonies and nonlactose-fermenters produce colorless colonies.

After isolation on EMB agar, differential screening media such as **triple sugar iron (TSI)** agar can be used to further categorize organisms. TSI contains:

0.1% Glucose

1.0% Lactose

1.0% Sucrose

0.02% Ferrous sulfate

Phenol red

Nutrient agar

As shown in Figure 61-1, if the organism ferments only glucose, the tube will turn yellow in a few hours. The bacteria quickly exhaust the limited supply of glucose and start oxidizing amino acids for energy, giving off ammonia as an end product. Oxidation of amino acids increases the pH and the indicator in the slanted portion of the tube will turn back to red. The butt will remain yellow. If the organism in the TSI slant ferments lactose and/or sucrose, the butt and slant will turn

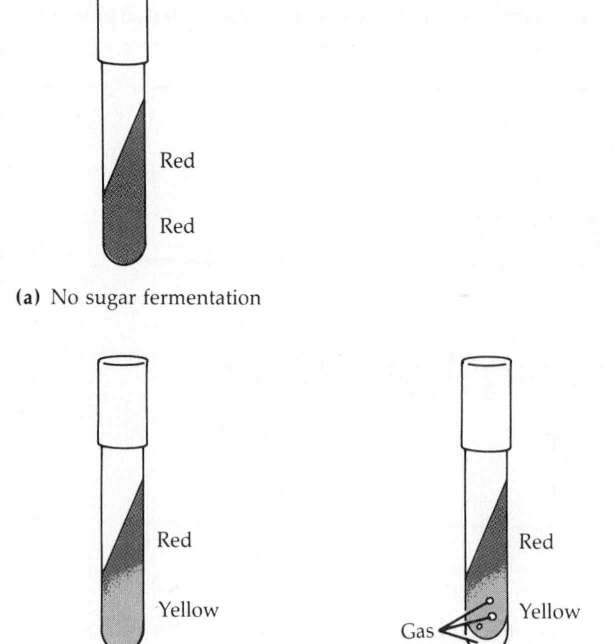

(a) No sugar fermentation

(b) Glucose fermented; lactose and sucrose not fermented

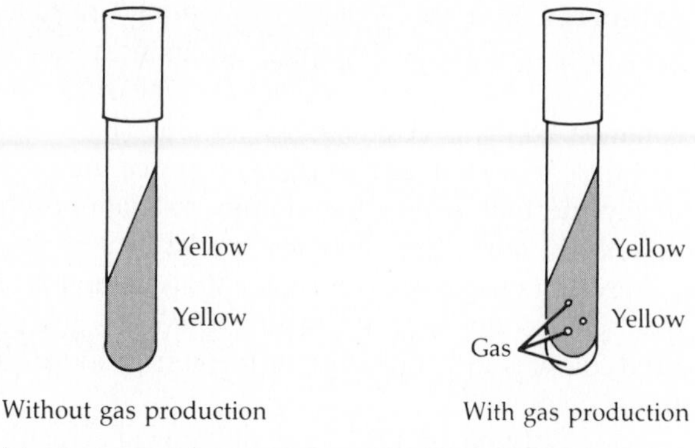

(c) Glucose and lactose and/or sucrose fermented

(d) H_2S production can occur in addition to (a), (b), and (c)

Figure 61-1.
Reactions in triple sugar iron (TSI) agar after 24 hours' incubation.

yellow and remain yellow for days due to the increased level of acid production. Gas production by an organism can be ascertained by the appearance of bubbles *in* the agar. TSI can also be used to indicate whether hydrogen sulfide (H_2S) (Exercise 21) has been produced due to the reduction of sulfur-containing compounds. H_2S reacts with ferrous sulfate in the medium producing ferric sulfide, a black precipitate. Table 61-1 shows how enteric bacteria can be identified using EMB, TSI, and additional tests.

The presence of enterococci can be used to indicate fecal contamination. Enterococci are fairly specific for their mammalian hosts. *Streptococcus faecalis* is found in humans, *S. bovis* in cattle, and *S. equinus* in horses. **SF (*Streptococcus faecalis*) broth** can be used to detect the presence of enterococci. SF broth contains sodium azide to inhibit the growth of gram-negative bacteria. Enterococci will grow and ferment the glucose in the broth. Growth and a change in the bromcresol purple indicator from purple (pH 6.6) to yellow (pH 5.2) may indicate the presence of enterococci.

Tomato juice agar is an enriched medium used to encourage the growth of lactobacilli. *Lactobacillus* does not grow well on ordinary media but its growth is enhanced by the addition of tomato juice and peptonized milk.

Materials

Petri plate containing EMB agar (1)

Petri plate containing tomato juice agar (1)

Tube containing SF broth (1)

TSI slant (1)

Sterile water

Brewer anaerobic jar and GasPak®

Tube containing sterile cotton swab (1)

10-ml pipette (1)

Fecal sample (animal or human feces; swab feces or anus with sterile swab and bring the inoculated swab to class in a sterile tube)

Gram stain reagents

Techniques Required

Exercises 7, 14, 18, 22, and 40

Procedure

1. If feces are hard, break apart in a sterile Petri plate and emulsify in sterile water. Swab a small area on the EMB with an inoculated swab, then streak for isolation with your loop (Figure 48-1*b*).
2. Incubate the plate inverted at 35°C for 24 to 48 hours.
3. Using the fecal swab, inoculate tomato juice agar. Incubate the plate inverted anaerobically (Exercise 40) at 35°C for 24 to 48 hours.

Table 61-1
Identification Scheme for Enteric Genera Primarily Using EMB and TSI

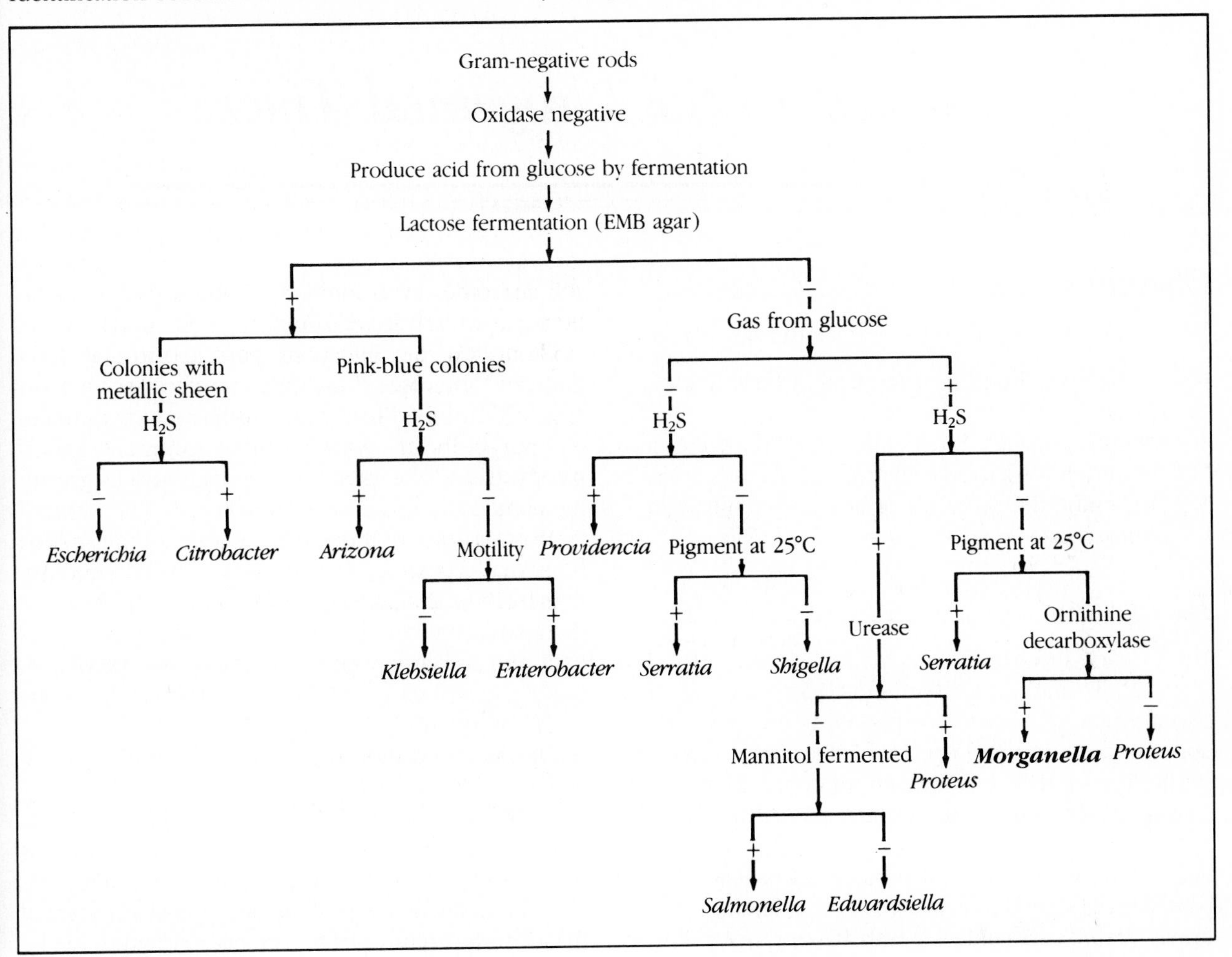

4. Inoculate SF broth by placing the fecal swab in the broth. Incubate the SF broth at 35°C for 24 to 48 hours.
5. Observe your EMB plate and those of other students. *Escherichia* and *Citrobacter* produce a green metallic sheen by reducing the dye. Did you see it? ______ Record the appearance and colors of the colonies. Prepare a Gram stain from an isolated colony.
6. Select one isolated colony using an inoculating needle and inoculate the TSI slant. Streak the surface of the slant, *then stab* into the butt of the agar. Incubate the tube at 35°C for 24 hours. Observe and record results from TSI slants inoculated from different appearing colonies.
7. Observe the tomato juice agar for the presence of *Lactobacillus.* Perform a Gram stain and a catalase test on isolated colonies. Is *Lactobacillus* catalase-positive or catalase-negative? ______
8. Observe the SF broth for the presence of enterococci. Gram stain.

Turn to the Laboratory Report for Exercise 61.

EXERCISE 62

Bacteria of the Urogenital Tract

Objectives

After completing this exercise you should be able to

1. List bacteria found in urine from a healthy individual.
2. Identify, through biochemical testing, bacteria commonly associated with urinary tract infections.
3. Determine the presence or absence of *Neisseria gonorrhoeae* in a G.C. smear.

Background

The urinary and genital systems are closely related anatomically, and some diseases that affect one system also affect the other system, especially in the male. The upper urinary tract and urinary bladder are normally sterile. The urethra does contain resident bacteria including *Streptococcus, Bacteroides, Mycobacterium, Neisseria,* and enterics. Most bacteria found in urine are due to contamination by skin flora during passage. The presence of bacteria in urine is not considered to be a urinary tract disease unless at least 10^5 cells per milliliter of urine are found.

Many infections of the urinary tract such as cystitis (inflammation of the urinary bladder) or pyelonephritis (inflammation of the kidney) are caused by opportunistic pathogens and are related to fecal contamination of the urethra and medical procedures such as catheterization.

Standard examination of urine consists of a plate count on blood agar for total number of organisms, coupled with a streak plate of undiluted urine on EMB agar (Exercise 48). Why? ________________
A "clean-catch" collection of a voided urine specimen depends on careful cleaning of the external urogenital surface with an antiseptic solution and collection of a midstream specimen in a sterile container.

You will examine normal urine in the first part of this exercise. In the second part, three gram-negative rods that commonly cause cystitis will be provided to you. *E. coli* and *Proteus* are enterics. *E. coli* is one of the coliforms (Exercise 25). *Proteus* is actively motile and exhibits "swarming" on solid media (Exercise 9) where the cells at the periphery move away from the main colony. *Pseudomonas* is a gram-negative aerobic rod. *Pseudomonas aeruginosa* is commonly found in the soil and other environments. Under the right conditions, particularly in weakened hosts, this organism can cause urinary tract infections, burn and wound infections, and abscesses. *P. aeruginosa* infections are characterized by blue-green pus. This bacterium produces an extracellular, water-soluble pigment called **pyocyanin** ("blue pus") that diffuses into its growth medium.

Most diseases of the genital system are transmitted by sexual activity and are therefore called **sexually transmitted diseases (STDs).** Most of the bacterial diseases can be readily cured with antibiotics if treated early and can largely be prevented by the use of condoms. Nevertheless, STDs are a major U.S. public health problem.

The most common reportable communicable disease in the United States is gonorrhea, an STD caused by the gram-negative diplococci *Neisseria gonorrhoeae* called gonococci or *G.C.* Diagnosis of gonorrhea is made by identifying the organism in the pus-filled discharges of patients, as demonstrated in the third part of the exercise. Cultures from patients' discharges can be made on Thayer-Martin medium and incubated in a CO_2 jar (Figure 59-2). An oxidase test is performed on characteristic colonies for confirmation.

Materials

Urine culture

Petri plate containing blood agar (1)

Petri plate containing EMB agar (1)

Sterile wide-mouth jar, approximately 50 to 250 ml

Sterile 1-ml pipettes (3)

Sterile 0.9-ml saline dilution blanks (2)

"Hockey stick" and alcohol

Cystitis

Petri plate containing EMB agar (1)

Petri plate containing Pseudomonas agar P (1)

Tubes containing OF-glucose medium (4)

Urea agar slants (2)

Vaspar

Oxidase reagent

Gram stain reagents

Cultures

Tube containing *Escherichia coli, Proteus vulgaris,* and *Pseudomonas aeruginosa* (1)

G.C. Smears

G.C. smears

One unknown smear # ____________

Techniques Required

Exercises 7, 13, 17, 20, 22, 61, and Appendix A

Procedure

Urine Culture

1. Collect a "clean-catch" urine specimen as described by your instructor, using the sterile jars available in the lab. Refrigerate until ready to perform step 2. Why? ____________
2. Add 0.1 ml urine to one dilution blank, pipette up and down three times to mix, and transfer 0.1 ml to the second tube. Mix and transfer 0.1 ml to the surface of the blood agar plate. This is a 1: ____________ dilution.
3. Dip the bent glass rod ("hockey stick") in alcohol and flame off the excess alcohol to disinfect the hockey stick. Spread the 0.1 ml diluted urine on the blood agar with the hockey stick. After spreading, dip the hockey stick in alcohol and ignite (Figure 26-2).
4. Streak a loopful of *undiluted* urine on EMB agar.
5. Incubate both plates inverted at 35°C for 24 to 48 hours.
6. Count the colonies on the blood agar plate and determine the number of bacteria per milliliter of urine.

$$\text{Bacteria per ml} = \frac{\text{number of colonies}}{\text{amount plated} \times \text{dilution}}$$

7. Examine the EMB plate for the presence of coliforms. Contact your instructor if alarmed by your results.

Cystitis

1. Inoculate an EMB plate and Pseudomonas agar P plate with the mixed culture of bacteria.
2. Incubate the plates inverted at 35°C for 24 to 48 hours.
3. Examine the EMB plate for lactose-fermenting colonies. Which of the three organisms ferments lactose? ____________
   Look for swarming. Which organism is actively motile on solid media? ____________
   Are any small, nonlactose-fermenting colonies present? ____________
4. Prepare a Gram stain from each different colony type.
5. Examine the Pseudomonas agar P plate. Can you identify *Pseudomonas?* ____________
6. Perform an oxidase test on each different organism (see Exercise 22).
7. Inoculate four tubes of OF-glucose medium: two with *Proteus* and two with *Pseudomonas.* Plug one tube of each organism with vaspar (Exercise 17). Inoculate a urea agar slant with each organism (Exercise 20). Incubate the tubes at 35°C for 24 to 48 hours.
8. Record the results of OF-glucose and urease production tests. Why isn't it necessary to perform these two biochemical tests on *E. coli?* ____________

G.C. Smears

1. Examine the G.C. smears provided.
2. Determine whether *N. gonorrhoeae* could be present in your unknown G.C. smear.

Turn to the Laboratory Report for Exercise 62.

EXERCISE 63

Identification of an Unknown from a Clinical Sample

*Unknown #3**

So simple did it seem to me
The day t'was first presented,
A Sherlock Holmes of lab technique
No effort be resented.

Now six Gram stains, three cultures later
The first step barely taken
Gram vari rods, both red and blue
My confidence was shaken.

Straight rods 'tis true, yet red and blue
In ones and twos and chain
Some spores would show the truth of it
But stains were all in vain.

Alas! No spores of vivid green
Would show in fields of red.
Ah Ha! Too soon! Spores take some time
Flashed through my seeking head.

With rod in hand, though young it be,
It surely must, said I,
Reveal some trait, another test
Than just a non-cocci.

So while I wait don't waste the time
The sugars show some trait.
So off to these, ferment or not
I can't just sit and wait.

Glucose a plus, Sucrose the same
No gas, and lactose zilch.
The spores again with no green spots
It's on to litmus milk.

So mannitol and nitrate too
I'll run, for these should show,
Though spores resist at now day six,
Some little cells should grow.

Day seven's gone and spores still hide
From vision sharply tooled.
But agents stained from tests complete
Have many others ruled.

Divorce now near, by child disowned
On job, the thought "to fire"
I'll take my stand and flunk or not
B. subtilis, *or tests are liars.*

**Source:* M. E. Davis. 1976. *ASM News* 42(3):164. With the permission of the American Society for Microbiology.

Objectives

After completing this exercise you should be able to

1. Separate and identify two bacteria from a sample.
2. Determine the unknown bacteria's sensitivity to chemotherapeutic agents.

Background

Clinical samples often contain many microorganisms, and a pathogen must be separated from resident normal flora. Differential and selective media are employed to facilitate isolation of the pathogen. After isolation, the pathogen must be identified. Also, antibiotic sensitivity tests are performed to aid the physician in prescribing treatment. Since effective treatment depends on identification of the pathogen, a clinical laboratory must provide results as quickly and as accurately as possible. Microbiology laboratory results are usually available within 48 hours.

You will be provided with a mixed culture containing two bacteria. After you have separated them in pure culture, prepare stock and working cultures. To obtain pure cultures, inoculate a trypticase soy agar plate *and* a differential medium. Select the differential medium using the type of sample as a clue. For in-

Table 63-1
Zone-Diameter Interpretive Standards

Antimicrobial Agent	Disk Content	Zone-diameter (mm)	
		Resistant	Susceptible
Ampicillin	10 µg		
Gram-negative enterics and enterococci		11 or less	14 or more
Staphylococci		20 or less	29 or more
Chloramphenicol	30 µg	12 or less	18 or more
Erythromycin	15 µg	13 or less	18 or more
Penicillin-G	10 U		
Staphylococci		20 or less	29 or more
Other bacteria		11 or less	22 or more
Streptomycin	10, 250, or 300 µg	11 or less	15 or more
Sulfonamides	300 µg	12 or less	17 or more
Tetracycline	30 µg	14 or less	19 or more

Adapted from Table 14-2 "Zone-Diameter Interpretive Standards and Approximate MIC Correlates." A. L. Barry, *The Antimicrobic Susceptibility Test: Principles and Practices.* Philadelphia, Lea & Febiger, 1976.

stance, a fecal sample should probably be inoculated onto ______________________ agar. Since differential media should give the information needed to isolate the organisms, why is a trypticase soy agar inoculated also? ______________________

In this exercise, you will be provided with a simulated clinical sample containing two bacteria. The procedure will differ from that used in a clinical laboratory in that you will be asked to isolate and identify both organisms in the sample.

Materials

Petri plates containing trypticase soy agar

Trypticase soy agar slants

All stains, reagents, and media previously used

Antimicrobial agents

Culture

Unknown sample # ______________________

Techniques Required

Exercises 24, 46, 58 through 62, and Appendix H

Procedure

1. Refer back to Exercise 24 for general information regarding identification of an unknown. Study the identification schemes in Exercises 24, 58, 60, 61 and Appendix H as you proceed.
2. You may be able to prepare a Gram stain directly from the unknown if it does not contain a high concentration of organic matter to interfere with the stain results.
3. Streak the unknown onto a trypticase soy agar plate and an appropriate differential medium. *Do not* be wasteful. Inoculate only the necessary media. Incubate the plates at 35°C for 24 to 48 hours.
4. Examine the plates and select colonies that differ from each other in appearance and are separated for easy isolation.
5. Streak each organism for isolation onto half of a trypticase soy agar plate. Incubate the plate at 35°C for 24 to 48 hours. This plate can be your first working culture. Prepare a stock culture of each organism.
6. Prepare a Gram stain of each organism. What should you do if you do not think you have a pure culture? ______________________
____
____

7. After determining the staining and morphologic characteristics of the two organisms, determine which biochemical tests you will need. Plan your work carefully. The same tests need not be performed on both bacteria.
8. Record which tests were performed and the results.
9. Determine whether your unknowns are resistant (R) or sensitive (S) to the antimicrobial agents listed in Table 63-1.

Turn to the Laboratory Report for Exercise 63.

EXERCISE 64

Animal Viruses

Objectives

After completing this exercise you should be able to

1. List the characteristics used to classify animal viruses.
2. Describe the chemical composition and structure of animal viruses.
3. Describe the method of classifying viruses.
4. Identify two shapes of viral capsids.

Background

Over the years, many systems of classifying viruses have been used, including classification according to hosts, tissue types, and diseases. These systems of classification are inadequate because some viruses can induce many types of diseases and can infect different tissues. Moreover, some symptoms can be caused by more than one virus. A classification system based on nucleic acid composition and sequence would be the most accurate, but that type of system would take a great deal of time to complete. Presently, viral classification is based on the virus's nucleic acid and capsid shape (symmetry), the presence of an envelope, and the number of capsomeres (see Table 64-1).

Animal viruses consist of a single piece of nucleic acid, either DNA or RNA, single-stranded or double-stranded, surrounded by a protein coat called a **capsid.** In some viruses, the capsid is covered by an **envelope** that is usually composed of lipids, proteins, and carbohydrates. The molecular organization of the envelope is generally unknown. Some viruses such as Rhabdovirus and Poxvirus have complex shapes due to the relationship between the envelope and the capsid.

Capsids may be either helical or icosahedral, depending on the arrangement of the protein subunits called **capsomeres.** A **helical virus,** which resembles a long rod, contains a strand of nucleic acid in a spiral or helical configuration. Capsomeres project out from the nucleic acid strand like beads on a string. The number of capsomeres depends on the length of the nucleic acid. The helix can be twisted into a spherical shape and have an envelope like the influenza virus.

An **icosahedron** is a regular polygon with 20 sides (faces) and 12 vertices (points). This is shown in Figure 64-1. This capsomeres consist of five- and six-protein units called **pentamers** and **hexamers,** respectively (Figure 64-2). There are very few different proteins in one virus because the virus has a limited amount of nucleic acid for storing genetic information for protein synthesis. Moreover, only a few different bonds are needed between similar protein units of the capsomeres; this facilitates the spontaneous formation of the capsid without complex biochemical reactions.

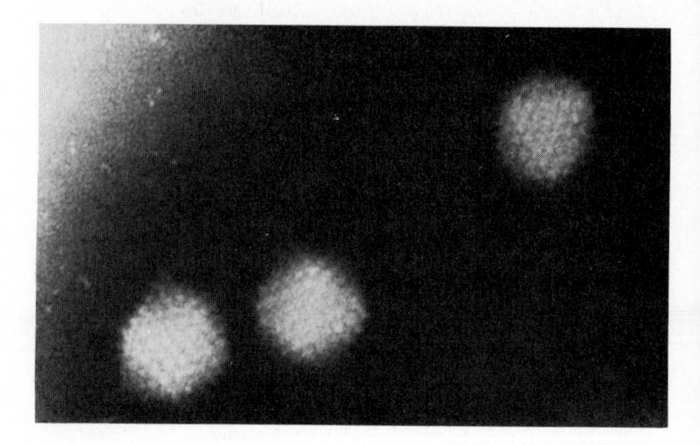

Figure 64-1.
Electron micograph of several icosahedral particles of adenovirus ($\times$267,000). Individual capsomeres in the protein coat are visible.

Table 64-1
Classification of Animal Viruses According to Morphologic, Chemical, and Physical Properties

Viral Group and Specific Examples	Morphologic Class	Nucleic Acid*	Dimensions of Capsid (Diameter, nm)	Number of Capsomeres
Parvoviruses (adenosatellite)	Naked icosahedron	ss DNA	18–26	12
Papovaviruses (papilloma, polyoma, simian virus 40)	Naked icosahedron	ds circular DNA	40–57	72
Adenoviruses	Naked icosahedron	ds DNA	70–80	252
Herpesviruses (herpes simplex, herpes varicella-zoster, Epstein-Barr)	Enveloped icosahedron	ds DNA	150–250	162
Poxviruses (variola, cowpox, vaccinia)	Enveloped	ds DNA	200–350	(brick-shaped)
Picornaviruses (poliovirus, rhinovirus)	Naked icosahedron	ss RNA	28–38	32
Togaviruses (arboviruses, rubella)	Enveloped icosahedron	ss RNA	40–60	32
Orthomyxoviruses (influenza)	Enveloped helical	ss segmented RNA	80–200	(coarse spikes)
Paramyxoviruses (measles, mumps)	Enveloped helical	ss segmented RNA	150–300	(fine spikes)
Rhabdoviruses (rabies)	Enveloped helical	ss RNA	70–180	(bullet-shaped)
Reoviruses	Naked icosahedron	ds segmented RNA	60–80	92

*ds = double-stranded, ss = single-stranded

EXERCISE 64

In addition to providing protection for the nucleic acid, some capsomeres also function as attachment sites to the target cells.

Icosahedral viruses have three axes of symmetry: **5:3:2.** These are shown in Figure 64-2. If a rod is put through a vertex, the capsid can be rotated in *five* identical positions. If a rod is put through the center of a face, the capsid can be rotated into *three* identical positions. And if a rod is placed through the midpoint of an edge, it divides the capsid into *two* identical parts, providing the third axis.

The smallest virus particle has 12 pentamers forming its capsid. There must be 12 pentamers to form the 12 vertices of an icosahedron. Larger viruses have hexamers attached to each edge of the pentamer.

Bacteriophages and plant viruses were cultured in Part 7. The shape of the nonenveloped or **naked animal viruses** will be studied in this exercise.

Materials

Scissors

Index cards or paper

Tape or glue

Colored pencils or felt tip markers

Newspaper

Techniques Required

None. Rereading Part 7 will be helpful.

Procedure

A. An icosahedron.* This will not produce an actual

*Source: M. J. Wenninger. 1971. *Polyhedron Models.* New York: Cambridge University Press, p. 17.

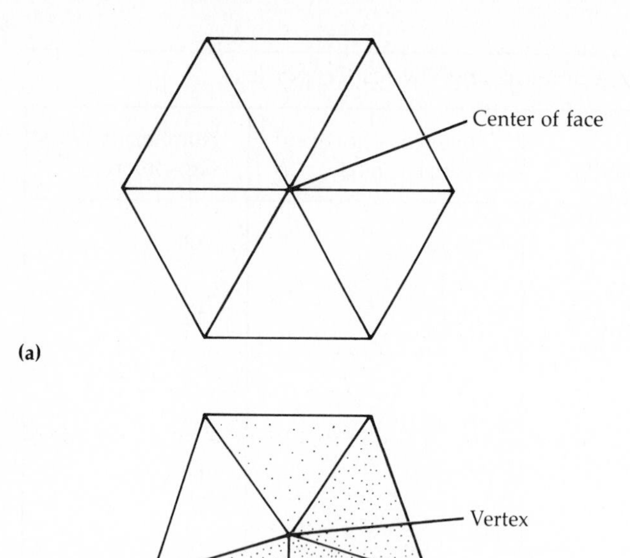

Figure 64-2. ______

Capsomeres. **(a)** Hexamers are composed of six proteins. **(b)** Pentamers consist of five proteins and form the vertices of an icosahedron.

model of a virus particle; however, it will show the shape and symmetry of an icosahedron.

1. Place the template for an equilateral triangle over a few cards or pieces of paper, and mark the vertices by poking a hole through the template with a pin or probe (Figure 64-3*a*). Connect each hole with a straight line to complete the triangles (Figure 64-3*b*) and cut out each triangle leaving a 5-mm margin (Figure 64-3*c*). Cut 20 triangles from the index cards or paper.
2. Color five triangles as follows: yellow, blue, orange, red, green. Glue them together as shown here:

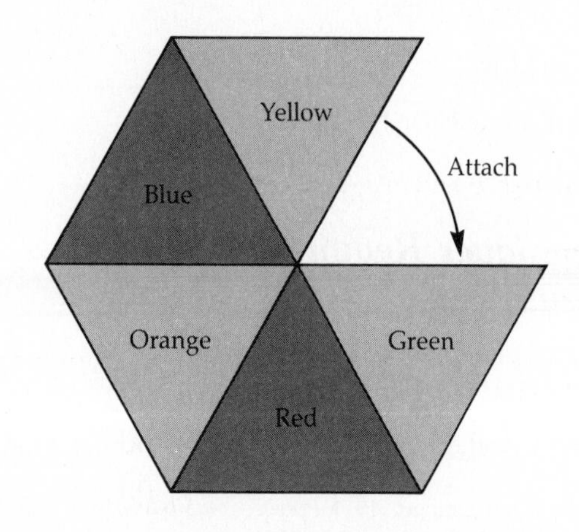

Fold the margin and glue margins on the inside (Figure 64-3*d*).
3. Color five more triangles: red, green, yellow, blue, orange. Attach them to the first five (red to yellow, green to blue, yellow to orange, blue to red, orange to green). This produces one-half of the icosahedron.
4. To make the other half, connect five triangles (yellow, blue, orange, red, green) together. Then attach these five triangles: orange, red, green, yellow, blue.
5. Assemble both halves of the triangle. The orange of the last row should fit between the red and green of the second row made in step 3.
6. What colors are around each vertex? ______

______

How many vertices are there on your model? ______

How many faces? ______

B. A pentamer. The purpose of this is to demonstrate the shape of each vertex formed by the pentamers. This will not be used in the virus model.
1. Cut out five equilateral triangles as described in part A, step 1.
2. Glue the triangles together to form a pentagonal pyramid.

C. A parvovirus model.
1. Cut out 12 pentamers as described in part A, step 1, using a pentamer template. Color six of them different colors and attach them together.
2. Color the remaining six and assemble them in the same order used in step 1. Attach the two halves together.
3. Each pentamer is composed of ___ protein units. How many faces does this model have?

______

D. A virus with 12 pentamers and 20 hexamers.
1. Cut out 12 pentamers and 20 hexamers as described in part A, step 1.
2. Color the pentamers one color and the hexamers a different color.
3. Begin by attaching a hexamer to each edge of one pentamer. Then add five pentamers at regular intervals, followed by hexamers, five more pentamers, hexamers, and the last pentamer.
4. What color are the pentamers? ______
What color are the vertices? ______
Mark the location of the vertices in black ink.

E. Helical virus
1. Obtain 3 full sheets of newspaper.
2. Cut at the crease.

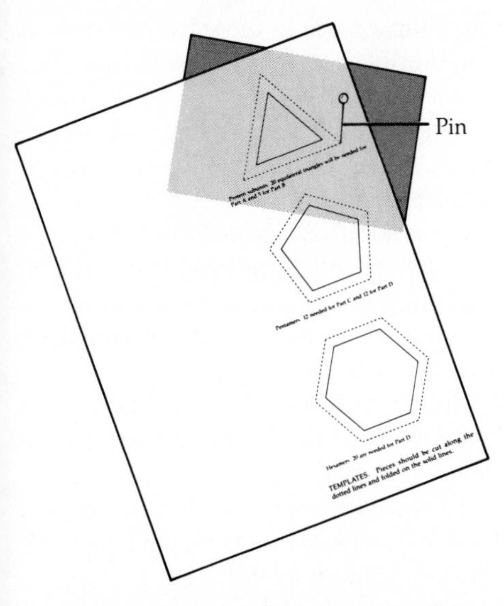

(a) Set the template over an index card or paper and mark each vertex using a pin.

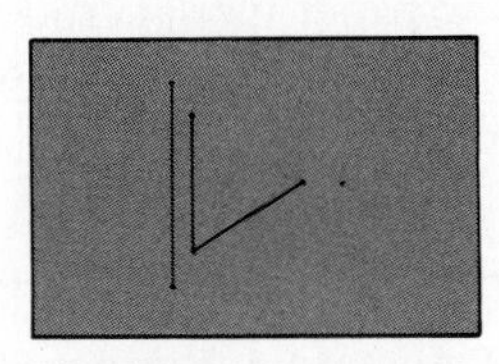

(b) Remove the template and connect each hole with a straight line.

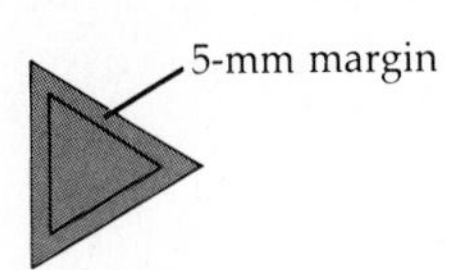

(c) Cut out the necessary number of pieces, leaving a 5-mm margin for folding.

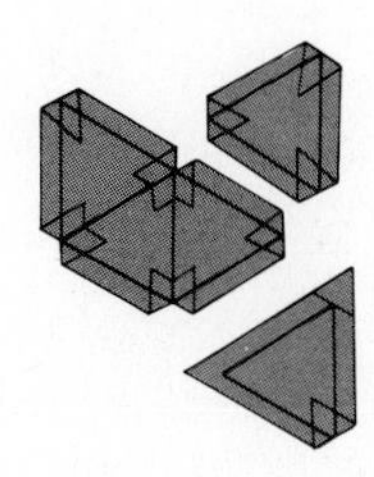

(d) To construct the models, fold along the solid line. The folded-in edges can be glued or taped to the folded edge of the next unit.

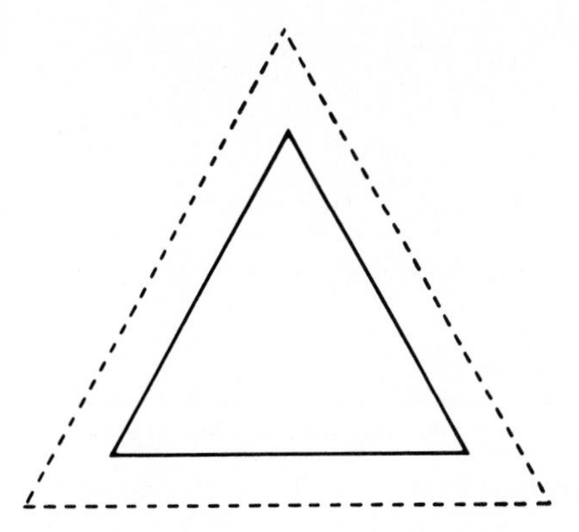

Protein subunits. 20 equilateral triangles will be needed for Part A and 5 for Part B.

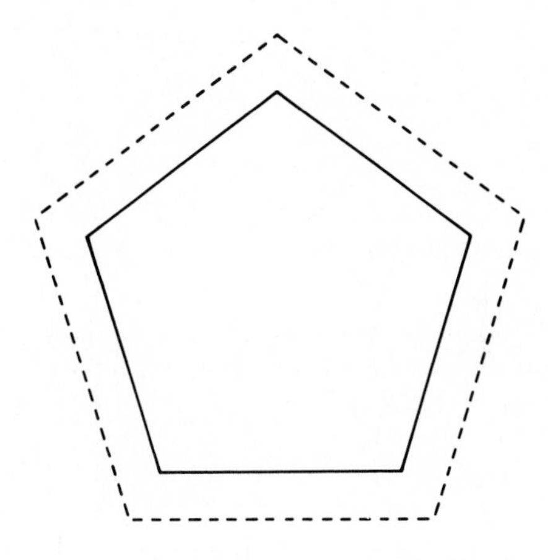

Pentamers. 12 needed for Part C and 12 for Part D.

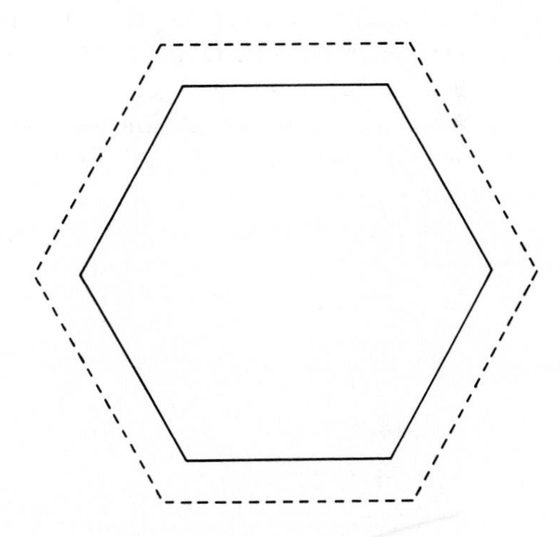

Hexamers. 20 are needed for Part D.

TEMPLATES. Pieces should be cut along the dotted lines and folded on the solid lines.

Figure 64-3.
Using the templates.

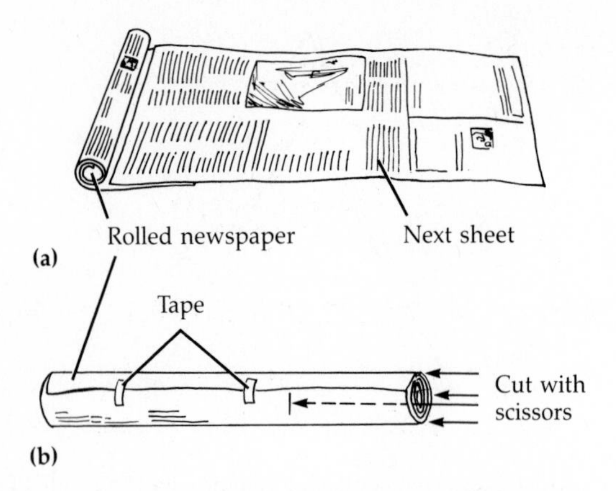

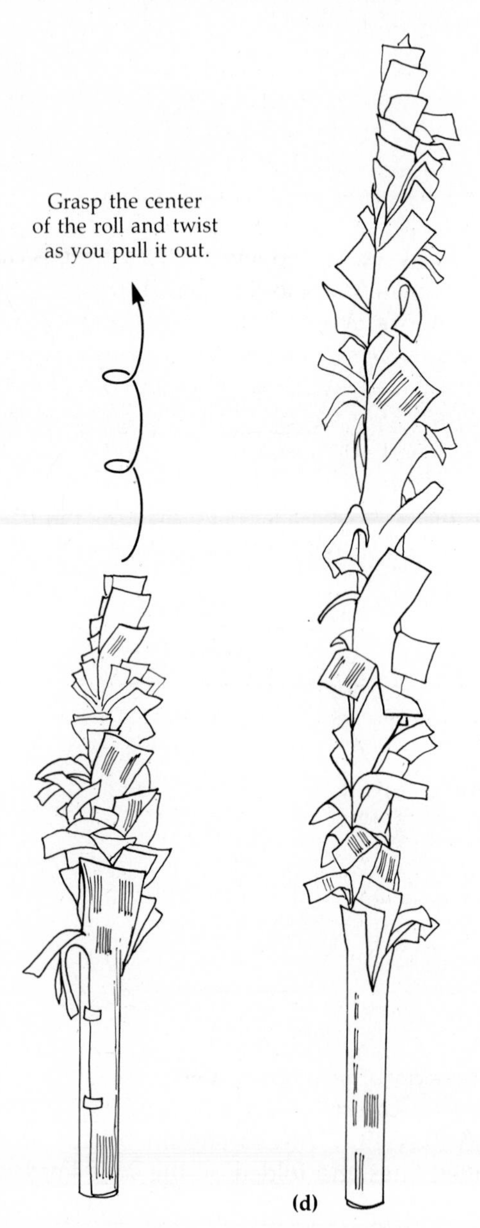

Figure 64-4.
Newspaper helical model.

3. Start rolling a piece in the same direction as the newsprint. Roll it tightly. Stop about 12 cm from the end.
4. Insert another sheet into the roll as shown in Figure 64-4*a*. Roll it tightly onto the roll completed in step 3.
5. Again stop about 12 cm from the end.
6. Repeat steps 4 and 5 until you have used all six half sheets.
7. Place two pieces of tape on the rolled up newspaper to keep the roll tight.
8. Make four cuts, each 11 to 12 cm long on one end to give 4 equal-sized pieces (Figure 64-4*b*).
9. Grasp the inner edge (inside of the cut pieces) and twist as you pull quickly on the newspaper (Figure 64-4*c*).
10. The newspaper should extend about 2 meters with the cut edges projecting out (Figure 64-4*d*).
11. This newspaper model crudely represents a helical virus with each cut strip of paper a capsomere.

Turn to the Laboratory Report for Exercise 64.

EXERCISE 65

Introduction to Medical Mycology

Objectives

After completing this exercise you should be able to

1. Identify asexual spores.
2. Explain dimorphism.
3. List major characteristics of the four phyla of fungi containing human pathogens.

Background

Seven years before Koch provided definitive proof that diseases of animals (tuberculosis and anthrax) were caused by bacteria (Exercise 57), Anton de Bary showed that the fungus *Phytophthora infestans* was responsible for the potato blight of 1845 in Europe. **Medical mycology** is the study of parasitic fungi. Most pathogenic fungi are soil saprophytes that cause infections when they gain access to a human host. Pathogenic fungi are found in four phyla: Zygomycota, Ascomycota, Basidiomycota, and Deuteromycota. Characteristics of the first three phyla were examined in Exercises 31 and 32.

The *Deuteromycota* are also called the Fungi Imperfecti. These fungi are "imperfect" because sexual spores have not been demonstrated as yet. The Deuteromycota have septate hyphae and produce a variety of asexual spores (see Table 65-1).

Most of the pathogenic fungi are, or once were, classified as Deuteromycota. When sexual spores are observed in members of a genus of Deuteromycota, the species is reclassified into a different genus (usually a new genus) and placed in the appropriate phylum.

Some pathogenic fungi exhibit **dimorphism;** that is, they have two growth forms. In pathogenic fungi, dimorphism is usually temperature-dependent. The fungus is yeastlike at 37°C and moldlike at 25°C.

In addition to the asexual spores studied in Exercise 32, pathogenic fungi may produce arthrospores, chlamydospores, and macroconidia. **Arthrospores** are formed by fragmentation of a hypha into single cells

Table 65-1
Characteristics of Some Parasitic Fungi

Phylum	Asexual Spore Types	Human Pathogens	Type of Mycosis
Zygomycota	Sporangiospores	*Rhizopus, Mucor*	Systemic (opportunistic)
Ascomycota	Conidiospores	*Allescheria boydii*	Subcutaneous
		Aspergillus	Systemic (opportunistic)
		Blastomyces dermatidis	Systemic
		Histoplasma capsulatum	Systemic
		Microsporum	Cutaneous
		Trichophyton	Cutaneous
Basidiomycota	Conidiospores	*Cryptococcus neoformans*	Systemic
Deuteromycota	Conidiospores	*Epidermophyton*	Cutaneous
		Sporothrix schenckii	Subcutaneous
		Phialophora, Fonsecaea	Subcutaneous
	Conidiospores, chlamydospores, arthrospores	*Paracoccidioides brasiliensis*	Systemic
	Chlamydospores	*Cladosporium werneckii*	Superficial
		Candida albicans	Cutaneous, systemic
	Arthrospores	*Coccidioides immitis*	Systemic
		Trichosporon	Superficial

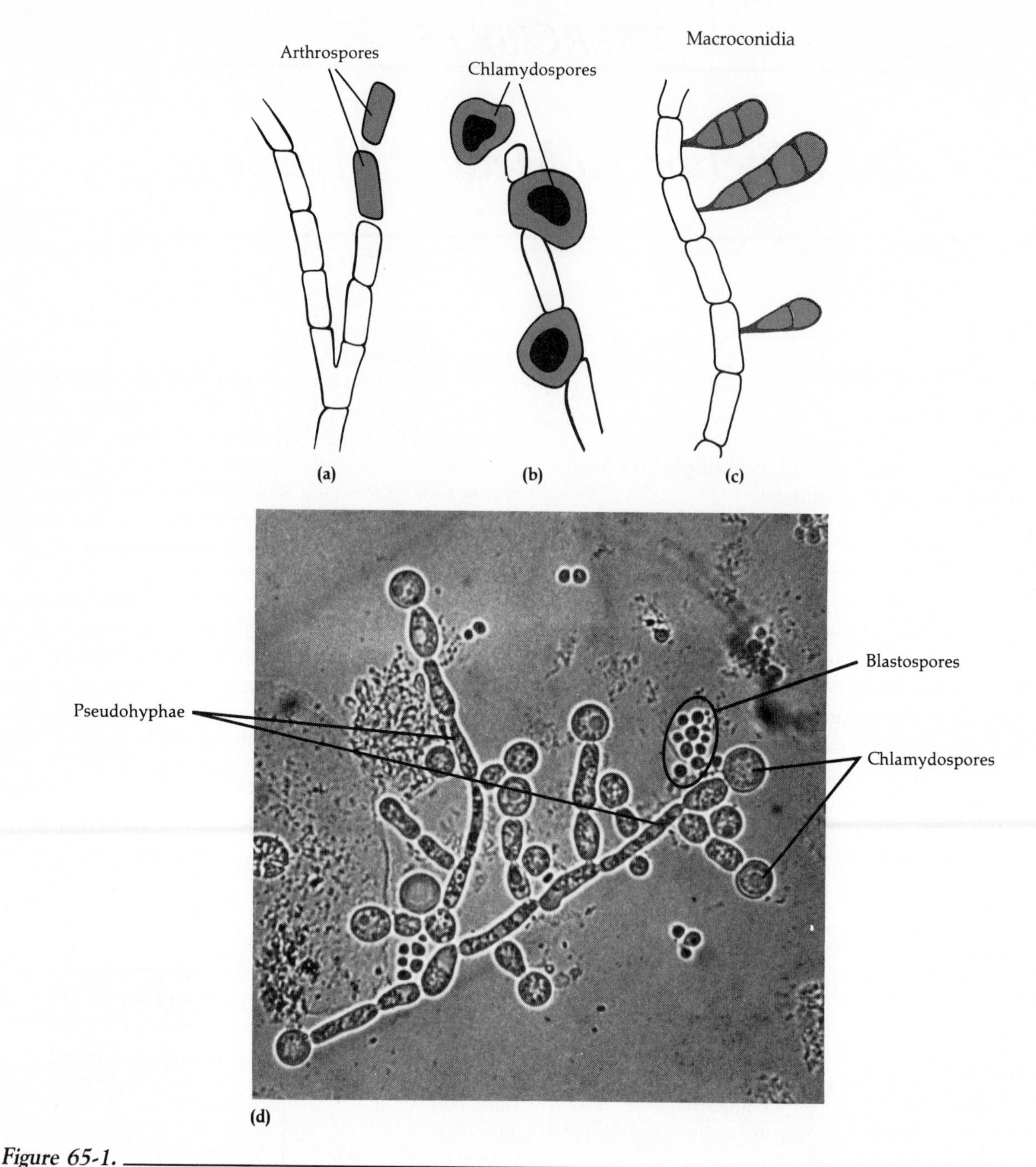

Figure 65-1.

Asexual spores. **(a)** Fragmentation of hyphae results in formation of arthrospores. **(b)** Chlamydospores are thick-walled cells within the hyphae. **(c)** Elongated spores divided into two or more cells are called macroconidia. **(d)** Photomicrograph of *Candida albicans,* an opportunistic fungal pathogen that causes candidiasis. Note the spherical chlamydospores, the smaller blastospores, and the pseudohyphae.

with thickened cell walls (Figure 65-1*a*). **Chlamydospores** are thick-walled spores formed within a hypha (Figure 65-1*b*). **Macroconidia** are elongated structures divided into two or more cells and produced along a hypha (Figure 65-1*c*). **Blastospores** are formed by budding from a parent cell (Figure 65-1*d*).

A summary of the medically important fungi is given in Table 65-1. Fungal infections are classified on the basis of the level of infected tissue and method of entry into the host. A **systemic mycosis** is deep within the patient and is most often acquired by inhalation of fungal spores from soil. **Subcutaneous mycoses** are localized beneath the skin and develop from implantation of the fungus in a puncture wound. A **cutaneous mycosis** is an infection of the hair, skin, or nails. The fungi that cause cutaneous mycoses are able to degrade keratin. These are the only fungal infections that are readily transmitted from one person to another. **Superficial mycoses** are local infections of hair shafts and surface epithelial cells. A few saprophytic fungi cause **opportunistic mycoses** in compromised (weakened) hosts.

Materials

Slide Culture

Sabouraud agar, melted

Petri plate, (1)

Bent glass rod (1)

Pasteur pipette (1)

Cultures (one of each)

Gymnoascus

Sepedonium

Dimorphic Gradient

Sabouraud agar, 5 ml, melted at 48°C

5-ml plastic microbeakers (2)

Razor blade

Tape

Cultures (one of each)

Mucor rouxii

Aspergillus niger

Penicillium notatum

Techniques Required

Exercises 2, 13, 31, and 32

Procedure

Slide Culture

1. Prepare a slide culture (Exercise 32) of *Gymnoascus* or *Sepedonium* and incubate at room temperature.
2. Observe the growth of both molds periodically for the next 7 days, and record your observations in the Laboratory Report. These fungi are saprophytes that produce asexual spores that are typical of pathogenic species. Why are saprophytes used in this laboratory exercise instead of pathogens? ____

Dimorphic Gradient*

1. Flick the fungal culture tube to resuspend the culture. Inoculate the melted agar with 2 or 3 loopfuls of the mold culture assigned to you. Mix the tube by rolling it between your hands, and quickly pour the contents into an empty beaker before it hardens. Why doesn't the beaker need to be sterile? ____

*Adapted from S. Bertnicki-Garcia, 1972, "The Dimorphic Gradient of *Mucor rouxii*: A Laboratory Exercise," *ASM News* 38(9):486–488. Permission granted by the American Society for Microbiology

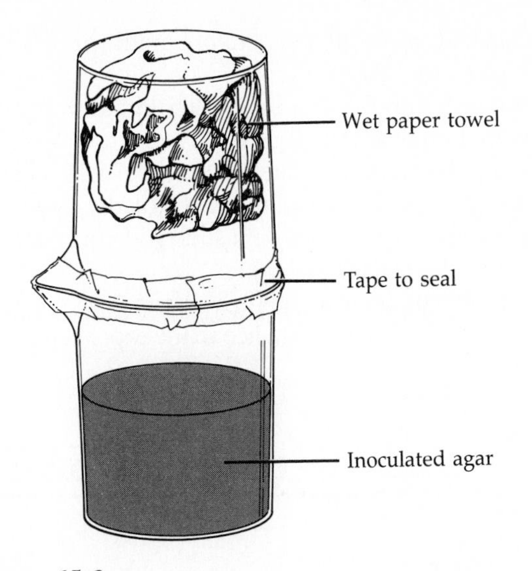

Figure 65-2.
Tape the two beakers together as shown.

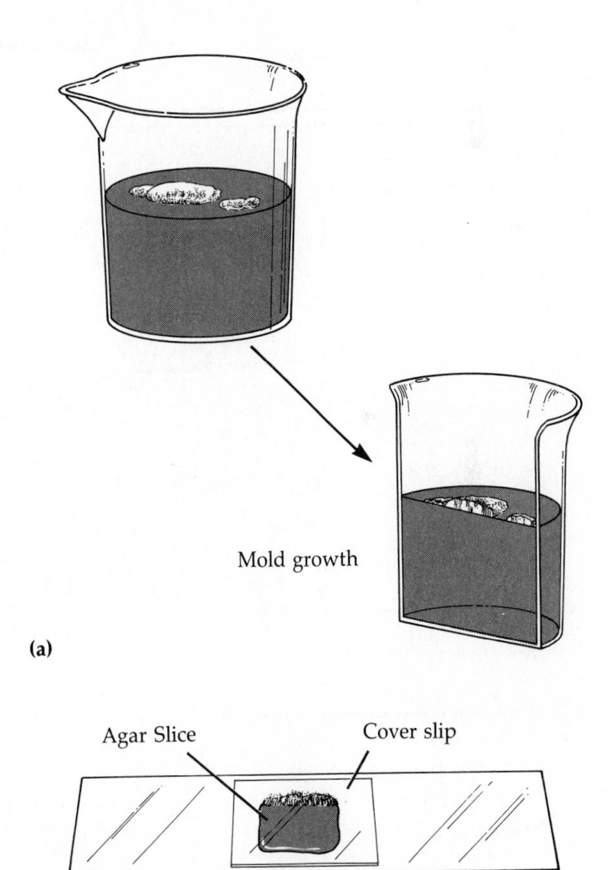

Figure 65-3.
To observe fungal growth microscopically **(a)** cut the bottom beaker in half with a razor blade and cut a vertical slice of agar. **(b)** Place the thin slice of agar on a slide and cover with a cover slip.

2. Place a piece of wet paper towel in the remaining empty beaker and invert over the beaker containing the agar (Figure 65-2). Tape the edges.

3. Incubate at room temperature until growth occurs (5 to 7 days).
4. Remove the tape. Cut the beaker in half vertically with a razor blade (Figure 65-3*a*).
5. Cut a thin (approximately 1 mm) vertical slice of the inoculated agar with a razor blade. Carefully place the slice of agar on a slide and cover with a cover slip (Figure 65-3*b*).
6. Observe under low and high power. Scan the agar from the bottom of the slice to the top. Many focal planes are present. Observe the slides of the other fungi.
7. Discard your beaker and disinfect your slide as instructed.

Turn to the Laboratory Report for Exercise 65.

EXERCISE 66

Introduction to Medical Protozoology

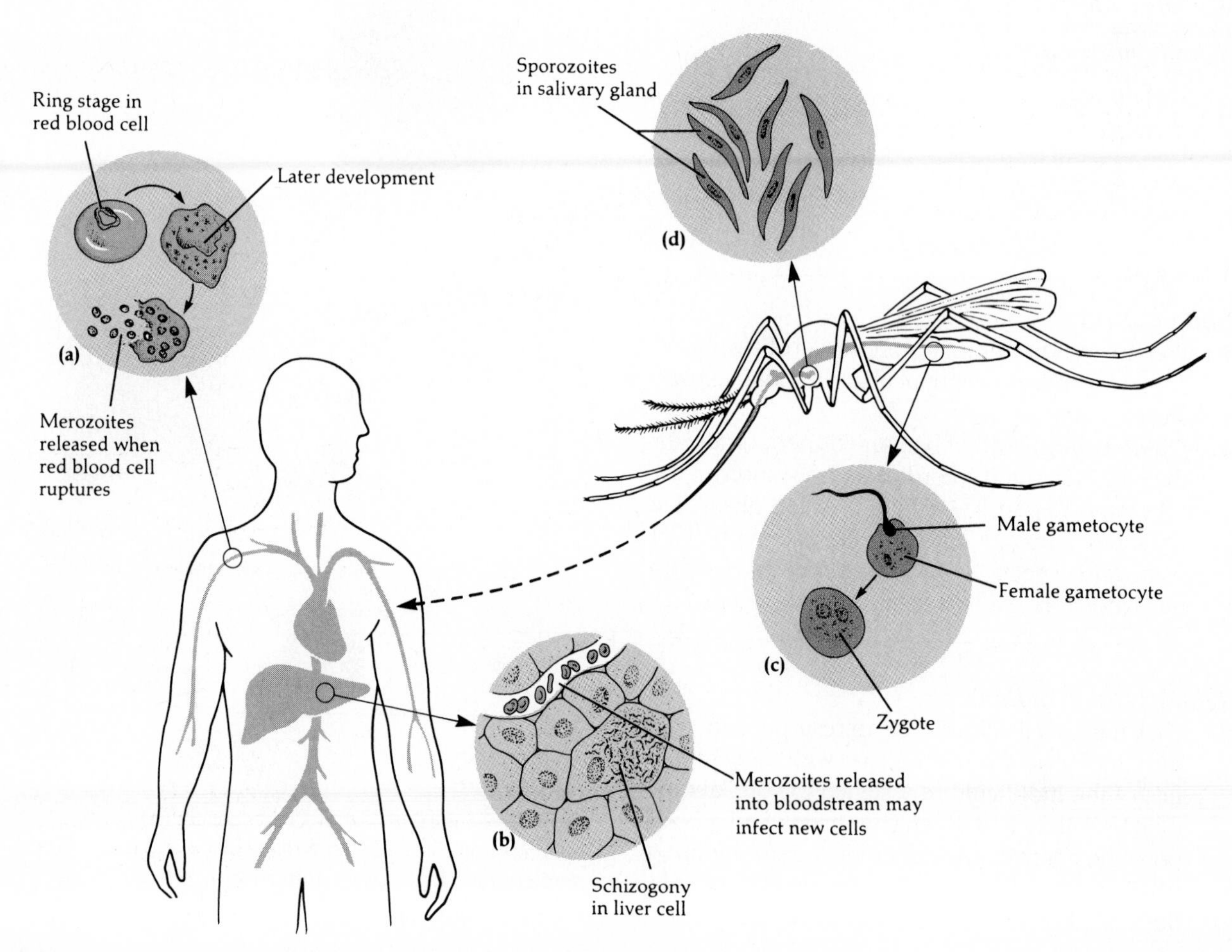

Figure 66-1.

Life cycle of *Plasmodium vivax.* Asexual reproduction, schizogony, of the merozoites takes place **(a)** in the red blood cells or **(b)** in the liver of a human host. **(c)** Sexual reproduction of the parasite occurs in the intestine of an *Anopheles* mosquito after ingestion of gametocytes by the mosquito. **(d)** Sporozoites resulting from sexual reproduction migrate to the salivary gland to be injected into the mosquito's next host.

Objectives

After completing this exercise you should be able to

1. Describe the four phyla of protozoans that include disease-causing species.
2. Compare and contrast free-living and parasitic protozoans.

Background

Disease-causing protozoans are found in four phyla: Sarcodina, Ciliata, Mastigophora, and Sporozoa. The vegetative forms, or **trophozoites,** of the phylum Sarcodina move by amoeboid movement. The Ciliata possess cilia. Trophozoites of the Mastigophora propel themselves by means of flagella. These three phyla were described in Exercise 34. Members of the fourth phylum, the Sporozoa, are incapable of movement.

Protozoans reproduce asexually by fission, budding, or schizogony. In **fission,** one cell divides into two by mitosis (nuclear division) and cytokinesis (cell division). **Budding** involves the formation of a small projection, a bud, from the parent cell. After nuclear division, one nucleus migrates to the growing bud. **Schizogony** is multiple fission. After many nuclear divisions, small portions of cytoplasm aggregate around each nucleus, which then separates into individual daughter cells.

Under certain conditions, some protozoans form a protective capsule called a **cyst.** A cyst permits the organism to survive when food, moisture, or oxygen are insufficient, when temperatures are not suitable, and when toxic chemicals are present.

Sporozoans are obligate intracellular parasites. They have complex life cycles that ensure their survival and transmission from host to host. An example is *Plasmodium,* the causative agent of malaria (Figure 66-1). Trophozoites of *Plasmodium* called **merozoites** are

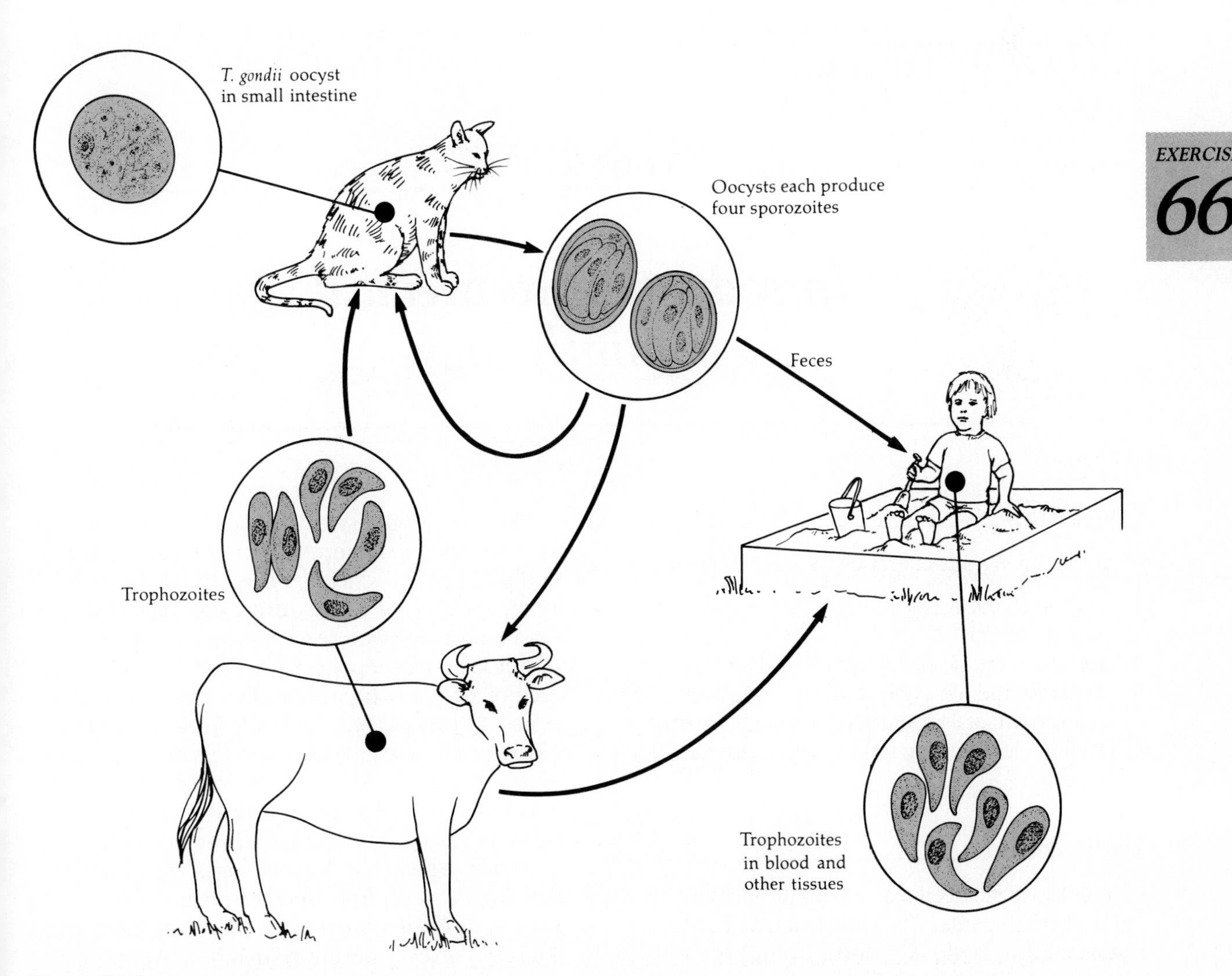

EXERCISE 66

Figure 66-2.

Life cycle of *Toxoplasma gondii.* The domestic cat is the definitive host in which the sporozoan reproduces sexually to produce oocysts, which are passed out in feces. Four sporozoites form within each oocyst. The sporozoites are then ingested by one of the intermediate hosts. They mature into trophozoites in blood and other tissues.

found in the red blood cells of a human host. The merozoites multiply by schizogony and cause lysis of the red blood cell. Most of the merozoites infect new red blood cells. However, some develop into male and female sexual forms called **gametocytes.** When gametocytes are ingested by an *Anopheles* mosquito, they produce sporozoites by sexual reproduction. The sporozoites are then injected into a new human host by the biting mosquito, and they undergo schizogony in the liver to produce infective merozoites.

Toxoplasma gondii is another sporozoan that infects humans. The life cycle (Figure 66-2) of this parasite is not well known but appears to involve sexual and asexual reproduction in domestic cats.

Techniques Required

Exercise 1. Rereading Exercise 34 may be helpful.

Materials

Prepared slides of each of the following:

Trichomonas vaginalis

Trypanosoma gambiense, blood smear

Balantidium coli, trophozoite and cyst

Toxoplasma gondii

Entamoeba histolytica

Plasmodium falciparum, ring stage

Procedure

1. Observe the prepared slides. Make accurate drawings of each parasite in the spaces provided in the Laboratory Report. Observe each slide at the same magnification and show the relative size of each parasite in your drawings. Which is the largest? ____________

Turn to the Laboratory Report for Exercise 66.

EXERCISE 67

Introduction to Medical Helminthology

Objectives

After completing this exercise you should be able to

1. List the characteristics of the three classes of parasitic helminths.
2. Identify parasitic helminths as to class.
3. Describe the life cycle of a fluke, tapeworm, and nematode that use humans as their definitive hosts.
4. Define the following terms: hermaphrodite, dioecious.

Background

Helminths are multicellular animals that belong to two phyla: Platyhelminthes (flatworms) and Aschelminthes (roundworms). **Medical helminthology** is the study of members of these phyla that spend part or all of their lives in humans.

The reproductive systems of helminths can be complex, resulting in the production of large numbers of fertilized eggs by which to infect suitable hosts. Adult helminths may be **hermaphroditic,** with male and female reproductive organs in one animal. Two hermaphrodites may copulate and simultaneously fertilize each other. A few hermaphrodites can fertilize themselves. Organisms in which the male reproductive organs are in one individual and female reproductive organs in another are called **dioecious.** In dioecious helminths, reproduction depends on two adults of opposite sex being in the same host.

Parasites spend one or more stages in their life cycle deriving nutrients from another organism called the **host.** A **definitive host** is the organism that harbors the adult, sexually mature helminth. However, one or more **intermediate hosts** may be necessary for each larval, or immature, stage of the parasite.

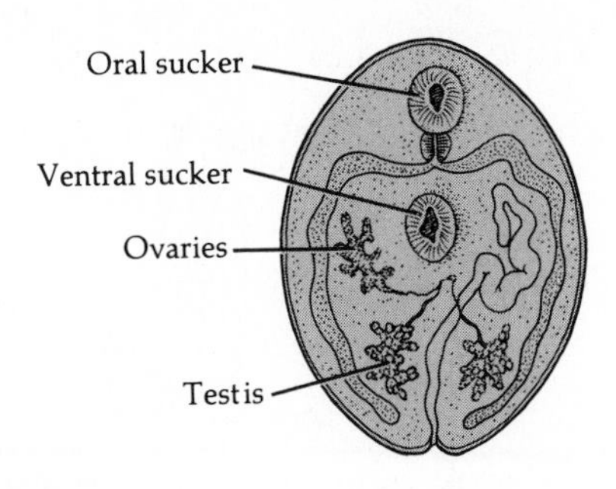

Figure 67-1.
This generalized diagram of the anatomy of an adult fluke shows the oral and ventral suckers. The suckers are used for attachment to the host. The mouth is located in the center of the oral sucker. Flukes are hermaphroditic, with testes and ovaries in each animal.

Platyhelminths are dorsoventrally flattened. They have an *incomplete* digestive system. In this type of digestive system, one opening functions to both take in food and eliminate wastes.

One class of parasitic flatworms is the **trematodes** or **flukes** (Figure 67-1). Flukes have leaf-shaped bodies with an oral sucker and a ventral sucker. Suckers hold the organism in place and may be used to suck fluids from the host. Flukes can also obtain food by absorption through their outer covering or cuticle.

The liver fluke, *Fasciola,* inhabits the liver and bile ducts of cattle, sheep, and occasionally humans (Figure 67-2). The hermaphroditic adults liberate eggs into the bile duct and small intestine, and the eggs must then be excreted into a body of water. A **miracidial larva** develops in the egg, and when it hatches, it enters a suitable snail. Inside the snail, the fluke reproduces asexually producing **rediae.** Each redia develops into a **cercaria** that bores out of the snail and encysts on aquatic vegetation such as watercress. The encysted stage is called a **metacercaria.** When the metacercaria is ingested by a mammal, the young fluke hatches out and burrows through the abdominal cavity to reach the liver.

The most important human disease caused by a helminth is **schistosomiasis.** Schistosomes or blood flukes (Figure 67-3) live in mesenteric or pelvic veins. The cercariae of schistosomes can penetrate directly through the skin. Schistosomes are dioecious and the males and females look quite different. The male is flattened around a central groove in which the cylindrical female permanently lies.

The second class of parasitic flatworms is the **cestodes** or **tapeworms** (Figure 67-4). The head, or **scolex,** has suckers and may have small hooks for attaching to the intestinal mucosa of the definitive host. Segments, called **proglottids,** are produced from the

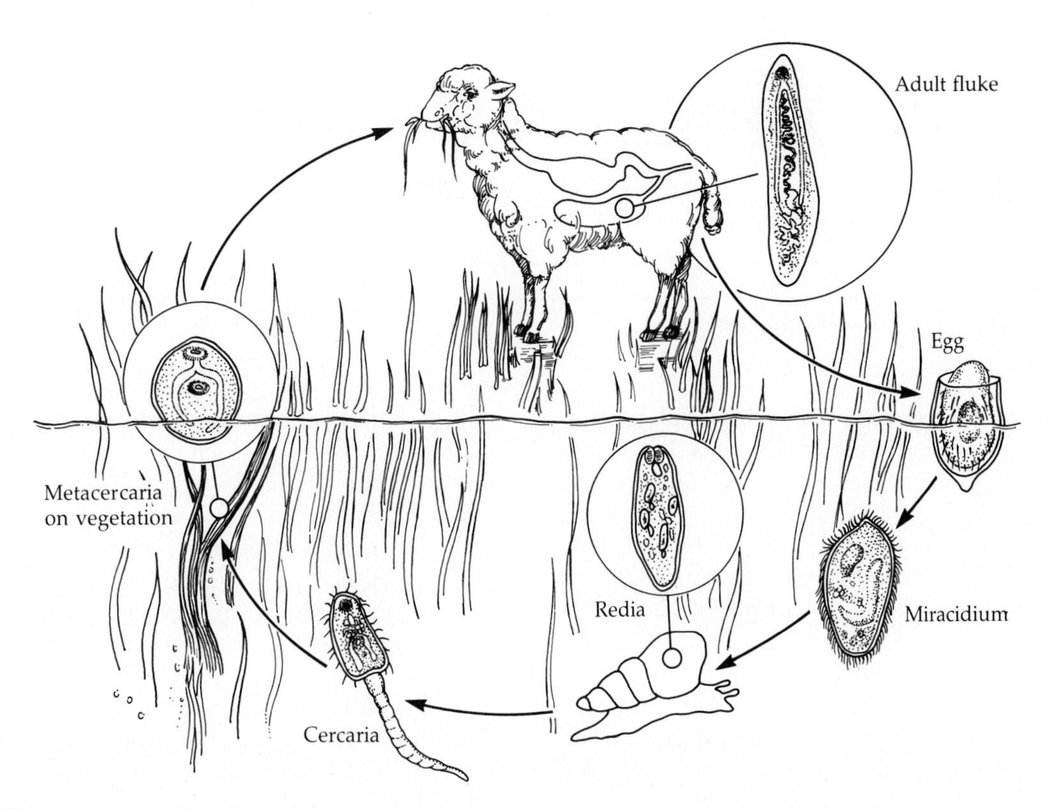

Figure 67-2.
Life cycle of *Fasciola.* The free-swimming miracidium invades a snail. Rediae develop within the snail, and the rediae give rise to cercariae. The cercariae leave the snail and encyst on vegetation as metacercariae. When a mammal ingests the vegetation, the young fluke hatches out and migrates to the liver to develop into an adult.

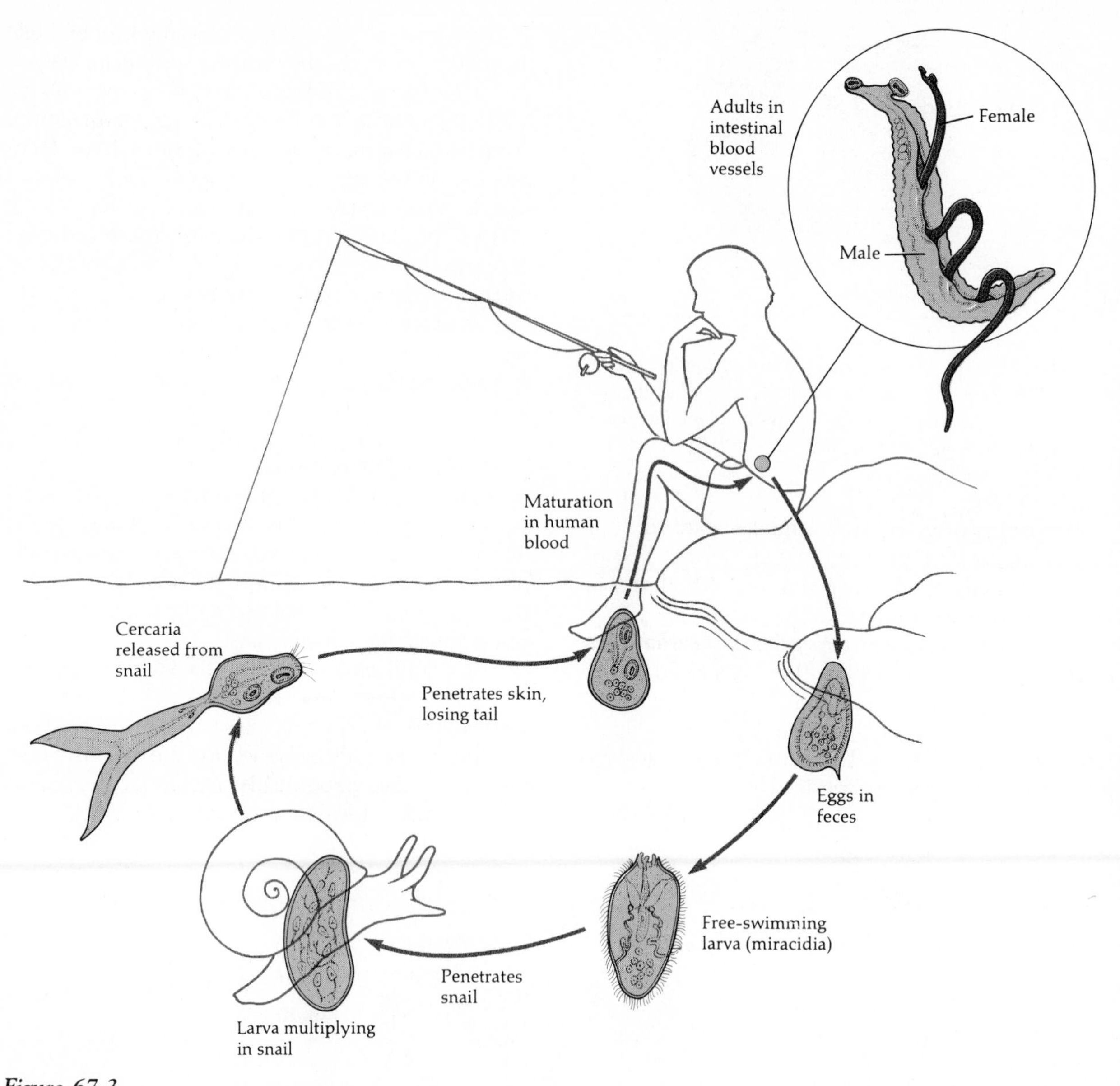

Figure 67-3.
Life cycle of *Schistosoma,* the causative agent of schistosomiasis.

neck region of the scolex. Each proglottid contains both male and female reproductive organs. The mature proglottids contain fertilized eggs. Proglottids are continuously produced as long as the scolex is attached and alive. Tapeworms obtain nutrients by absorbing food from the small intestine through their cuticle.

Taenia saginata (beef tapeworm) adults live in humans. When mature proglottids are excreted they wiggle away from the fecal material. Upon ingestion by grazing cattle, the larvae hatch and bore through the intestine. The larvae migrate to muscle (meat) and encyst as **cysticerci.** The cysticercus contains a scolex that is released when the infected meat is eaten.

Aschelminths, called roundworms, are cylindrical in shape. Roundworms have a *complete* digestive system, consisting of a mouth, an intestine, and an anus (Figure 67-5). Parasitic roundworms belong to the class of **nematodes.**

The pinworm *Enterobius vermicularis* is an example of a nematode in which the eggs are infective for humans. Pinworms are dioecious. Adult pinworms are found in the large intestine. The female pinworm migrates to the anus to deposit her eggs in the perianal skin. The eggs can be ingested by the same person or another person. Pinworm infections are diagnosed by picking up eggs from the perianal skin on a piece of sticky tape and examining the tape microscopically.

The larvae of hookworms *(Necator americanus)* are also infective for humans. The larvae hatch from eggs in the soil. The larva penetrates skin, enters a blood vessel, and is carried to the lungs. It is then swallowed and reaches the small intestine. The hookworm matures into an adult nematode in the small intestine.

Humans serve as an intermediate host for *Trichinella spiralis,* the causative agent of trichinosis. Trichinosis is acquired by ingesting encysted larvae.

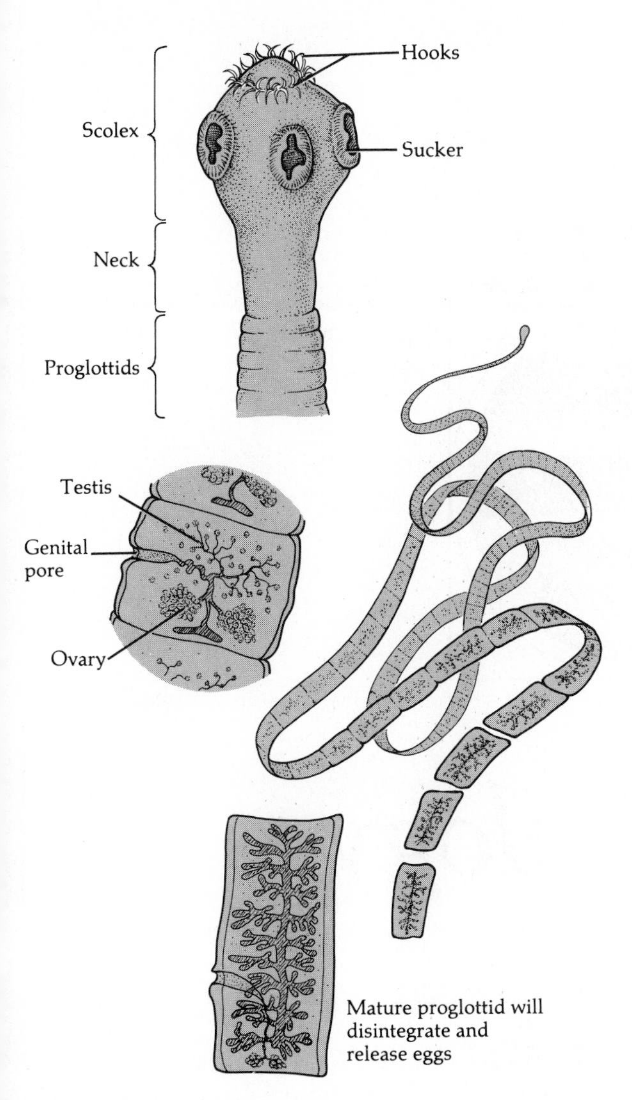

Figure 67-4.
The general anatomy of an adult tapeworm. The scolex consists of suckers and hooks that attach to host tissues. The body lengthens as new proglottids form at the neck. Each proglottid contains both testes and ovaries.

The larvae mature in the intestine and reproduce sexually. Eggs mature in the female. When the larvae are born, they migrate throughout the body and encyst in muscles and other tissues.

Materials

Planaria

Vinegar eels

Prepared slides of the following:

Fasciola hepatica egg, miracidium, metacercaria, redia

Clonorchis sinensis adult

Schistosoma in copula

Necator americanus adult, filariform (infectious) larva

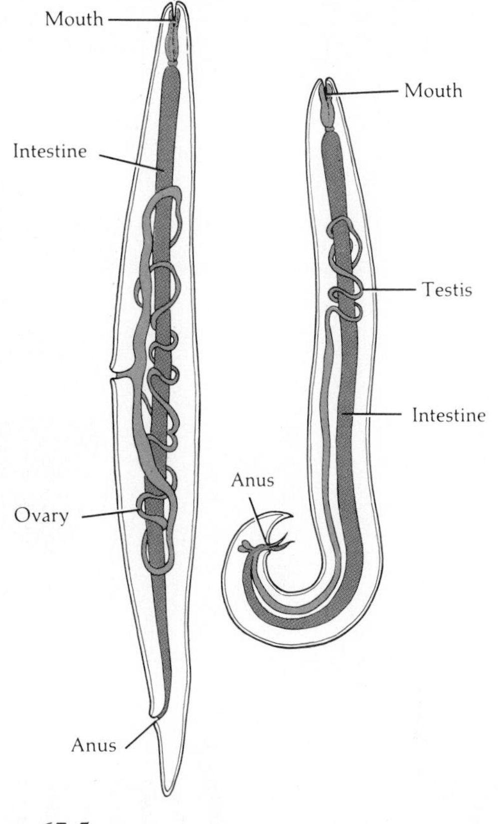

Figure 67-5.
Ascaris lumbricoides. Roundworms have a complete digestive system, with a mouth, intestine, and anus. Most roundworms are dioecious, and the female (left) is often distinctly larger than the male (right).

Enterobius vermicularis adult, eggs

Taenia scolex, proglottids

Trichinella spiralis, in muscle

Techniques Required

Exercise 1 and Appendix E

Procedure

1. Free-living helminths. Observe the free-living examples of a flatworm and a roundworm. Observe their movement using a dissecting microscope. Compare and contrast the movement of each.
2. Parasitic helminths. Observe the prepared slides using the low and high-dry objectives. Carefully draw each of the specimens you observed. Look for similarities and differences between each of the parasitic helminths.

Turn to the Laboratory Report for Exercise 67.

PART 12

Immunology

EXERCISES

In Part 11, we looked at microorganisms that are present on and in our bodies. In this part, we will examine our defenses against invasion by microorganisms. In 1883, Elie Metchnikoff began his investigations into the body's defenses with this entry into his diary:

> *These wandering cells in the body of the larva of a starfish, these cells eat food, they gobble up carmine granules—but they must eat up microbes too! Of course—the wandering cells are what protect the starfish from microbes! Our wandering cells, the white cells of blood—they must be what protects us from invading germs—*

These white blood cells (see the figure) and certain chemicals are considered nonspecific immunity because they will combat any microorganism that invades the body. Factors involved in nonspecific defenses are the topics of Exercises 68, 69, and 70.

Specific defenses or **immunity** was demonstrated when Emil von Behring developed what he called diphtheria "antitoxin" and, in 1894, successfully treated humans with it. Immunity involves the production of specific proteins called **antibodies.** Antibodies are directed against distinct microorganisms. If microorganism A invades the body, antibodies are produced against A; if microorganism B invades the body, antibodies are produced against B. The chemical substance that induces antibody production is called an **antigen.** In the examples just mentioned, microorganisms A and B possess antigens on their surfaces.

Specific and nonspecific immunity involves blood. **Blood** consists of a fluid called **plasma** and formed elements, that is, cells and cell fragments. The cells that are of immunologic importance are the white blood cells (Table 69-1). **Serum** is the fluid portion that remains after blood has clotted. We will study the techniques and mechanisms of **serology,** that is antigen–antibody reactions in vitro, in Exercises 71 through 75.

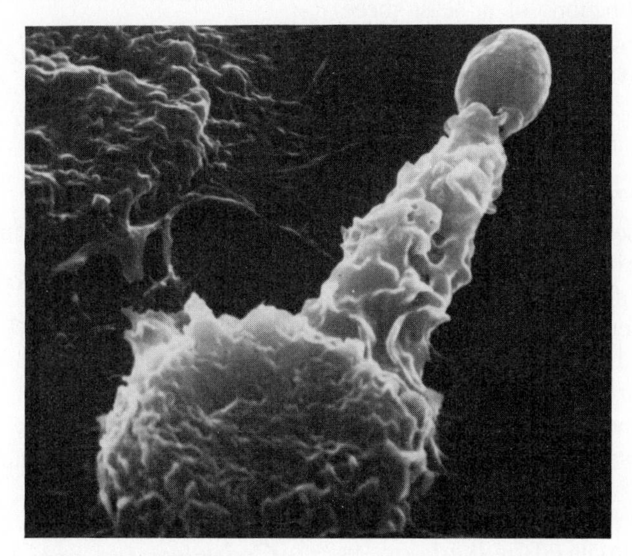

Early stage of phagocytosis by an alveolar macrophage (a type of macrophage found in the alveoli of the lungs). The "wrinkled" macrophage has contacted and adhered to a smooth, roughly spherical yeast cell by means of a pseudopod (×3100).

EXERCISE 68

Nonspecific Resistance

Objectives

After completing this exercise you should be able to

1. Differentiate between specific and nonspecific immunity.
2. List and discuss the functions of at least three chemicals involved in nonspecific resistance.
3. Determine the effects of normal serum bactericidins.

Background

Our ability to ward off diseases through our body's defenses is called **immunity.** Immunity can be divided into two kinds: specific and nonspecific. **Specific immunity** is the defense against a specific microorganism (Exercises 71 and 74). **Nonspecific immunity** or **resistance** refers to all our defenses that protect us from invasion by any microorganism. Physical barriers such as the skin and mucous membranes are part of nonspecific immunity. Inflammation and phagocytosis play major roles in nonspecific immunity, as do chemicals produced by the body. As early as 1888, Metchnikoff noted that bacteria *"show degenerative changes"* when inoculated into samples of mammalian blood.

Lysozyme and complement are examples of chemicals involved in nonspecific immunity. **Lysozyme** is an enzyme found in body fluids that is capable of breaking down the cell walls of gram-positive bacteria and a few gram-negative bacteria. **Complement** is a group of proteins found in normal serum. Another group of serum proteins, the **properdin system,** works with complement to attack and destroy invading microorganisms. Complement and properdin proteins can be activated by bacterial cell wall polysaccharides and antigen–antibody reactions (Exercises 71, 73, and 74).

In this exercise we will examine chemicals involved in nonspecific immunity.

Materials

Lysozyme Activity

Lysozyme buffer, 4.5 ml

Lysozyme, 1:10^5 dilution, 5 ml

Spectrophotometer tubes (2)

Pipettes, 1.0 ml and 5.0 ml (1 each)

Spectrophotometer

Micrococcus lysodeikticus suspension, 5 ml

Normal Serum Bactericidins

Nutrient agar, melted and cooled to 45°C

Sterile Petri plates (9)

Sterile pipettes, 1.0 ml (10)

Sterile 99-ml dilution blanks (3)

Sterile serological tubes (9)

Serological test tube rack

0.85% saline solution

Normal serum

Serum heated to 56°C for 30 minutes

Cultures (as assigned)

Escherichia coli

Staphylococcus aureus

Techniques Required

Exercises 13, 14, pp. 109–110, and Appendices A, B, and D

Procedure

Lysozyme Activity

1. Collect tears or saliva in a Petri plate as described by your instructor.
2. Prepare a 1:10 dilution of tears or saliva by adding 4.5 ml lysozyme buffer to 0.5 ml tears or saliva. What is the purpose of the buffer? ____________ ____________
3. Mix 5 ml of the tear or saliva preparation (step 2) with 5 ml *Micrococcus lysodeikticus* suspension in a spectrophotometer tube. Carefully pipette the contents of the tube up and down three times to mix.

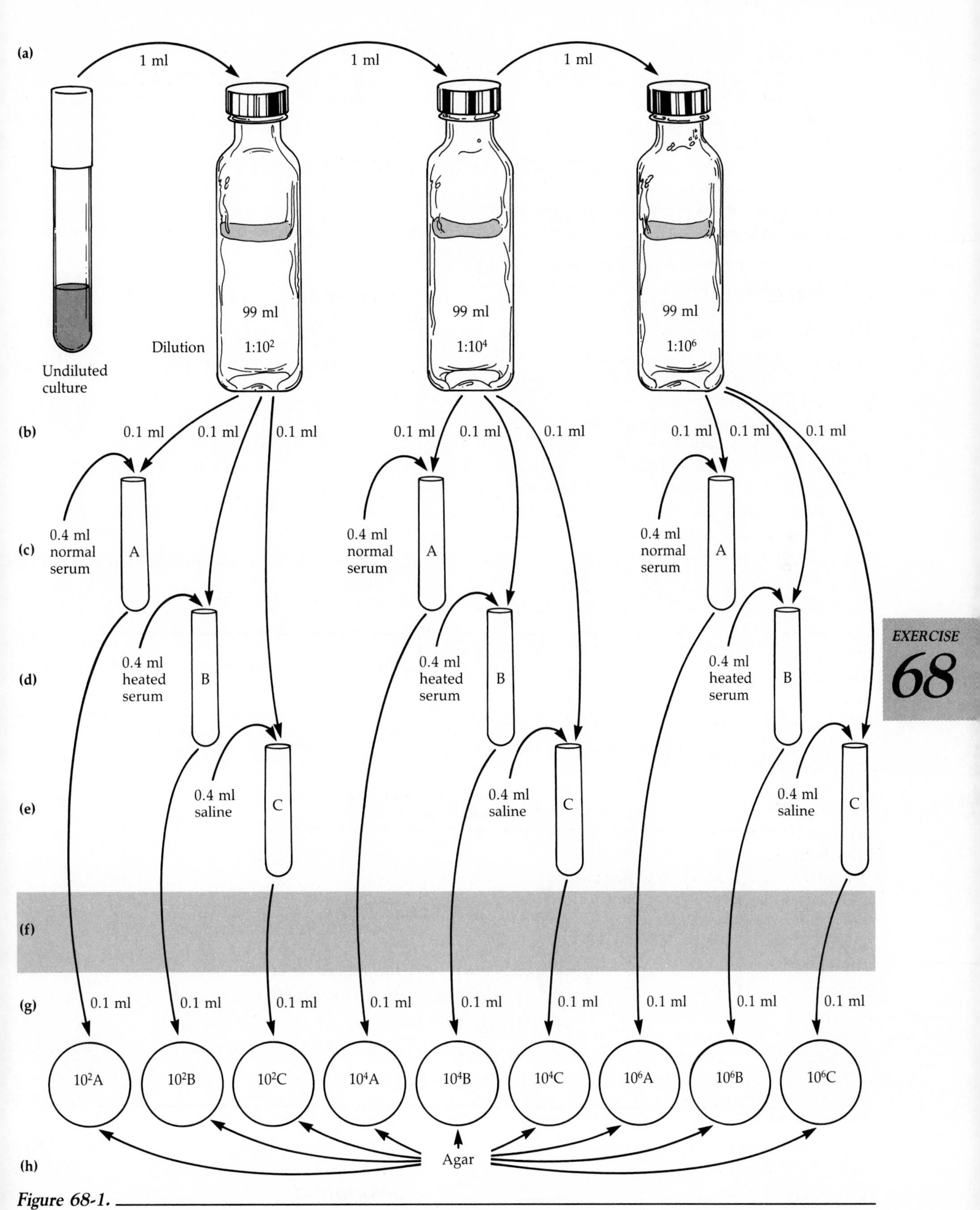

Figure 68-1.
Procedure to determine bactericidal activity of serum.

4. Record the optical density at 540 nm at 30-second intervals for 5 minutes and at 5-minute intervals for the next 15 minutes.
5. Repeat steps 2 and 3 using egg white lysozyme.
6. Plot your data in the Laboratory Report.

Normal Serum Bactericidins

(Work in groups as assigned.) Refer to the outline of the procedure in Figure 68-1.

1. Aseptically prepare $1{:}10^2$, $1{:}10^4$, and $1{:}10^6$ dilutions of the culture (*S. aureus* or *E. coli*) assigned to you (see Figure 68-1).
2. Aseptically transfer 0.1 ml of culture from the highest dilution (1: ______) to each of three sterile serological tubes. Label the tubes A, B, and C and the dilution. Repeat with the other dilutions using the same pipette.
3. To each A tube, add 0.4 ml of unheated normal serum; to each B tube, add 0.4 ml serum heated to 56°C for 30 minutes; to the C tubes, add 0.4 ml saline. What is the purpose of the C tubes?

____
4. Mix all tubes and incubate at 35°C for 1 hour.
5. Label nine sterile Petri plates in the same manner as the tubes: "$1{:}10^2$ A", "$1{:}10^2$ B", "$1{:}10^2$ C", "$1{:}10^4$ A", "$1{:}10^4$ B", and so on.
6. Remove 0.1 ml from the A tube of the highest dilution and place it in the appropriate Petri plate. Using the same pipette, transfer 0.1 ml from the "$1{:}10^4$ A" to the corresponding Petri plate. Repeat this procedure with the "$1{:}10^2$ A" tube. Why are the samples transferred from the highest dilution first?______________________
7. Using another pipette, repeat step 6 with the B tubes. With your remaining pipette, repeat step 6 with the C tubes.
8. Pour melted, cooled nutrient agar into each plate to a depth of approximately 5 mm. Gently swirl to mix the contents and allow the agar to solidify.
9. Incubate the plates for 24 to 48 hours at 35°C. Record the number of colonies on each plate.
10. Calculate the percent increase or decrease in bacterial numbers due to exposure to heated and unheated serum as compared to the control.

$$\begin{array}{l} \text{Control} \\ \underline{-\,\text{Experiment}} \\ \text{Difference} \end{array}$$

$$\frac{\text{Difference}}{\text{Control}} \times 100 = \%$$

11. Compare your data with those of a group using the other bacterial species.

Turn to the Laboratory Report for Exercise 68.

EXERCISE 69

Inflammation

The outcome of any serious research can only be to make two questions grow where only one grew before.

THORSTEIN VEBLEU

Objectives

After completing this exercise you should be able to

1. Describe the process of inflammation.
2. Perform a white blood cell differential count.
3. Document the process of inflammation.

Background

When injury to tissue occurs, the body forms an immediate response to the injury called **inflammation.** Tissue damage may be due to infection, burning, a cut or abrasion, or chemicals. Damaged tissues release mediators such as histamine and leukocytosis-promoting factor, which initiate the inflammatory response from the circulatory system. **Leukocytosis-promoting factor** causes an eventual increase in the total number of white blood cells in the blood. During the active stage of bacterial infection, the white blood cell number may triple or quadruple (although a few infections cause a decrease in white blood cells). The type of cells involved in the change in number can be determined by a **differential count,** which calculates the percentage of each type of white blood cell in a sample of 100 white cells. The percentages for a normal differential count are shown in Table 69-1. The classification of the different white blood cells is determined by their morphology and reaction when stained by Wright's stain.

Histamine causes vasodilation and circulation increases in the damaged area, causing redness and heat. Fluids leak from capillaries into the damaged area, causing swelling. Certain white blood cells become sticky and adhere to capillary walls, eventually squeezing between the cells that line the blood vessel by a process called **diapedesis** and entering the damaged area. The white blood cells are monocytes and neutrophils, which can **phagocytize** or ingest particulate material and bacteria. Neutrophils arrive first and are followed by monocytes, which are called **macrophages** when they are outside of the circulatory system. Neutrophils and macrophages engulf bacteria and then die, leaving a mixture of fluids and dead cells called **pus.** Pus may be discharged to the surface or body cavity or slowly absorbed by the body.

Platelets form clots that prevent further leakage and spread of the damage. Granular tissue fibroblasts replace damaged tissue, and healing or repair occurs. The purpose of inflammation is to localize the damage in order to prevent the spread of an infection and to enhance tissue repair.

In this exercise a differential white blood cell count will be performed, and a simple procedure called *Rebuck skin window* will be used to examine the inflammatory process. The skin window can be used to study in situ the morphology, incidence, and changes in cells involved in the inflammatory response.

Materials

Differential Count

Cotton moistened with 70% ethyl alcohol

Sterile cotton ball and bandage

Clean microscope slides

Sterile lancet

Staining rack (make sure it's level)

Wright's stain

Wright's buffer

Wash bottle of water

Skin Window

Razor

Razor blade

Nonallergenic tape

Cover slips (10)

Autoclaving wrap and tape

Cardboard

Scissors

Grease pencil

Wright's stain

Wright's buffer

Bandage

Corks

Pie tin

Slides

Small envelopes (10)

Slide mounting medium

EXERCISE 69

Techniques Required

Exercise 1

Procedure

Differential Count

1. Disinfect your middle finger with cotton saturated in alcohol (Figure 69-1), then dry and pierce it with a sterile lancet.
2. Let a small drop of blood fall onto a microscope slide, about 2 cm from the end of the slide. Let

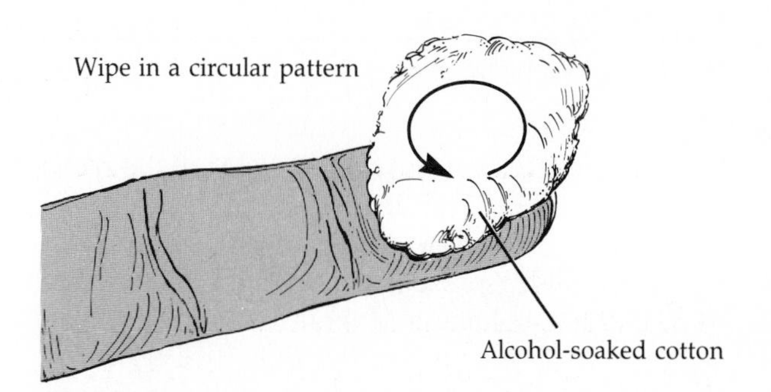

Figure 69-1.
To disinfect skin: Rub with 70% ethyl alcohol and wipe dry.

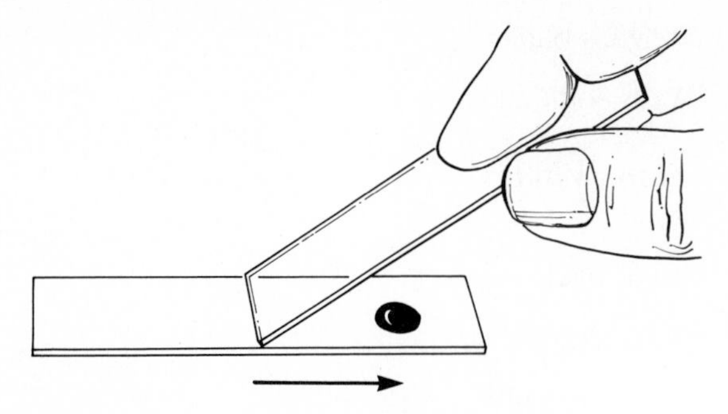

(a) Draw another slide toward the drop of blood.

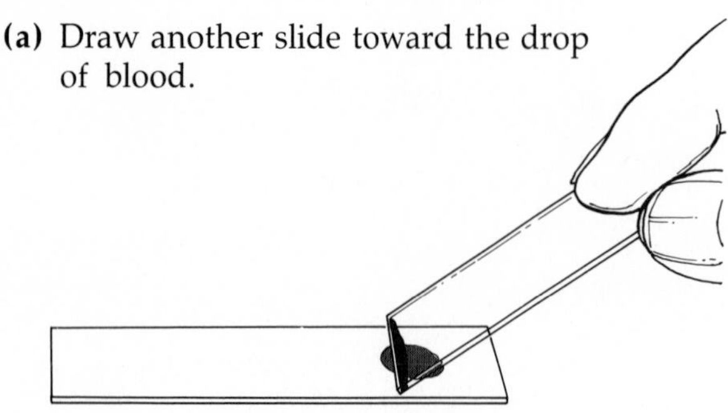

(b) Let the blood flow along the edge of the spreader slide.

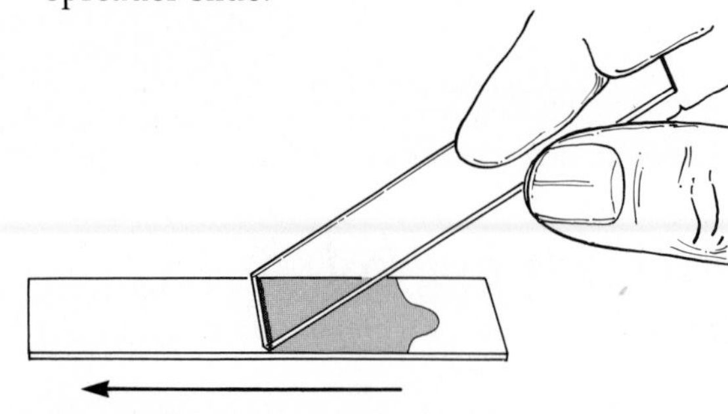

(c) Push the spreader slide and blood across the first slide.

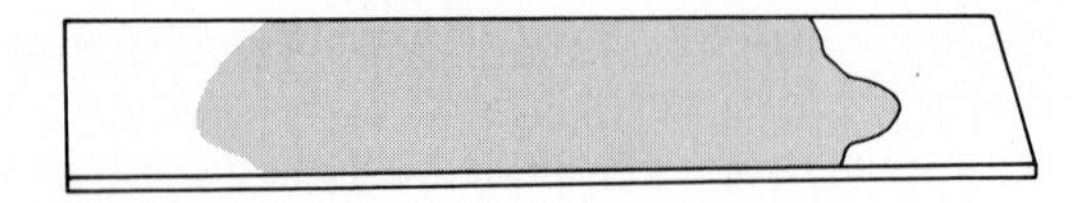

(d) A thin smear is produced.

Figure 69-2.
Preparation of a blood smear.

drops fall onto two other slides. Stop the bleeding with a sterile dry cotton ball and apply an adhesive bandage. Quickly place the end of a clean slide (spreader slide) against the surface of the first, holding it at about a 45° angle (Figure 69-2*a*). Draw the spreader slide back against the drop of blood (Figure 69-2*a*) and wait until the blood spreads the width of the spreader slide (Figure 69-2*b*). Then spread the blood by keeping the spreader slide at a constant angle and constant pressure and pushing the spreader slide smoothly and rapidly across the slide (Figure 69-2*c*). The blood will follow the spreader slide, making an even film (Figure 69-2*d*). A smaller angle of the spreader slide to the blood will produce a thick smear. Therefore, if the smear is thin, increase the angle; if the blood smear is thick, decrease the angle. If your smear was a disaster, try again using the remaining slides. When you are satisifed with your smear, let it air dry.

3. When the smear is dry, write your initials and date in pencil in the heaviest area of the smear.
4. Place the slide on a level staining rack and *flood* with Wright's stain. Leave the stain on the smear for 5 minutes.
5. Carefully add about an equal amount of buffer to the Wright's stain on the slide. Gently blow on the slide to mix the buffer and stain. A brassy, green sheen will appear as the solutions mix. Leave the mixture on for 10 minutes.
6. Wash off the staining mixture with distilled water by spraying water at a low angle so the staining mixture floats off.
7. Remove excess stain from the back of the slide by wiping with alcohol.
8. Stand the slide on end and let it air dry.
9. When the stain is dry, place it on the microscope stage and focus on the feathery area of the slide.
10. Using the oil immersion lens, and moving through the slide widthwise, count and differentiate a hundred white blood cells. Use Table 69-1 as well as your textbook or other reference. Record on the Laboratory Report each type of white blood cell as you see it.

Skin Window

1. Preparing materials
   a. Obtain 10 new cover slips. Handle the cover slips by the edges only. Cut 10 cardboard pieces slightly larger than a cover slip. Write a B on each cover slip.
   b. Wrap a razor blade in autoclaving paper. Carefully wrap one cover slip with a piece of cardboard for autoclaving. Wrap the remaining cover slips and cardboard in a similar manner.
   c. Your instructor will sterilize the cover slips and razor blades.
2. This technique must be done *carefully,* and if done properly, no pain or infection will result.
   a. Shave a 5-cm-by-5-cm square on the inside of your left arm near your elbow (if right-handed). Wash well with soap and rinse. Dry with a clean towel.
   b. Carefully scrape your partner's skin (in the shaved area) with the sterile razor blade,

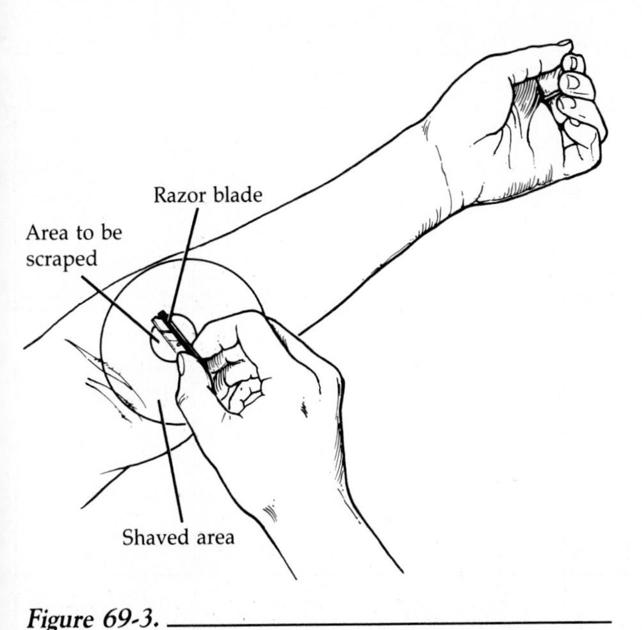

Figure 69-3.
Holding the razor blade perpendicular to the skin, carefully scrape the skin.

holding the blade perpendicular to the skin as shown in Figure 69-3. Scrape rapidly and lightly until the skin develops some small red dots, then scrape a few more times. The scraped area should appear moist. Be careful that you do not scratch your partner with the corners of the razor blade. *Do not* overdo it. Let your partner do the same for you.

3. Carefully and aseptically, unwrap a cover slip and cardboard. Place the sterile cover slip next to your abrasion, "B" side away from your skin, and place cardboard on top (Figure 69-4).

4. Carefully tape the cover slip and cardboard to your skin using the nonallergenic tape. Make note of the time. Time: ______________________
Bring the remaining cover slips, cardboard, tape, and envelopes home.

5. In 3 hours, remove the patch and cover slip. Place a new sterile cover slip on the abrasion as before. Let the old cover slip dry, then place in an envelope and mark the time on the envelope.

6. Cover slips should be changed every 3 hours for 24 hours if at all possible. Some students could use different time intervals between 1 and 4 hours to give a more complete experiment. Time may be adjusted to facilitate sleeping. Stop if your skin dries out.

7. When done, wash your skin and let it heal. *Report any problems.*

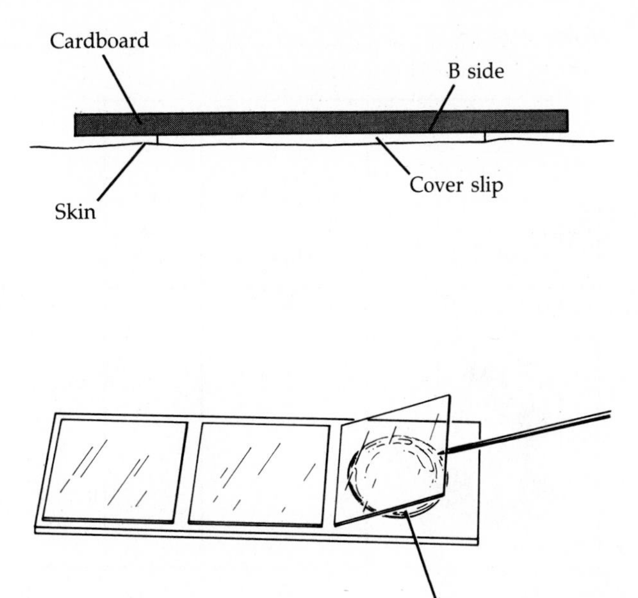

Figure 69-5.
Place a cover slip in a drop of mounting medium on a slide. Two or three cover slips can be mounted on one slide.

8. Back in the laboratory:

a. Place 10 small corks (or a number equal to the number of cover slips) on a pie tin, and place a small drop of water on each cork.

b. Put one of your cover slips on each cork in a manner that will allow you to keep track of them. Place them cell-side up, with the "B" next to the corks.

c. Carefully stain for 5 minutes with Wright's stain (see part A). Add a small amount of buffer until the metallic sheen forms. Buffer for 10 minutes.

d. Rinse with distilled water and let dry.

e. Place a drop of mounting medium on a slide and place the dry, stained cover slip cell-side down on the mounting medium. Several cover slips can be mounted on one slide (Figure 69-5).

f. Examine each cover slip, determine the type of white blood cells present (Table 69-1), and estimate the number of each type. Record your data in the Laboratory Report.

Table 69-1
Types of White Blood Cells

Cell Type	Percent of Total White Blood Cells	Size (μm)	Nucleus: Shape	Nucleus: Chromatin	Cytoplasm: Ground Substance	Cytoplasm: Granules
Neutrophils	60–70	9–12	2–4 lobes united by thin filaments or broad bands.	Irregularly stained. Dark blue–deep purple.	Pale purple.	Deeper purple. Equal sized, fine.
Eosinophils	2–4	9–12	Usually 2 lobes.	Reddish-blue.	Pale purple.	Red. Equal sized, large, coarse refractive.
Basophils	0.5–1	9–12	Rosette shaped or slightly polymorph.	Cloudy. Pale blue.	More pinkish.	Round, coarse, equal sized, few in number. Deep bluish-purple.
Monocytes	3–8	14–18	Kidney or lobular shape.	Loosely meshed interlacing strands. Pale blue.	Gray-blue.	Muddy blue, nonspecific granules. "Ground glass" appearance.
Lymphocytes	20–25	7–13	Round or oval, shallow or deeply indented.	Very coarse, regular dark blue.	Small amount clear blue, light area around nucleus.	May contain granules that are nonspecific.

Turn to the Laboratory Report for Exercise 69.

EXERCISE 70

Phagocytosis

Science doesn't prescribe, it describes.
AUTHOR UNKNOWN

Objectives

After completing this exercise you should be able to

1. Define phagocytosis and describe the mechanism.
2. Determine a viability index and phagocytic index.

Background

As we observed in Exercise 69, inflammation results in neutrophils and monocytes coming to the site of injury. These white blood cells, referred to as **phagocytes,** will engulf any microorganisms or particulates at the site by phagocytosis.

The phagocyte is attracted to a microbe by **chemotaxis,** that is, attraction to certain chemicals. The attachment of a microbial cell surface to the surface of the phagocyte is called **adherence.**

The phagocyte engulfs the microbe by invagination of the cell membrane and formation of a **phagosome.** In neutrophils, organelles containing digestive enzymes called **lysosomes** fuse with the phagocytic vacuole, forming a **phagolysosome.** Most microbial cells are destroyed in the phagolysosome. In most cases the phagocyte dies after several phagocytic events. The mechanism of phagocytosis is illustrated in Figure 70-1.

We will evaluate phagocytosis by observing the ability of phagocytes to engulf either India ink or yeast.

Materials

Tube containing fresh heparinized blood (3–5 ml)

India ink or yeast suspension (work in groups—one use ink and the other, yeast)

Wright's stain

Wright's buffer

Staining rack

Pasteur pipette

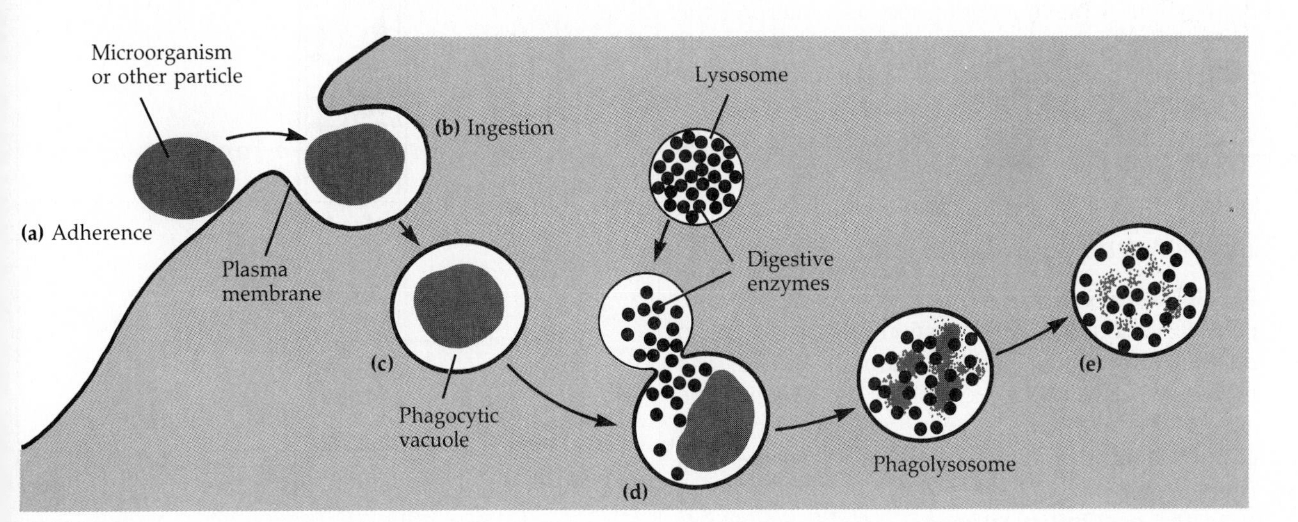

Figure 70-1.
Phagocytosis. **(a)** Adherence. **(b)** Ingestion. **(c)** Formation of the phagosome. **(d)** Fusion of the phagosome with lysosome. **(e)** Destruction of the ingested microorganism.

Techniques Required

Exercise 69

Procedure

1. Mix your blood sample. Why doesn't it clot? ________________
2. Add either 1 drop of India ink or 1 drop of yeast suspension to the blood.
3. Incubate the mixture at 35°C for approximately 1 hour.
4. Centrifuge at 1000 rpm for 10 minutes; balance the centrifuge by placing your tube opposite another tube of blood, or a similar tube filled with water.
5. Carefully remove your tube from the centrifuge. Forceps may help remove it. Look at the tube without shaking it; the white cells are on top of the red cells. The layer of white cells is called a buffy coat.
6. With a Pasteur pipette, remove some of the buffy coat.
7. Make two smears with the buffy coat (see Figure 69-2).
8. Stain with Wright's stain as described in Exercise 69, Part A.
9. Students using India ink, scan 25 neutrophils (refer to Table 69-1). How many had engulfed India ink? ________________
10. Students using yeast, count the number of yeast particles in each of 25 neutrophils. How will you recognize neutrophils? ________________

Turn to the Laboratory Report for Exercise 70.

EXERCISE 71

Precipitation Reactions

Objectives

After completing this exercise you should be able to

1. Define precipitation reactions.
2. Perform a precipitin ring test and interpret the results.
3. Perform an Ouchterlony test and interpret the results.
4. Provide two examples of applications of precipitation reactions.

Background

Precipitation reactions occur between *soluble* antigens and antibodies. Precipitation reactions depend on the formation of antigen–antibody lattices. The maximum amount of precipitation occurs when the antigen and antibody concentrations are optimal; excessive amounts of either component interfere with lattice formation.

In the **precipitin ring test,** soluble antigen is overlayed onto a solution of antibodies in a small test tube. A white precipitation ring forms at the antigen–antibody interface (Figure 71-1).

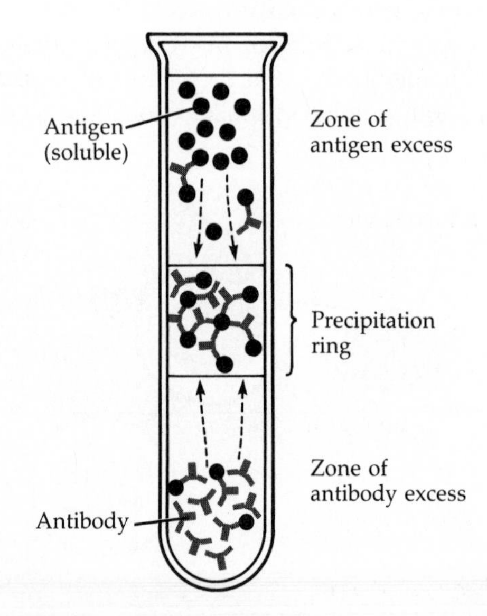

Figure 71-1.
Diagrammatic representation of a precipitin ring test.

Immunodiffusion tests such as the **Ouchterlony test** can be used to detect the presence of multiple antigens and antibodies. The test is performed in a Petri plate containing a gel medium. The gel used in this exercise is Noble Agar. Noble Agar is highly purified agar with no nutritional contaminates that might interfere with reactions or on which bacteria can

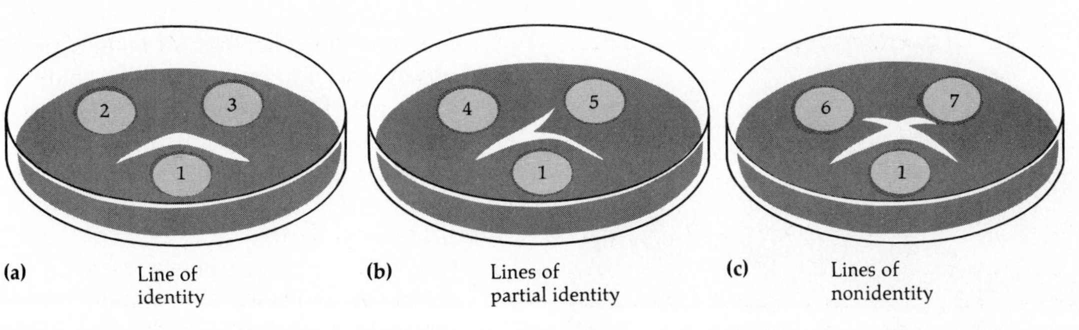

Figure 71-2.
Immunodiffusion (Ouchterlony) precipitation test. Well 1 contains antibodies to antigens 2 through 7. **(a)** Wells 2 and 3 contain identical antigens. **(b)** Wells 4 and 5 contain closely related antigens. **(c)** Wells 6 and 7 contain different, unrelated antigens.

grow. The agar may be further preserved from microbial growth by the addition of antimicrobial chemicals. Antibodies and antigens are added to wells (holes) in the agar. The reactants diffuse toward each other and a precipitate forms where the ratio of antigen to antibody is optimal. If two antigen preparations contain identical antigens, the precipitation lines come together to form **lines of identity** (Figure 71-2a). If the antigens are not identical but are closely related, **lines of partial identity** form (Figure 71-2b). The precipitation bands cross to form **lines of nonidentity** when the antigens are different (Figure 71-2c).

Materials

Precipitin Ring Test

Precipitin tubes (5)

Test tube rack

Pasteur pipettes (6)

Antisera (1 of each):

antibovine serum

anticat serum

antihuman serum

antisheep serum

Unknown antigen, diluted

$1{:}10^3$, # ______

Saline, 0.85% NaCl

Ouchterlony Test

Petri plate containing Noble agar (1)

Cork borer, #1 to #5 (1)

Pasteur pipettes (4)

Parafilm or transparent tape

Forceps (1)

Probe (1)

Bovine serum

Human serum

Human serum albumin

Antihuman serum

Techniques Required

None

Procedure

EXERCISE 71

Precipitin Ring Test

1. Arrange the precipitin tubes and label the tubes or the rack "bovine," "cat," "human," "sheep," and "control."
2. Using a Pasteur pipette, place the unknown antigen ($1{:}10^3$) in each tube to a depth of approximately 7 mm.
3. Hold the control tube nearly horizontally and, using a clean Pasteur pipette, let saline run down the side of the tube (Figure 71-3). The final depth should be approximately 14 mm. A *sharp interface* between the two solutions is required. Why?

   ____

   ____

   Replace the tube without mixing the contents.

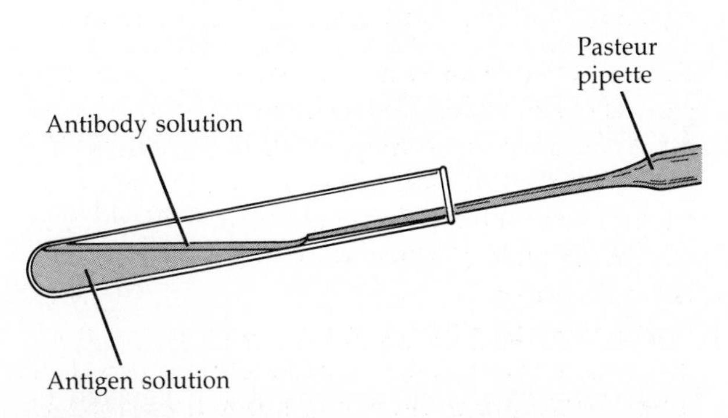

Figure 71-3.
Hold the tube nearly horizontal so that the new solution lies on top of the antigen.

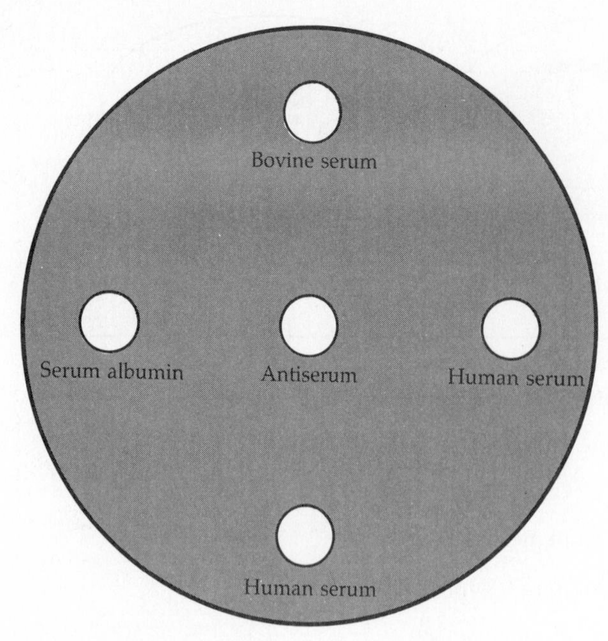

Figure 71-4.
Ouchterlony test. Five wells are made in the agar. Antiserum is placed in the center well, and solutions of antigens in the surrounding wells.

4. Using a different Pasteur pipette for each antiserum, *carefully* overlay the antigen in each tube with the appropriate antiserum. Why is a different pipette used for each antiserum? ______________

5. Incubate the tubes for 30 minutes at 35°C and observe for formation of a precipitin ring. The tubes may be kept in the refrigerator until the next laboratory period. By that time, the precipitates may fall to the bottom of the tubes.

Ouchterlony Test

1. Using the cork borer, punch five holes in the agar as shown in Figure 71-4. Carefully remove the agar from each well using forceps and a probe. Why should the wells have straight, even sides? ______________
2. Fill each well just to the top with the appropriate solution as shown in Figure 71-4.
3. Seal the opening between the top and bottom of the Petri plate with Parafilm or transparent tape, being careful not to spill the contents of the wells.
4. Incubate the plate *right side up* at room temperature for 2 to 4 days.
5. Diagram the appearance of the precipitate bands in your Laboratory Report.

Turn to the Laboratory Report for Exercise 71.

EXERCISE 72

Immunoelectrophoresis

Objectives

After completing this exercise you should be able to

1. Define electrophoresis.
2. Differentiate between electrophoresis and immunoelectrophoresis.
3. Use immunoelectrophoresis to separate and identify serum proteins.

Background

Electrophoresis is the movement or migration of charged particles in solution under the influence of an electric field. The rate of migration of different ions depends mainly on four factors: (1) nature of the charged particle itself, (2) composition and pH of the buffer, (3) electrical field (voltage), and (4) time of electrophoresis. Difference in migration rate is used to separate the components of a mixture. Usually the particles separated are colloids and macromolecular ions, but materials such as viruses, subcellular organelles, and amino acids can also be separated by electrophoresis.

There are many types of electrophoresis, but all are based on the same principles. In immunology, serum proteins are the most frequently studied substances.

Proteins are amphoteric. **Amphoteric** substances can possess a positive or negative charge. The relative amounts of the amphoteric components in a protein determine the charge and magnitude of the charge of the molecule. The charge can be varied by changing the pH of the buffer. In an alkaline solution, the charge is negative, and the protein migrates toward the *anode* (positive pole); if the pH is lower (below pH 7)—that is, if the solution is acidic—the protein migrates toward the *cathode* (negative pole). The pH at which the protein will have a net charge of zero is its **isoelectric point;** at this point it will not migrate.

Protein mixtures are made up of different charges and may be readily separated and purified by electrophoresis. When serum is electrophoresed (pH 8.6), *albumin* migrates the fastest toward the anode, followed by *alpha, beta,* and *gamma globulins* (Figure 72-1*a*).

Immunoelectrophoresis is a very accurate method of analyzing biologic fluids. This technique essentially separates sample components by electrophoresis and subsequently resolves the separated components by **immunodiffusion** (Exercise 71). This technique can separate proteins with similar electrophoretic mobilities.

Generally, the initial electrophoretic separation is done on an agar gel medium poured onto a microscope slide (Figure 72-1*a*). After electrophoresis, an antiserum is added that contains antibodies known to precipitate the components studied. The antiserum is added so that diffusion will occur perpendicular to the axis of electrophoretic migration (Figure 72-1*b*). The antibodies precipitate at equivalence with the radially diffusing components, forming arc-shaped lines in the agar (Figure 72-1*c* and *d*).

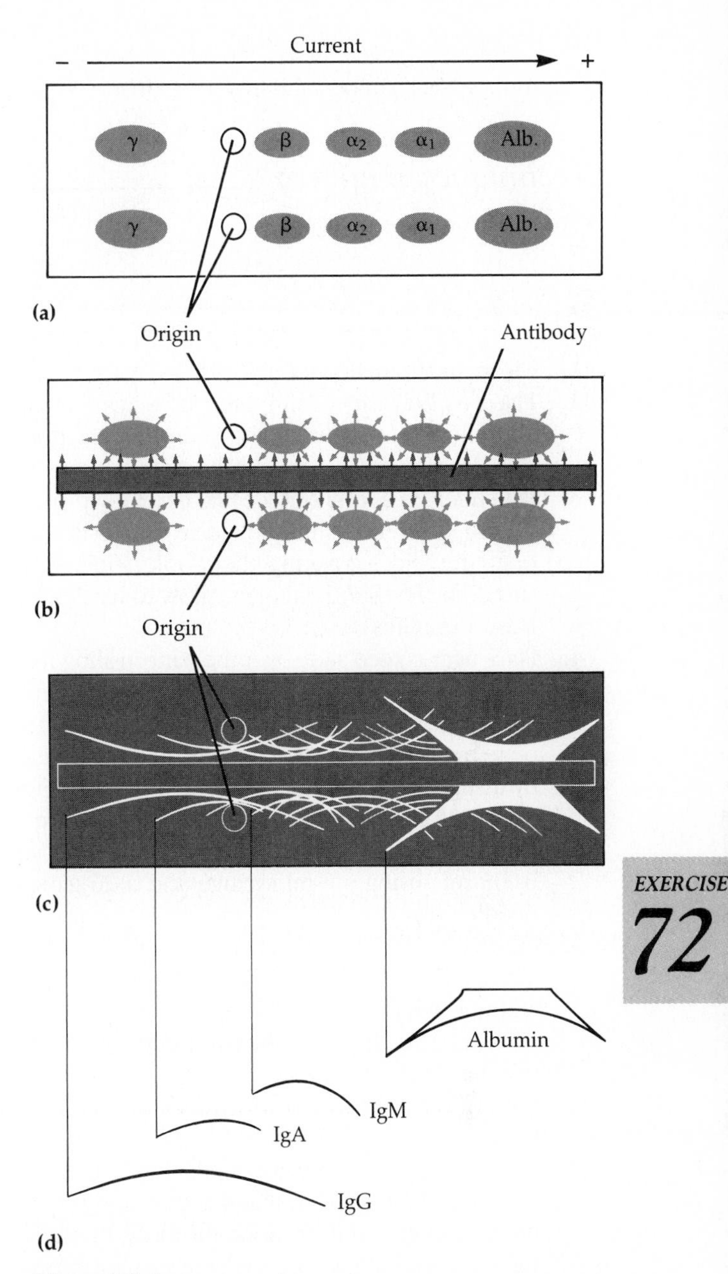

Figure 72-1.
Immunoelectrophoresis. **(a)** Electrophoresis separation. **(b)** Immunodiffusion. **(c)** Resulting pattern. **(d)** Interpreting the pattern.

Materials

Impregnation agar

Immunoelectrophoresis agar

Electrophoresis buffer

Cold saline (0.85%)

Distilled water

Ponceau S stain

Washing solution, 5% acetic acid

Human serum

Antihuman rabbit serum

Filter paper strips (No. 4, 2.5 cm by 20 cm) (3)

Petri plate with filter paper

Large beaker or bowl

Microscope slides

Razor blade

Pasteur pipettes (2)

5-ml pipette (1)

1-ml syringe, 26 gauge needle (2)

22–23 gauge needle (1)

Coplin jars (4)

Vacuum source with rubber hose

Electrophoresis chamber and power source

Techniques Required

Exercise 71 and Appendix A

Procedure

1. Clean slides in alcohol, and dry.
2. Place in hot, melted impregnation agar. Drain and allow to dry overnight in an upright position, leaning against a test tube rack.
3. Put slides on a *level* surface and carefully pipette 2.0 ml of melted immunoelectrophoresis agar onto the slide, covering the whole slide. Agar should be level and uniform. Allow to harden (at least 5 minutes).
4. Place agar-coated slide over the pattern shown in Figure 72-2*a*. Attach a bulb to a Pasteur pipette, apply pressure to the bulb, place tip of the pipette over one circle in the pattern. Remove agar by a slight release of pressure on the bulb (Figure 72-2*b*). Remove agar over the other circle.
5. Fill the holes with the human serum (about 0.005 ml) using a 1-ml syringe and a 26 gauge needle.
6. Fill the electrophoresis chamber with cold buffer. Each buffer vessel should have an equal amount. Why? ____________________
   Place the slide between the two buffer vessels so that the holes containing the serum are nearer to the cathode (−) than to the anode (+) (Figure 72-3).
7. Gently attach buffer-dampened filter paper on one end of the slide so that it covers about one-half centimeter of the agar on the slide. Immerse the other end of the strip in the buffer, forming a bridge from the buffer to the slide (Figure 72-3). Make a bridge from the other end of the slide to the buffer.
8. Connect to the power supply, applying 150 volts for 45 to 90 minutes.
9. Turn off the power supply and remove the slide.
10. Using the pattern shown in Figure 72-2*a*, cut the center trough by cutting over the two parallel lines. Aspirate out the agar with a 23 gauge needle (bevel up) connected to a vacuum source (Figure 72-2*c*).
11. With a 1-ml syringe and a 26 gauge needle, fill the trough with about 0.07 ml of antihuman serum. Put the slide in a Petri plate lined with wet filter paper.
12. Place the slide at 5°C for 24 to 48 hours until the pattern develops.

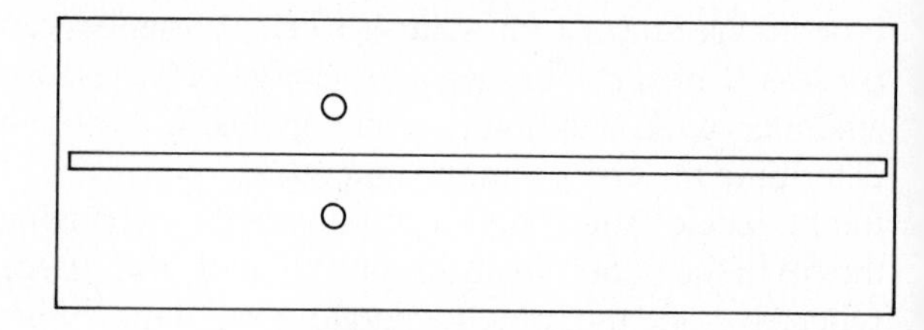

(a)

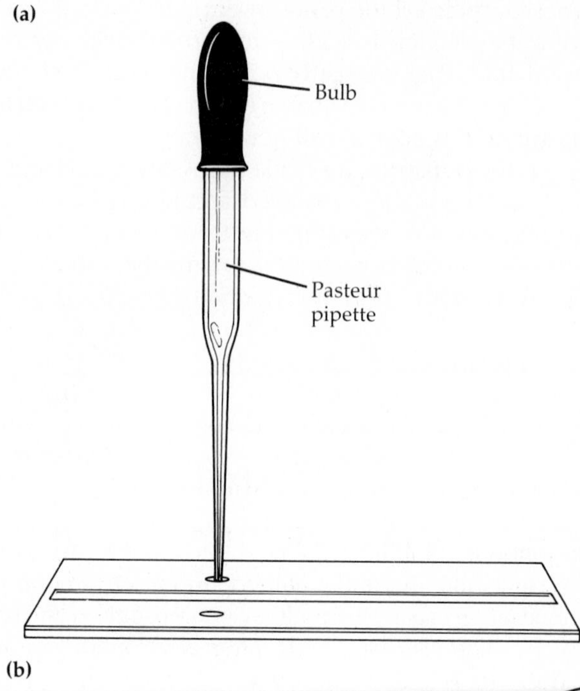

(b)

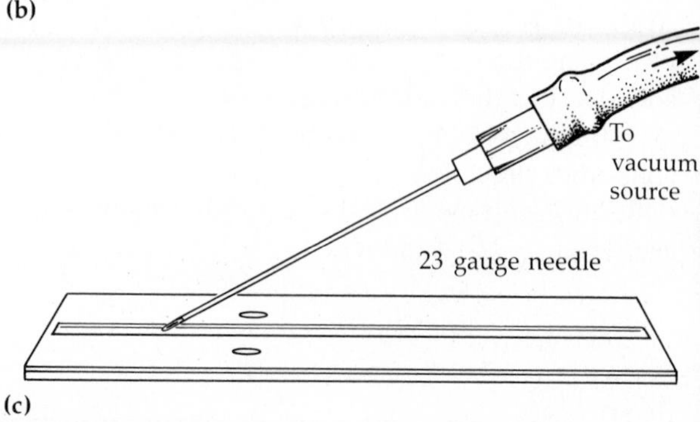

(c)

Figure 72-2. ____________________
Immunoelectrophoresis agar slide. **(a)** Pattern. **(b)** Cutting the wells (holes). **(c)** Cutting the trough.

13. Record patterns. Add 100 ml cold saline to a large beaker or bowl. Carefully place the slide in the bowl. After 10 minutes, decant the saline. Add 100 ml fresh saline. Place at 5°C for 24 hours. What is the purpose of the saline wash?
   ____
   ____
14. Discard saline and fill the container with distilled water. Wash carefully in the distilled water for 15 to 30 minutes.
15. Saturate a piece of filter paper the size of the slide with distilled water and place over the gel

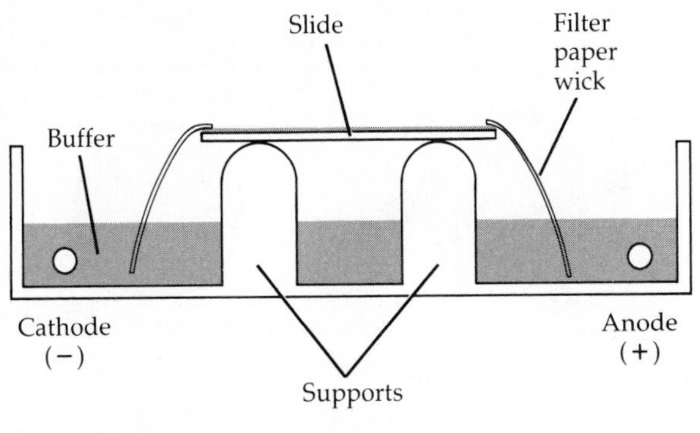

Figure 72-3.
Electrophoresis. Place agar-coated slide on the inner bridge between the two buffer vessels. Attach buffer-dampened filter paper strips to each end on the slide, and immerse the other end of the strip in the buffer vessel. Connect electrodes.

surface of the slide. Place on a rack in the incubator for 24 hours.

16. Remove paper when it's dry.
17. Stain the dry agar film by placing slide for 10 minutes in a Coplin jar containing Ponceau S, which stains proteins.
18. Remove excess stain by placing in Coplin jar of washing solution. Change to fresh washing solutions two or three times during a period of 15–20 minutes.
19. Remove slide, air dry in a vertical position.
20. Examine and record your results.

Turn to the Laboratory Report for Exercise 72.

EXERCISE 73

Blood Group Determination: Slide Agglutination

Objectives

After completing this exercise you should be able to

1. Compare and contrast the terms agglutination and hemagglutination.
2. Determine ABO and Rh blood types.
3. Determine compatible transfusions.

Background

Agglutination reactions occur between high-molecular-weight, particulate antigens and multivalent antibodies. Since many antigens are on cells, agglutination reactions lead to the clumping or **agglutination** of cells. When the cells involved are red blood cells, the reaction is called **hemagglutination.**

Hemagglutination reactions are used in the typing of blood. The presence or absence of two very similar carbohydrate antigens (designated **A** and **B**) located on the surface of red blood cells is determined using specific antisera. These antigens are called *isoantigens* (Figure 73-1). When anti-A antiserum is mixed with type B red blood cells, no hemagglutination occurs. Persons with type AB blood possess both A and B antigens on their red blood cells and persons with type O blood lack A and B isoantigens (see Figure 73-1).

Many other blood antigen series exist on human red blood cells. Another surface antigen on red blood cells is designated the Rh factor. The **Rh factor** is a complex of many antigens. The Rh factor that is used routinely in blood typing is the $\mathbf{Rh_o}$ antigen or *D* antigen. Individuals are Rh-positive when Rh_o antigen is present. The presence of the Rh factor is determined by a hemagglutination reaction between anti-D antiserum and red blood cells.

Isoantibodies are present in human serum. An individual possesses isoantibodies to the opposite isoantigen. Thus, persons of blood type A will have isoantibodies to the B antigen (see Figure 73-1) in their sera. Rh-negative individuals do not naturally have anti-Rh antibodies in their sera. Anti-Rh antibodies are produced when red blood cells with Rh antigen are introduced into Rh-negative individuals.

The ABO and Rh systems place restrictions on how blood may be transfused from one person to another. An incompatible transfusion results when the antigens of the donor red blood cells react with the isoantibodies in the recipient's serum or induce the formation of isoantibodies.

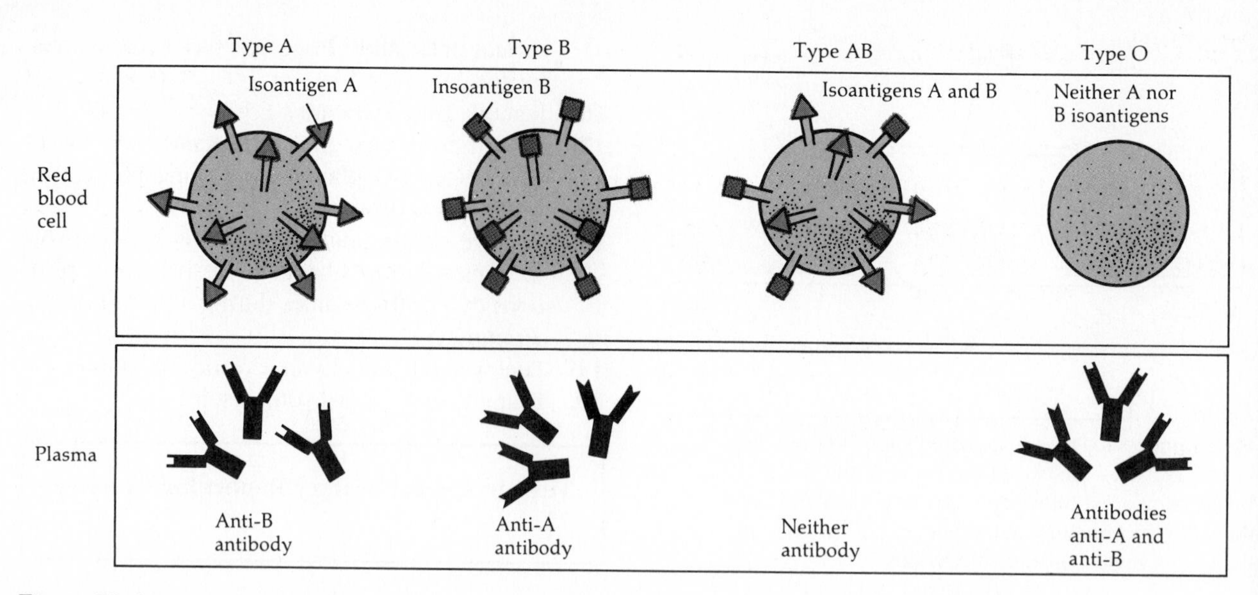

Figure 73-1.
Relationship of isoantigens and isoantibodies involved in the ABO blood group system.

Table 73-1
The ABO and Rh Blood Group Systems

	Blood Type					
Characteristic	A	B	AB	O	Rh+	Rh−
Isoantigen present on the red blood cells	A	B	Both A and B	Neither A nor B	D	No D
Isoantibody normally present in the serum	anti-B	anti-A	Neither anti-A nor anti-B	Both anti-A and anti-B	No anti-D	No anti-D*
Serum causes agglutination of red blood cells of these types	B, AB	A, AB	None	A, B, AB	Neither Rh+ nor Rh−	Neither Rh+ nor Rh−*
Cells agglutinated by serum of red blood cells of these types	B, O	A, O	A, B, O	None	Neither Rh+ nor Rh−	Neither Rh+ nor Rh−
Percent occurrence in a mixed Caucasian population	41	10	4	45	85	15
Percent occurrence in a mixed Black population	27	20	7	46	85	15

Source: Adapted from G. J. Tortora, B. R. Funke, and C. L. Case. 1982. *Microbiology: An Introduction.* Menlo Park, Ca.: Benjamin/Cummings.
*Anti-D antibodies are not naturally present in the serum of Rh− persons. Anti-D antibodies can be produced upon exposure to the D antigen through blood transfusions or pregnancy.

A summary of the major characteristics of the ABO and Rh blood groups is presented in Table 73-1.

Materials

Grease pencil

Cotton moistened with 70% ethyl alcohol

Sterile cotton ball and bandage

Sterile lancet

Anti-A, anti-B, and anti-D antisera

Glass slide (2)

Toothpicks

Techniques Required

None

Procedure

1. With a grease pencil, draw two circles on a clean glass slide, and label one A and the other B. Draw a circle on the second slide and label it D.
2. Disinfect your finger with 70% alcohol (Figure 69-1). Use any finger except your thumb. Pierce the disinfected finger with a sterile lancet.
3. Let a drop of blood fall into each circle. Stop the bleeding with a sterile cotton ball and apply an adhesive bandage.
4. To the A circle, add 1 drop of anti-A antiserum. Add 1 drop of anti-B antiserum to the B circle and 1 drop of anti-D antiserum to the D circle.
5. Mix each suspension with toothpicks. Use a different toothpick for each antiserum. Why? ______

   ______
6. Observe for agglutination and determine your blood type. A dissecting microscope may help you see hemagglutination (Appendix E).

Turn to the Laboratory Report for Exercise 73.

EXERCISE 74

Agglutination Reactions: Tube Agglutination

Objectives

After completing this exercise you should be able to

1. Define agglutination and titer.
2. Determine the titer of antiserum by the tube agglutination method.
3. Provide two applications for agglutination reactions.

Background

The outer surfaces of bacterial cells contain antigens that can be used in agglutination reactions. **Agglutination reactions** occur between *particulate* antigens, such as cell walls, flagella, or capsules *bound* to cells, and antibodies (Figure 74-1). **Agglutination** or clumping of bacteria by antibodies is a useful laboratory diagnostic technique.

In a slide agglutination test (Exercise 73), an unknown bacterium is suspended in a saline solution on a slide and mixed with a drop of known antiserum. This test is done with different antisera on different slides. The bacteria will agglutinate when mixed with antibodies produced against the same strain and species. A positive test is therefore determined by agglutination and can be used to identify the bacterium.

A **tube agglutination test** can be used to estimate the concentration **(titer)** of antibody in serum. This test is used to determine whether a particular organism is causing the patient's symptoms. If an increase in titer is shown in successive daily tests, the patient probably has an infection caused by the organism used in the test.

In the tube agglutination test, bacterial cells are mixed with dilutions of serum (e.g., 1:5, 1:10, 1:20, 1:40, and so on). The *end point* is the greatest dilution of serum showing an agglutination reaction. The *reciprocal* of this dilution is the titer. For example, in the illustration in Figure 74-1, the titer is 40.

A test used to diagnose typhoid fever, the **Widal test,** will be performed in this exercise. *Salmonella typhi* is the causative agent of typhoid fever. *Be very careful when handling this pathogenic bacterium.*

Materials

0.85% saline solution

Patient's serum

Serological tubes (10) and rack

1-ml pipettes (3)

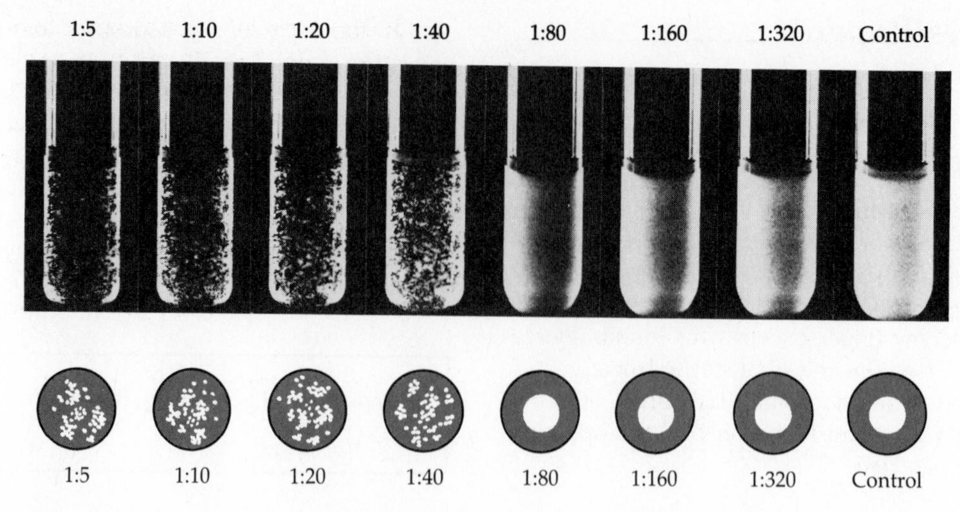

Figure 74-1.
Tube agglutination test for determining antibody titer. Here the titer is 40 since there is no agglutination in the next tube in the dilution series (1:80). Nowadays, such tests are rarely performed in tubes, but rather on plates (microtitration plates) that have a large number of small wells. Inset shows appearance of tubes from the underside after refrigeration for 24 hours.

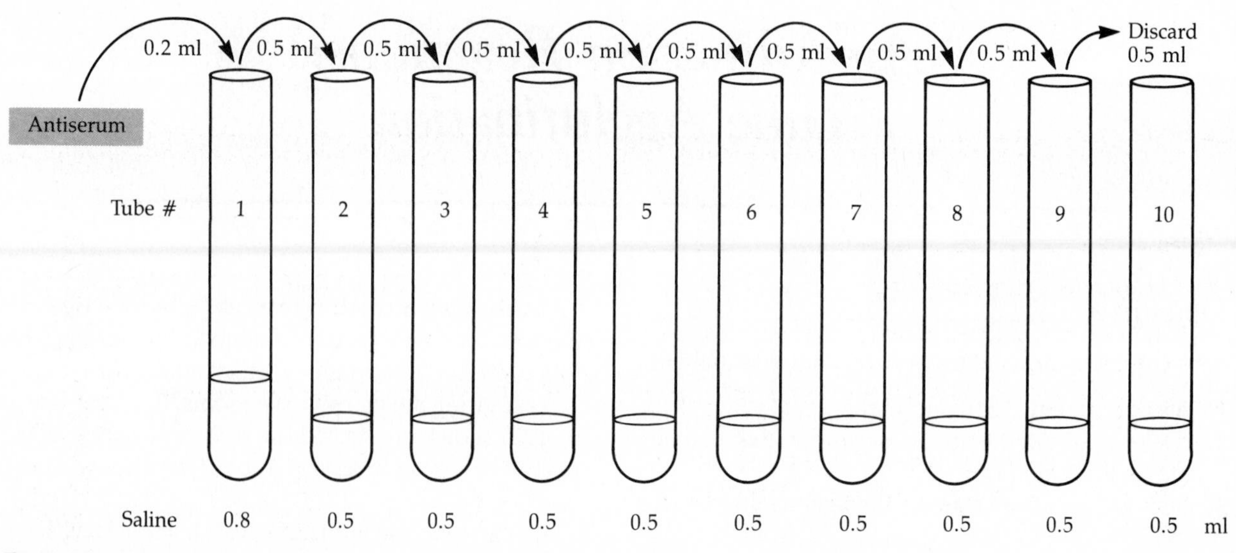

Figure 74-2.
Dilution of antiserum. 0.2 ml patient's serum is mixed with 0.8 ml saline in tube 1. Serial dilutions are made by transferring 0.5 ml from one tube to the next, through tube 9. 0.5 ml is discarded from tube 9.

Culture

Salmonella typhi (heated for 30 minutes at 56°C in water bath)

Techniques Required

Appendix A

Procedure

1. Label the serological tubes 1 through 10 and place in the rack (Figure 74-2).
2. Add 0.8 ml saline to the first tube and 0.5 ml to each of the remaining tubes.
3. Add 0.2 ml patient's serum to the first tube. Mix up and down 3 times and transfer 0.5 ml to the second tube. Mix and transfer 0.5 ml to the third tube, and so on. Continue until you have reached the ninth tube. Discard the 0.5 ml from that tube as shown in Figure 74-2. The patient's serum has now been diluted. The dilutions are as follows:

   Tube #1 1:5

   Tube #2 1:10

   Tube #3 1:20

   Tube #4 1: ______

   Tube #5 1: ______

Tube #6 1: ____

Tube #7 1: ____

Tube #8 1: ____

Tube #9 1: ____

Tube #10 no serum

Complete the missing dilutions.

4. Carefully add 0.5 ml of the antigen to each tube. What is the antigen? ____
How does the addition of 0.5 ml of antigen affect the dilutions? ____
5. Shake the rack to mix the contents in each tube. Place the rack in a 35°C water bath or incubator for 60 minutes.
6. Refrigerate until the next laboratory period.
7. Observe the bottom of each tube for agglutination. Which tube serves as the control? ____
What should occur in this tube? ____

____

Determine the end point and the titer.

Turn to the Laboratory Report for Exercise 74.

EXERCISE 75

Fluorescent-Antibody Technique

Objectives

After completing this exercise you should be able to

1. Explain how fluorescent-antibody tests can be used to diagnose diseases.
2. Differentiate between direct and indirect fluorescent-antibody tests.
3. Compare and contrast direct fluorescent-antibody testing with simple staining and brightfield microscopy.

Background

Fluorescent-antibody (F.A.) techniques are used to identify microorganisms or to detect the presence of specific antibodies in a patient's serum. F.A. tests employ **antibodies** that have been combined **(conjugated)** with fluorescent dyes such as fluoresceine isothiocyanate. **Fluorescent dyes** emit visible light when they absorb ultraviolet or near ultraviolet wavelengths.

There are two types of fluorescent-antibody tests: direct and indirect. **Direct F.A.** tests are used to determine the identity of a microorganism (antigen). In this procedure, the specimen containing the antigen is fixed onto a slide. Fluorescent-labeled antibody (F.A.) is then added to the slide. After a brief incubation, the slide is washed and examined under a fluorescent microscope. The presence of fluorescing bacterial cells (Figure 75-1) is a positive test.

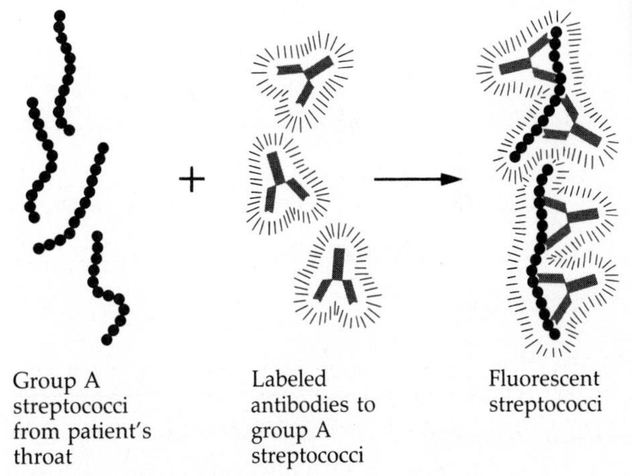

Figure 75-1.
Use of direct fluorescent-antibody technique for the identification of group A streptococci.

Indirect F.A. tests are used to detect the presence of a specific antibody in a patient's serum. In this procedure, a known species (or serotype) of microorganism is fixed onto a slide. Serum is then added, and if antibody specific to the microorganism is present, it reacts with the antigens forming a bound complex. To observe this antigen–antibody complex, fluorescent-labeled antihuman gamma globulin *(anti-antibody)* is added to the slide. After incubation and washing, the slide is observed for fluorescing antigens under a fluorescent microscope (Figure 75-2).

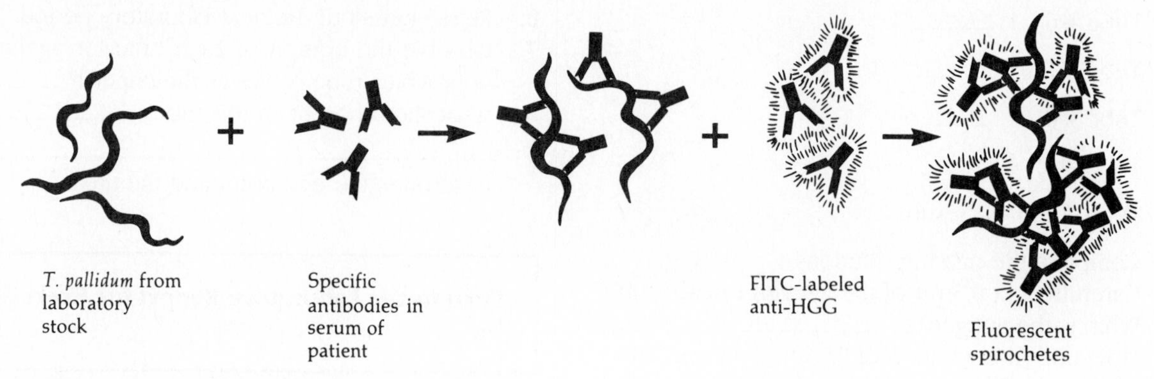

Figure 75-2.
Use of the indirect fluorescent-antibody technique for the diagnosis of syphilis (fluorescent treponemal antibody test).

Materials

E. coli polyvalent fluorescent antibody

F.A. buffer

F.A. mounting medium

95% ethyl alcohol

Petri plate with filter paper liner (2)

Applicator sticks (2)

Gram staining reagents

Cover slips (2)

Cultures

Unknown bacterial cultures (2)

Techniques Required

Exercises 3 and 7

Procedure

Gram Staining

Prepare Gram stains of each unknown culture. Record your observations. Can you identify these organisms from the Gram stain? ____________

Can you differentiate these organisms based on the Gram stain? ____________

F.A. Test

1. Prepare a smear of each culture on clean glass slides and allow the smears to air dry.
2. Fix the smears by immersing in 95% ethyl alcohol for 2 minutes.
3. Drain off excess alcohol and quickly rinse the slides in F.A. buffer.
4. Apply 2 drops of fluorescent antibody to each smear and spread the antibody by tilting the slide.
5. Place the slides in a Petri plate containing moistened filter paper for 30 minutes. The moistened filter paper will prevent evaporation during incubation.
6. Remove the slides and air dry.
7. Place a drop of F.A. mounting medium on each smear and set a cover slip on the mounting medium.
8. Examine each slide using a brightfield microscope and record your observations.
9. Examine each slide using a fluorescent microscope and look for fluorescence. Record your observations.

Turn to the Laboratory Report for Exercise 75.

Appendices

APPENDICES

APPENDIX A

Pipetting

Pipettes are used for measuring small volumes of fluid. They are usually coded according to the total volume and graduation units (Figure A-1). To fill a pipette use a bulb or other mechanical device as shown in Figure A-2. *Do not use your mouth.* Draw the desired amount of fluid into the pipette. The volume is read at the bottom of the meniscus (Figure A-3). Fill the pipette to above the zero mark, then allow it to drain just to the zero. The desired amount can then be dispensed.

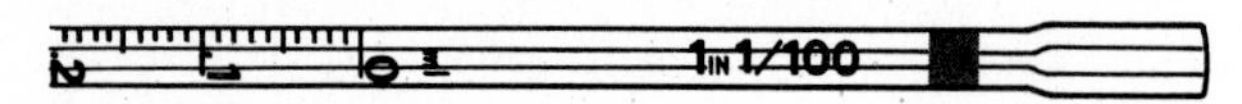

Figure A-1.
This pipette holds a total volume of 1 milliliter when filled to the zero mark. It is graduated in 0.01-milliliter units.

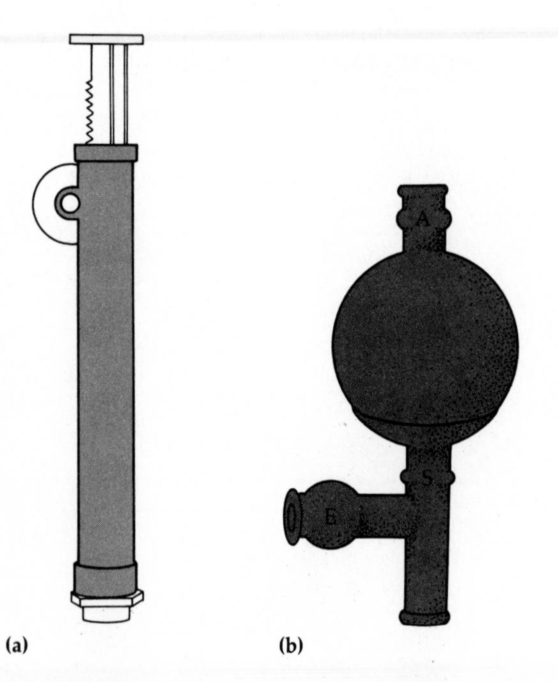

Figure A-2.
Two types of pipette aspirators. **(a)** This plastic pump is attached to the pipette, and the wheel is turned to draw fluid into the pipette. Turning the wheel in the other direction will release the fluid. **(b)** A pipette is inserted into this bulb; while pressing the A valve, squeeze the bulb, and it will remain collapsed. To draw fluid into the pipette, press the S valve, and to release fluid, press the E valve.

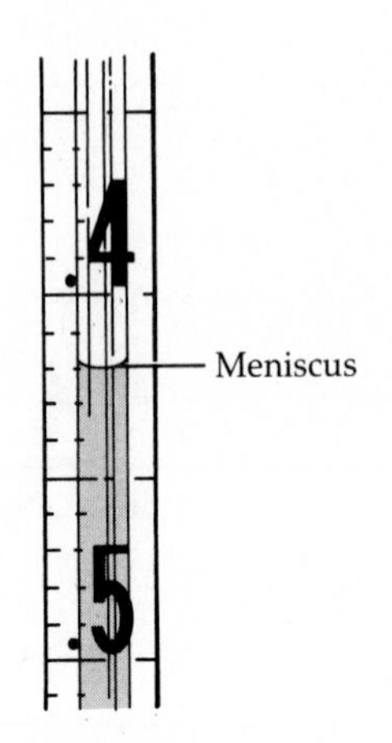

Figure A-3.
The volume is read at the lowest level of the meniscus.

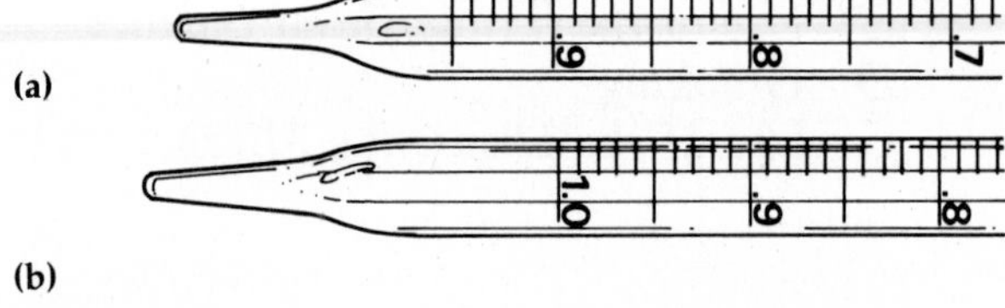

Figure A-4.
(a) A serological pipette. **(b)** A measuring pipette.

Microbiologists use two types of pipettes. The **serological pipette** is meant to be emptied to deliver the total volume (Figure A-4*a*). Note that the graduations stop above the tip. The **measuring pipette** delivers the volume read on the graduations (Figure A-4b). This pipette is not emptied, but the flow must be stopped when the meniscus reaches the desired level.

Aseptic use of a pipette is often required in microbiology. Bring the entire closed pipette container to your work area. If the pipettes are wrapped in paper, open the wrapper at the end opposite the delivery end; in a canister, the delivery end will be at the bottom of the canister. Do not touch the delivery end of a sterile pipette. Remove a pipette and, with your other hand, pick up the sample to be pipetted. Remove the cap from the sample with the little finger of the hand holding the pipette. Fill the pipette and replace the cap on the sample. After pipetting, place the contaminated pipette in the appropriate container of disinfectant.

Exponents, Exponential Notation, and Logarithms

Very large and very small numbers—such as 4,650,000,000 and 0.00000032—are cumbersome to work with. It is more convenient to express such numbers in exponential notation, that is, as a power of 10. For example, 4.65×10^9 is written in **standard exponential notation,** or **scientific notation.** 4.65 is the **coefficient,** and 9 is the power or **exponent.** In standard exponential notation, the coefficient is a number between 1 and 10 and the exponent is a positive or negative number.

To change a number into exponential notation, follow two steps. First, determine the coefficient by moving the decimal point so you leave only one nonzero digit to the left of it. For example,

$$0.00000032$$

The coefficient is 3.2. Second, determine the exponent by counting the number of places you moved the decimal point. If you moved it to the left, the exponent is a positive number. If you moved it to the right, the exponent is negative. In the example, you moved the decimal point 7 places to the right, so the exponent is -7. Thus

$$0.00000032 = 3.2 \times 10^{-7}$$

Now suppose we are working with a large number instead of a very small number. The same rules apply, but our exponential value will be positive rather than negative. For example,

$$\begin{aligned} 4{,}650{,}000{,}000. &= 4.65 \times 10^{+9} \\ &= 4.65 \times 10^{9} \end{aligned}$$

To multiply numbers written in exponential notation, multiply the coefficients and *add* the exponents. For example,

$$\begin{aligned} (3 \times 10^4) \times (2 \times 10^3) &= \\ (3 \times 2) \times 10^{4+3} &= 6 \times 10^7 \end{aligned}$$

To divide, divide the coefficients and *subtract* the exponents. For example,

$$\frac{3 \times 10^4}{2 \times 10^3} = \frac{3}{2} \times 10^{4-3} = 1.5 \times 10^1$$

Microbiologists use exponential notation in many kinds of situations. For instance, exponential notation is used to describe the number of microorganisms in a population. Such numbers are often very large. Another application of exponential notation is to express concentrations of chemicals in a solution—chemicals such as media components, disinfectants, or antibiotics. Such numbers are often very small. Converting from one unit of measurement to another in the metric system requires multiplying or dividing by a power of 10, which is easiest to carry out in exponential notation.

A **logarithm** is the power to which a base number is raised to produce a given number. Usually we work with logarithms to the base 10, abbreviated $\mathbf{log_{10}}$. The first step in finding the $\log_{10}$ of a number is to write the number in standard exponential notation. If the coefficient is exactly 1, the $\log_{10}$ is simply equal to the exponent. For example,

$$\begin{aligned} \log_{10} 0.00001 &= \log_{10}(1 \times 10^{-5}) \\ &= -5 \end{aligned}$$

If the coefficient is not 1, as is often the case, a logarithm table or calculator must be used to determine the logarithm.

Microbiologists use logs for pH calculations and for graphing the growth of microbial populations in culture.

Source: G. J. Tortora, B. R. Funke, and C. L. Case. 1982. *Microbiology: An Introduction.* Menlo Park, Ca.: Benjamin/Cummings.

APPENDIX B

Dilution Techniques and Calculations*

Bacteria, under good growing conditions, will multiply into such large populations that it is often necessary to dilute them to isolate single colonies or to obtain estimates of their numbers. This requires mixing a small, accurately measured sample with a large volume of sterile water or saline called the **diluent** or **dilution blank.** Accurate dilutions of a sample are obtained through the use of pipettes. For convenience, dilutions are usually made in multiples of ten.

A single dilution is calculated as follows:

$$\text{Dilution} = \frac{\text{Volume of the sample}}{\text{Total volume of the sample and the diluent}}$$

For example, the dilution of 1 milliliter into 9 milliliters equals

$$\frac{1}{1+9} \text{ which is } \frac{1}{10} \text{ and is written } 1{:}10$$

Experience has shown that better accuracy is obtained with very large dilutions if the total dilution is made out of a series of smaller dilutions rather than one large dilution. This series is called a **serial dilution,** and the total dilution is the product of each dilution in the series. For example, if 1 milliliter is diluted with 9 milliliters, and then 1 milliliter of that dilution is put into a second 9-milliliter diluent, the final dilution will be

$$\frac{1}{10} \times \frac{1}{10} = \frac{1}{100} \quad \text{or} \quad 1{:}100$$

To facilitate calculations, the dilution is written in exponential notation. In the example just given, the final dilution 1:100 would be written 10^{-2}. (Refer to the box on exponential notation.) A serial dilution is illustrated in Figure B-1.

Procedure

1. Aseptically pipette 1 ml of sample into a dilution blank.
   a. If the dilution is into a tube, mix the contents by rolling the tube back and forth between your hands (Figure 14-4).
   b. If the dilution is into a 99-ml blank, hold the cap in place with your index finger and shake the bottle up and down through a 35-cm arc (Figure 51-3*a*).
2. It is necessary to use a fresh pipette for each dilution in a series, but it is permissible to use the

*Adapted from C. W. Brady. "Dilutions and dilution calculations." Unpublished. University of Wisconsin, Whitewater.

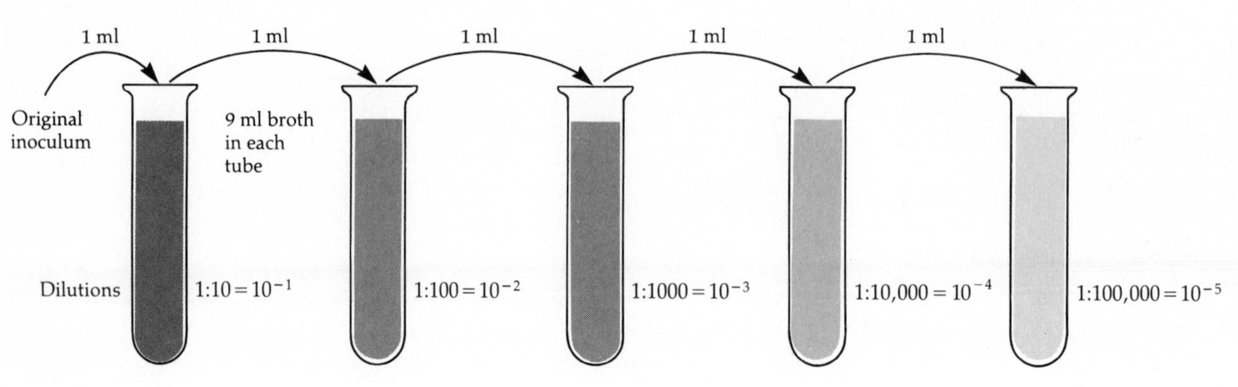

Figure B-1.
A 1-milliliter sample from the first tube will contain 1/10 the number of cells present in 1 milliliter of the original sample. A 1-milliliter sample from the last tube will contain 1/100,000 the number of cells present in 1 milliliter of the original sample.

same pipette to remove several samples from the same bottle, as when plating out samples from a series of dilutions.

Problems

Practice calculating serial dilutions using the following problems.

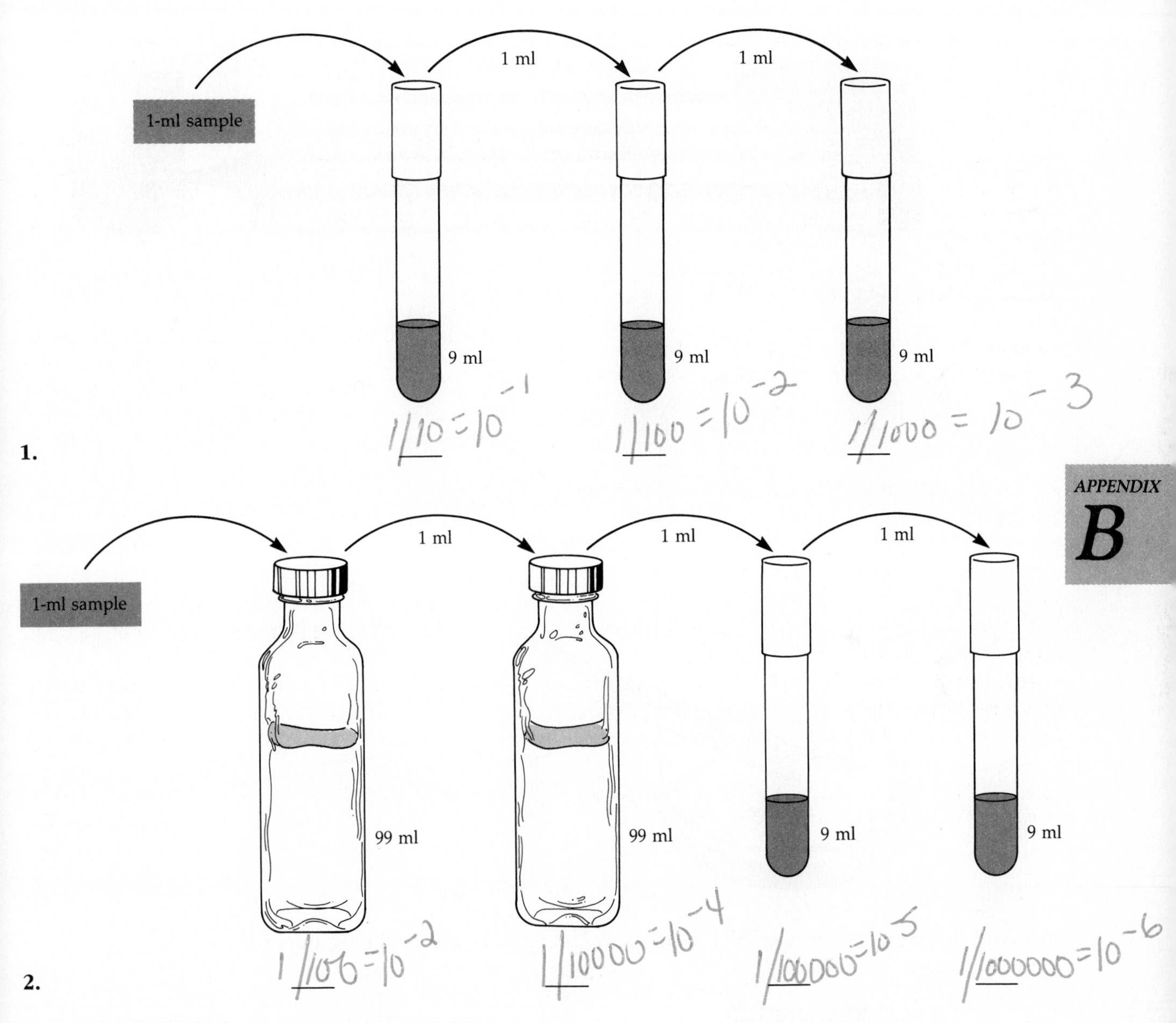

3. Design a serial dilution to achieve a final dilution of 10^{-8}.

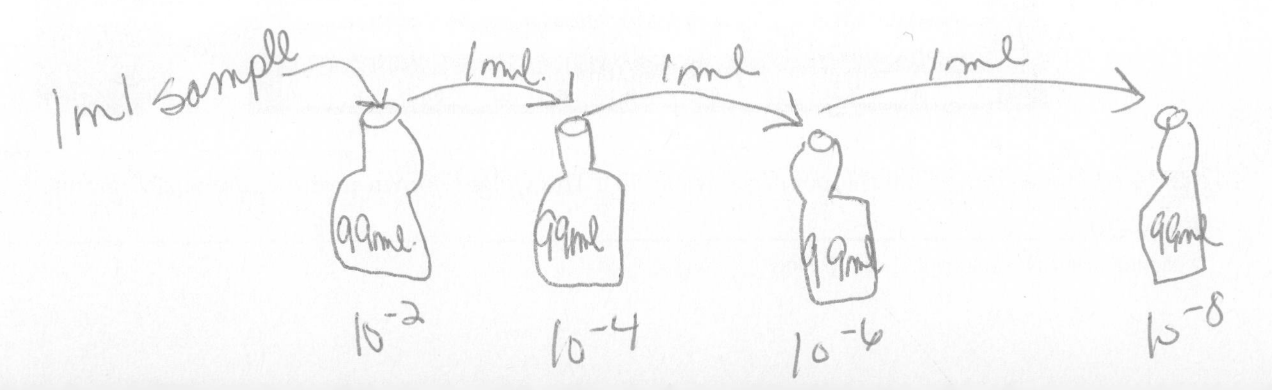

APPENDIX C

Use of the Triple Beam Balance

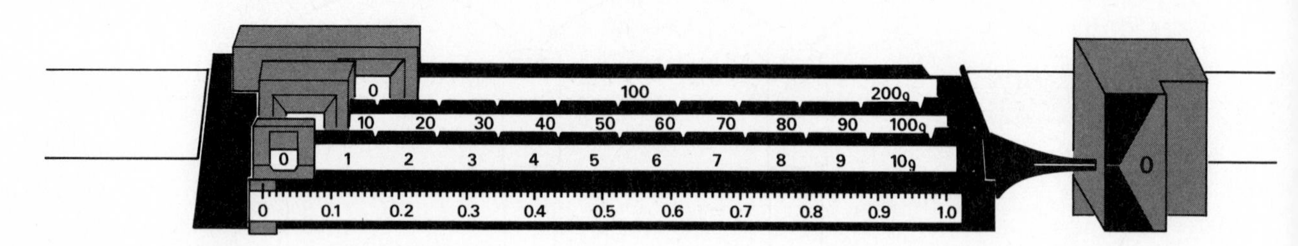

(a) Begin with the pointer at "0".

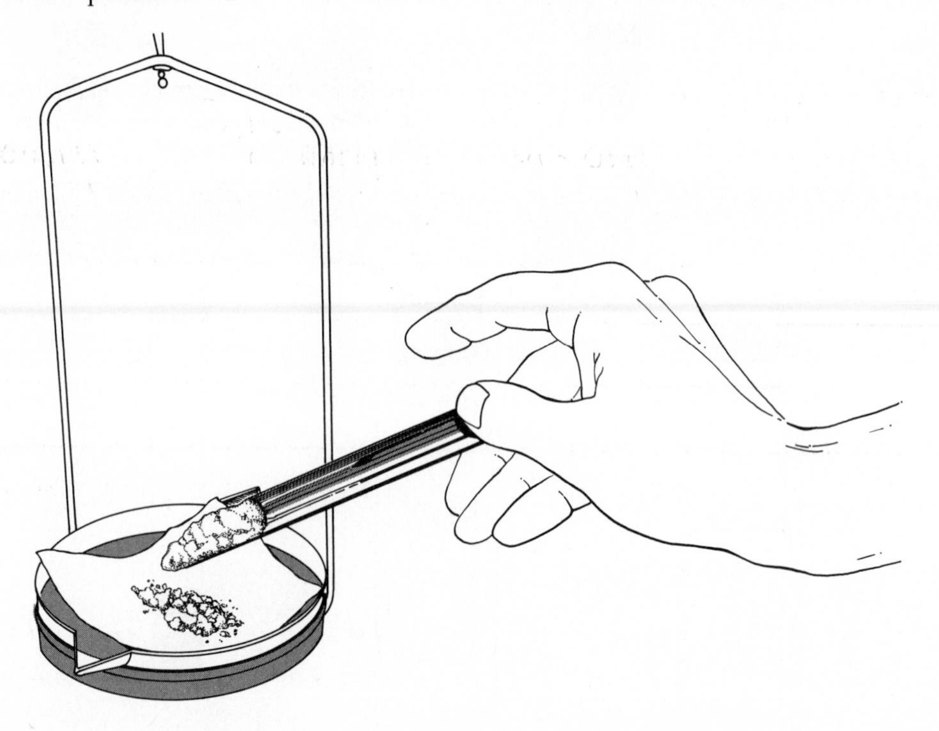

(b) Add the sample to be weighed to the pan. Use spatula and weighing paper.

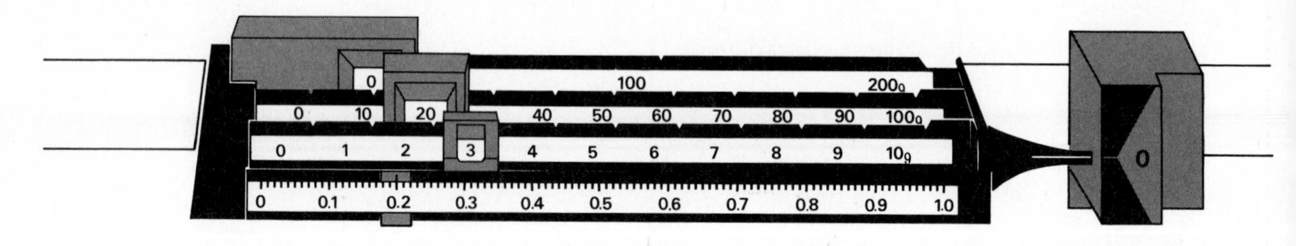

(c) Move the weights until the pointer stays at "0". The weight shown in the figure is 23.2 grams.

Figure C-1.
Weighing materials on a triple beam balance.

Materials to be weighed are placed on a piece of weighing paper, in a weighing boat, or in another suitable container. Reagents are never placed directly on the pan. Powders and crystals can be lost if you attempt to pour them to and from the pan. Use the smallest possible weighing container to minimize error.

Weigh the weighing paper or container first; this weight is called the **tare.** Then add the sample and weigh again. Subtract the weight of the container from the total weight to get the sample weight.

To weigh materials, begin with an empty pan and check to see that the needle points to zero (Figure C-1*a*). Place the material to be weighed on the pan and move the weights until the needle again points to zero (Figure C-1*b*). Add the weights together for the total weight of the material on the pan.

APPENDIX D

Graphing

A **graph** is a pictorial representation of a relationship between two variables. Whenever one variable changes in a definite way in relation to another variable, this relationship may be graphed.

Microbiologists work with large populations of bacteria and frequently use graphs to illustrate the activity of these populations. Graphs dealing with populations of cells are drawn on semi-log graph paper. The hori-

Table D-1
Numbers of *E. coli* in Three Cultures Incubated at Different Temperatures

Time (hours)	Temperature (°C)		
	20	15	35
0	5.00×10^5	1.50×10^5	5.40×10^5
4	—	1.50×10^5	—
5	4.60×10^5	—	—
6	—	—	5.20×10^5
10	5.78×10^5	5.62×10^5	6.00×10^5
11	8.80×10^5	—	1.23×10^6
12	1.05×10^6	—	2.58×10^6
13	1.16×10^6	—	4.93×10^6
14	2.32×10^6	—	9.00×10^6
15	3.80×10^6	1.62×10^6	—
16	7.00×10^6	—	—
17	8.10×10^6	—	—
18	9.60×10^6	—	—
19	1.95×10^7	—	—
20	—	2.70×10^6	—

zontal (x) axis is a linear scale. The x-axis is used for the **independent variable,** that is, the variable not being tested. The vertical (y) axis is a logarithmic scale where each cycle represents a power of 10. The dependent variable is marked off along the y-axis. The **dependent variable** changes in relation to the independent variable. The intersection point of the x-axis and y-axis is called the **origin,** and all units are marked off from this point.

To draw a graph, first establish the two variables you want to graph—for example, time and number of bacteria (Table D-1). The independent variable is time and the dependent variable is the number of bacteria. Mark the appropriate units of measure along the axes (Figure D-1).

We will graph the numbers of bacteria in the culture incubated at 20°C. The first point on the graph is ($0, 5.00 \times 10^5$). Locate 0 on the x-axis, measure directly above it the distance 5.00×10^5, and mark this point. After a sufficient number of points have been plotted, a smooth curve can be drawn connecting these points. The line through the points should be a

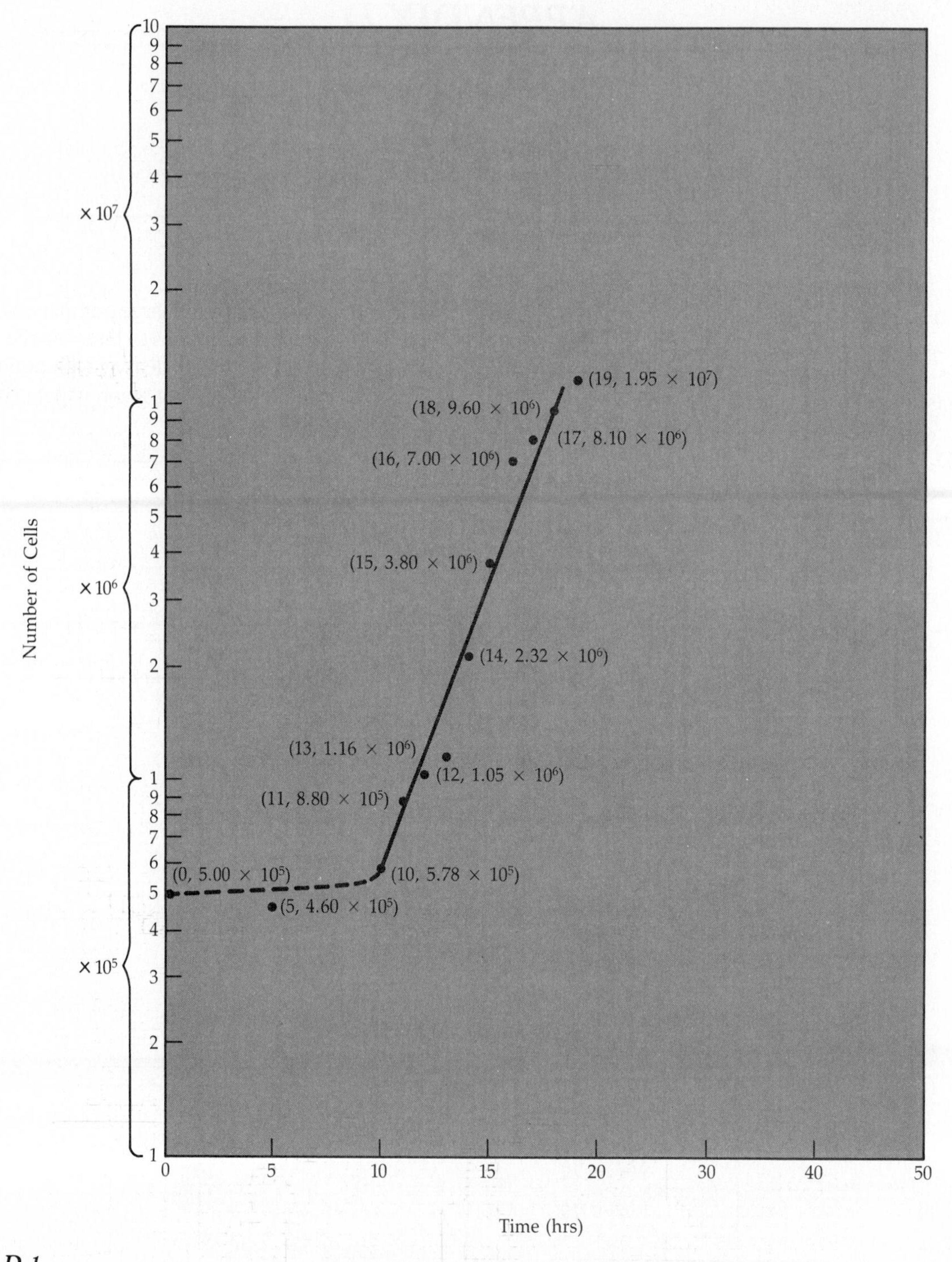

Figure D-1.
Growth of *E. coli* at 20°C.

straight line and does not have to connect all of the points. When the points do not fall in a straight line, a line can be drawn between the points leaving an equal number of points above and below the line.

When graphs are drawn that compare the same variables under different conditions (for example, bacterial growth rates at different temperatures), it is best to draw all the graphs on the same paper so comparisons are obvious and easily made. Graphs should be titled to explain the data presented.

The data for cultures incubated at 15°C and 35°C (Table D-1) can be plotted on the graph in Figure D-1. It is not necessary to have bacterial numbers for the same "time points" in each graph. The missing points will be filled in by the line. This is possible because the line between each point is an interpretation of what was happening between measurements.

Activity

Using the data provided in Table D-1, graph the growth curves on the graph paper and compare the effect of temperature on the rate of growth. Since *rate* is a change over time, the rate can be interpreted by the slope of the line. What will you title this graph?

____

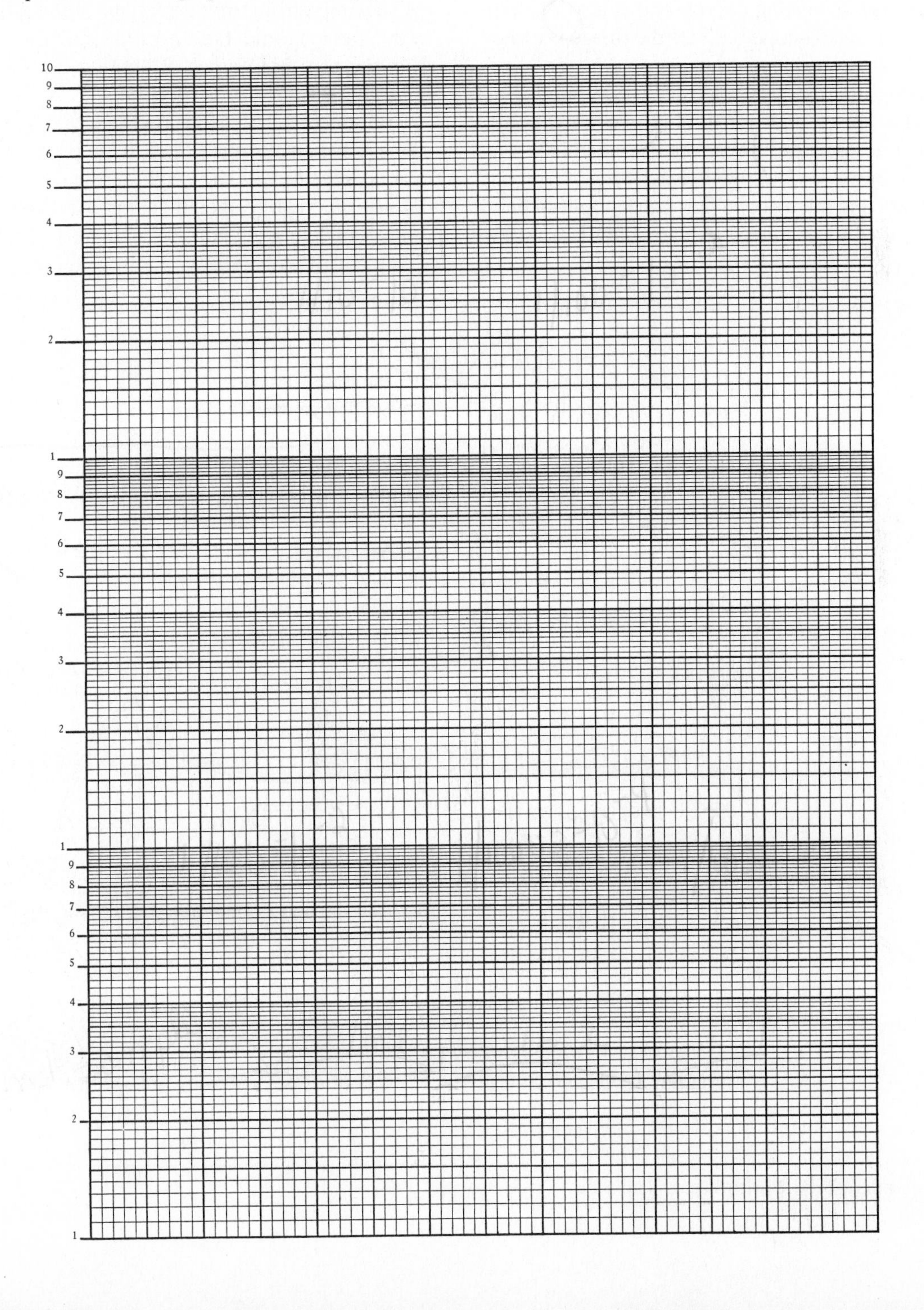

APPENDIX E

Use of the Dissecting Microscope

The **dissecting microscope** (Figure E-1) is useful for observing specimens that are too large for the compound microscope and too small to be seen with the naked eye. The dissecting microscope is also called a **stereoscopic** microscope because the image is three-dimensional.

Dissecting microscopes are usually equipped with 10× ocular lenses and a 1× objective lens. Some dissecting microscopes have a rotating nosepiece that houses 1× and 2× objective lenses.

The dissecting microscope may have a built-in light source or a substage mirror and auxiliary lamp. Illumination can be adjusted on microscopes with two built-in lamps and/or by using a mirror. For transparent specimens, the light should be directed through the specimen from under the stage. For opaque specimens, light should be directed onto the specimen from the top. Both types of lighting can be used for observing a translucent specimen.

The specimen is placed on the stage. Some microscopes have an alternate black stage plate for use with opaque specimens. The stage clips can be moved to the side when large specimens such as Petri plates are used. While looking through the microscope, adjust

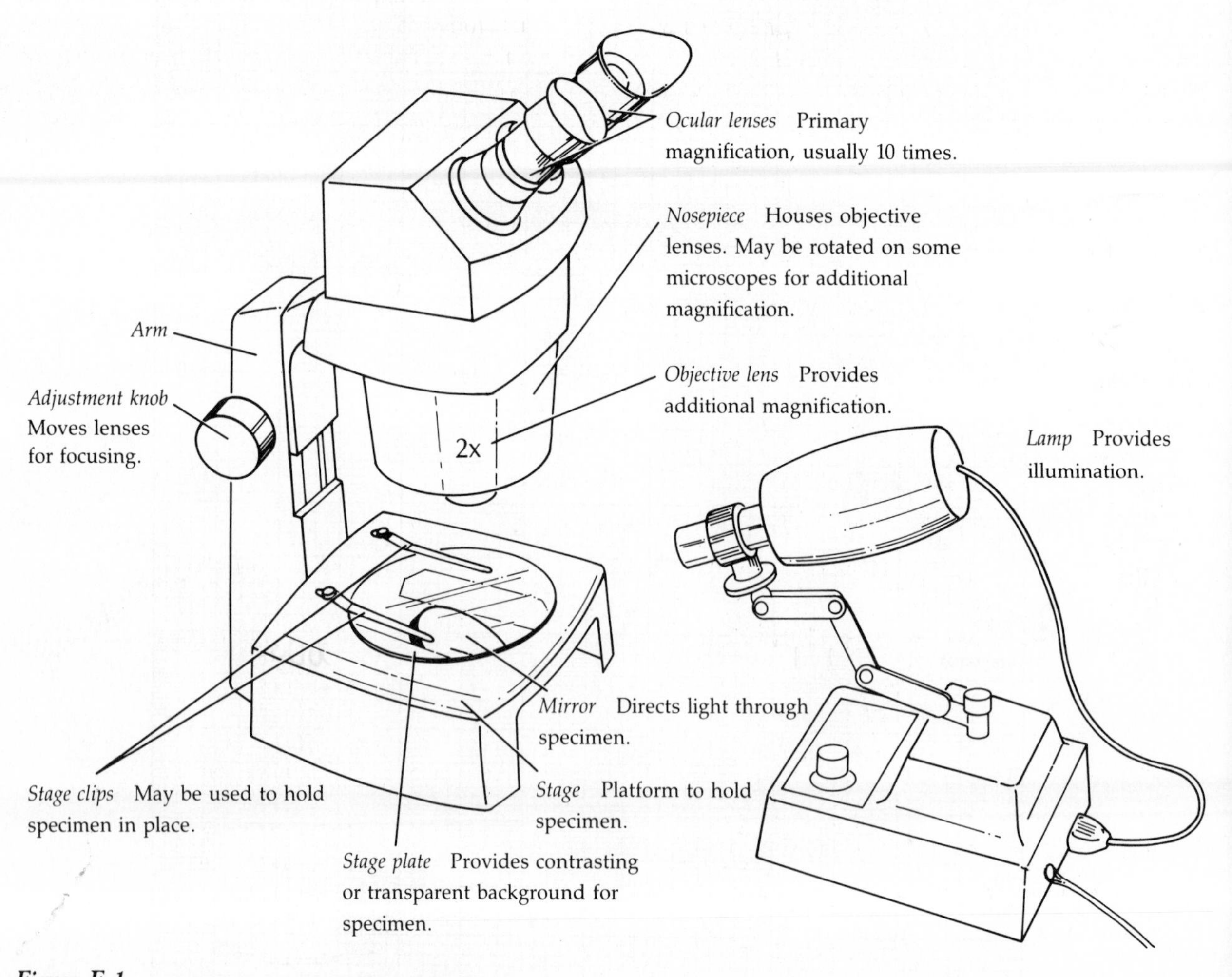

Figure E-1.
A dissecting microscope. Principal parts and their functions.

the width of the eyepieces so you see one field of vision. If two circles or fields are visible, the ocular lenses are too far apart; if the two fields overlap, the ocular lenses are too close together.

To focus, turn the adjustment knob.

Questions

1. What is the magnification of your dissecting microscope? ____________
2. Using a ruler, measure the size of the field of vision at low power. ______ mm. At high power. ______ mm.
3. How can you adjust the direction of illumination on your microscope? ____________

APPENDIX F

Use of the Membrane Filter

Membrane filtration can be used to separate bacteria, algae, yeasts, and molds from solutions. Cellulose or polyvinyl membrane filters with 0.45 μm pores are commonly used to trap microorganisms. In Exercise 35, membrane filtration can be used to separate viruses from their host cells because the viruses will pass through the filter. In Exercise 50, membrane filters are used to trap bacteria found in water in order to count the bacteria.

Membrane filters have many uses in industrial microbiology. They are used to filter wine, soft drinks, air, and water to detect microorganisms and particu-

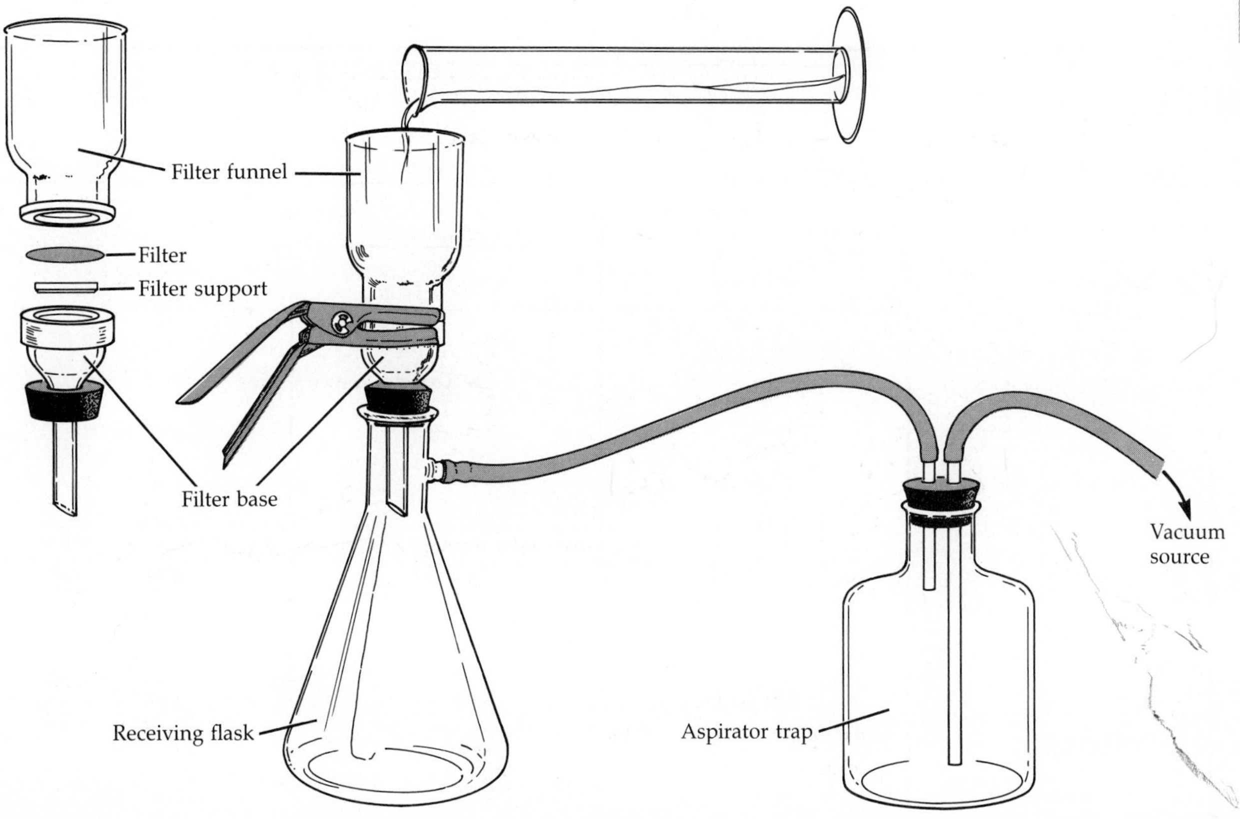

Figure F-1.
Membrane filtration setup.

late matter. Additionally, membrane filters can be used to separate large molecules such as DNA from solutions.

The receiving flask, filter base, and filter support can be wrapped in paper and sterilized by autoclaving (Figure F-1). Membrane filters can also be sterilized by autoclaving. For use, the filter base must be unwrapped aseptically. Using sterile, flat forceps, a sterile filter is placed on the filter support and the filter funnel is clamped or screwed into place.

Gravity alone will not pull a sample through the small filter pores, and the filtering flask is attached to a vacuum source (Figure F-1). Since air pressure is then greater outside the flask, the sample is pushed into the vacuum inside the flask. A vacuum pump or aspirator connected to a water faucet is usually used to provide the necessary vacuum. An aspirator trap is placed between the vacuum source and the receiving flask to prevent the flow of solution into the pump or sink. When small filters and small volumes of fluid are used, a syringe can be used to supply the needed vacuum.

Following filtration, the filtrate in the flask is free of all microorganisms larger than viruses. To observe cells trapped on the filter, the filter can be dipped into immersion oil to make it translucent and then mounted on a microscope slide. Microorganisms can be cultured from the filter by placing the filter on a solid nutrient medium (Exercise 50).

APPENDIX G

pH Indicators Used in Microbiology

Indicator	pH Range	Color	
		Acidic	Alkaline
Brilliant Green	0.0 to 2.6	0.0 Yellow 2.6 Green	Green
Bromcresol Green	3.8 to 5.4	Yellow	Blue-green
Bromcresol Purple	5.2 to 6.8	Yellow	Purple
Bromthymol Blue	6.0 to 7.6	Yellow	Blue
Congo Red	3.0 to 5.0	Blue-violet	Red
Cresol Red	2-3 to 8.8	Orange	7.2 Yellow 8.8 Red
Litmus	4.5 to 8.3	Red	Blue
Methyl Red	4.4 to 6.2	Red	Yellow
Phenol Red	6.8 to 8.4	Yellow	Red

APPENDIX H

Key to the Identification of Selected Heterotrophic Bacteria

Part, family, genus, and species names *and* numbers are taken from *Bergey's Manual.**

I. Cells are cocci.

A. Gram-positive, aerobic, or facultatively anaerobic, **Part 14.**

Family I. **Micrococcaceae**

Cells divide in more than one plane to form irregular clusters, packets, or tetrads. Catalase-positive. Tolerate 6.5% NaCl.

Genus I. *Micrococcus*

Does not ferment glucose. Often pigmented.

1. *M. luteus:* Yellow pigment; hydrolyzes lipids.
2. *M. roseus:* Red pigment; does not hydrolyze lipids.

Genus II. *Staphylococcus*

Ferments glucose (no gas). V-P$^+$.

1. *S. aureus:* Acid from mannitol.
2. *S. epidermidis:* No acid from mannitol. Acid from fructose.
3. *S. saprophyticus:* No acid from mannitol or fructose.

Family II. **Streptococcaceae**

Catalase-negative. Glucose fermented. MR$^+$V-P$^-$.

Genus I. *Streptococcus*

Cells divide to give pairs or chains.

Does not grow in 6.5% NaCl or at pH 9.6. No growth at 10° or 45°C.

1. *S. pyogenes:* Does not ferment glycerol or sorbitol.
3. *S. zooepidemicus:* Ferments glycerol and sorbitol.

Does not grow in 6.5% NaCl or at pH 9.6. Growth at 45° but not at 10°C.

11. *S. salivarius:* Does not hydrolyze starch.
13. *S. bovis:* Hydrolyzes starch.

Grows in 6.5% NaCl and at pH 9.6. Grows at 10° and 45°C.

17. *S. faecalis:* Ferments sorbitol but not arabinose.
18. *S. avium:* Ferments sorbitol and arabinose.

Does not grow in 6.5% NaCl or at pH 9.6. Grows at 10° but not at 45°C.

20. *S. Lactis*

B. Gram-negative, catalase-positive, aerobic or facultatively anaerobic, oxidase-positive, frequently in pairs, **Part 10.**

Family I. **Neisseriaceae**

Genus I. *Neisseria*

Nitrates not reduced.

3. *N. sicca:* Ferments glucose.
5. *N. flavescens:* Does not ferment glucose.

Nitrates reduced.

6. *N. mucosa:* Ferments glucose.

II. Cells are rods.

A. Gram-positive, endospores present, **Part 15.**

Family I. **Bacillaceae**

Genus I. *Bacillus*

Aerobic or facultatively anaerobic, catalase-positive.

Nitrates reduced.

1. *B. subtilis:* Slow hydrolysis of gelatin, acid from mannitol, V-P$^+$.
4. *B. cereus:* Rapid hydrolysis of gelatin, no acid from mannitol, V-P$^+$.

Nitrates not reduced.

7. *B. megaterium:* Slow hydrolysis of gelatin, acid from mannitol, V-P$^-$.

Genus II. *Clostridium*

Anaerobic, catalase-negative.

B. Gram-positive, no endospores present, catalase-negative, **Part 16.**

Family I. **Lactobacillaceae**

Most strains are nonmotile, aerotolerant anaerobes.

*R. E. Buchanan and N. E. Gibbons, eds. 1974. *Bergey's Manual of Determinative Bacteriology,* 8th ed. Baltimore: Williams and Wilkins.

Genus I. *Lactobacillus*

Ferments glucose (no gas).

7. *L. acidophilus:* No acid from mannitol.
9. *L. casei:* Acid from mannitol.

Acid and gas from glucose.

15. *L. fermentum*

C. Gram-positive, no endospores present, catalase-positive, **Part 17.**

Coryneform group

Genus I. *Corynebacterium*

Aerobic, nonmotile.

3. *C. xerosis:* Acid from glucose, urease not produced.
6. *C. pseudodiphtheriticum:* No acid from glucose, urease produced.

D. Gram-negative, oxidase-positive, **Part 7.**

Family I. **Pseudomonadaceae**

Aerobic, oxidize glucose.

Genus I. *Pseudomonas*

Many species are pigmented. Liquefies gelatin.

1. *P. aeruginosa:* Litmus milk coagulated, peptonized, and reduced. Good growth at 37°C.
3. *P. fluorescens:* Litmus milk alkaline. Optimum temperature, 22° to 25°C.

Genera of Uncertain Affiliation

Genus. *Alcaligenes*

Aerobic. Does not ferment or oxidize glucose. Gelatin not usually hydrolyzed.

1. *A. faecalis:* Starch not hydrolyzed. Nitrates reduced.
2. *A. aquamarinus:* Starch hydrolyzed. Nitrates not reduced.

E. Gram-negative, oxidase-negative, **Part 8.**

Family I. **Enterobacteriaceae**

Ferments glucose. Facultatively anaerobic. Nitrates reduced.

Acid and gas from lactose, usually within 24 to 48 hours.

Genus I. *Escherichia*

$MR^+V\text{-}P^-$. Citrate not utilized.

1. *E. coli:* H_2S not produced, indole produced.

Genus III. *Citrobacter*

$MR^+V\text{-}P^-$. Citrate utilized.

1. *C. freundii:* H_2S produced, indole not produced.

Genus VII. *Enterobacter*

$MR^-V\text{-}P^+$. Citrate utilized. Indole not produced.

1. *E. cloacae:* Acid from glycerol (no gas).
2. *E. aerogenes:* Acid and gas from glycerol.

Lactose not fermented.

Genus IX. *Serratia*

Urease not produced.

1. *S. marcescens:* Red pigment at 25°C.

Genus X. *Proteus*

No red pigment. Urease produced.

1. *P. vulgaris:* Indole and H_2S produced.
2. *P. mirabilis:* Indole not produced, H_2S produced.

EXERCISE 1 — *LABORATORY REPORT*

Use and Care of the Microscope

Name Julie Cavallucci

Date 2/19/92

Lab. Section ____________

Purpose Demonstrate correct use of a microscope.

Data

Microscope number 1083 Monocular or binocular? Monocular

Eyepiece adjustment notes: ____________

Draw a few representative cells from each slide and show how they appeared at each magnification. Note the differences in size at each magnification.

Algae

Oscillatoria

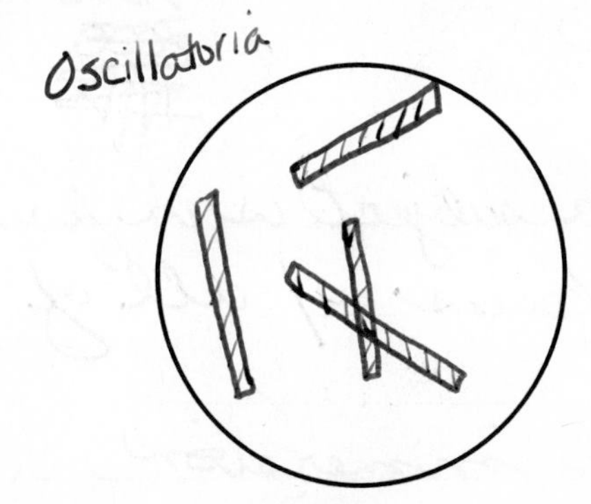

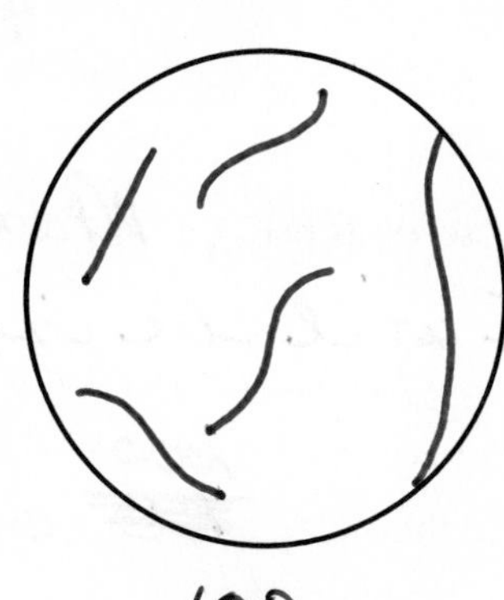

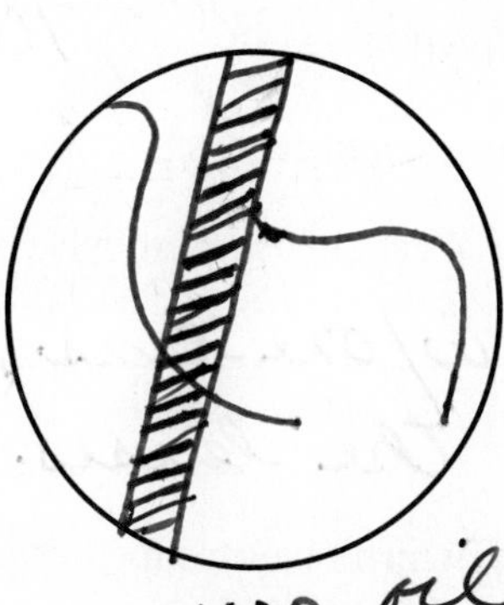

Magnification

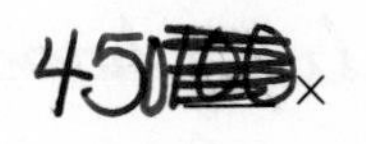

450 ~~[illegible]~~×

100×

1000× oil immergent

Fungi

hizopus (bread mold)

green

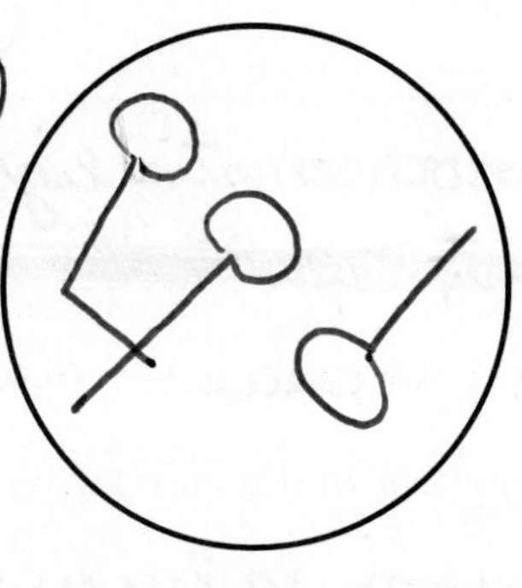

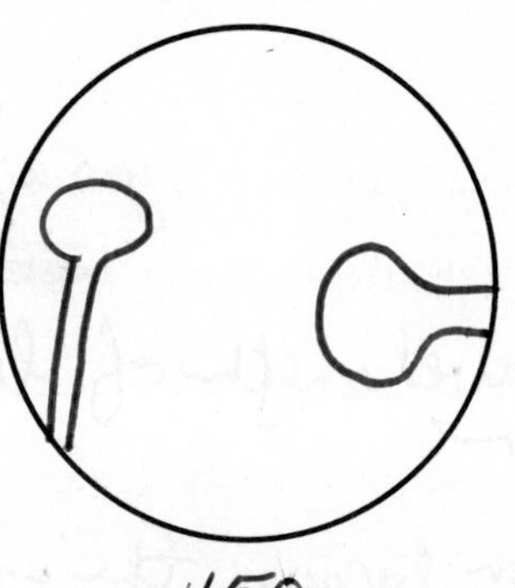

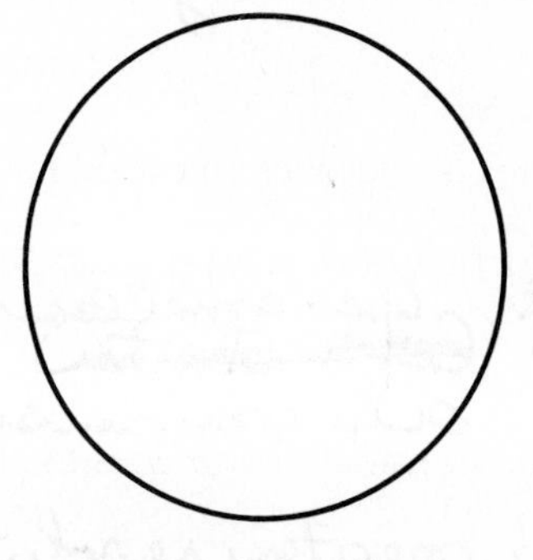

Magnification 100× 450× N/A

EXERCISE 1

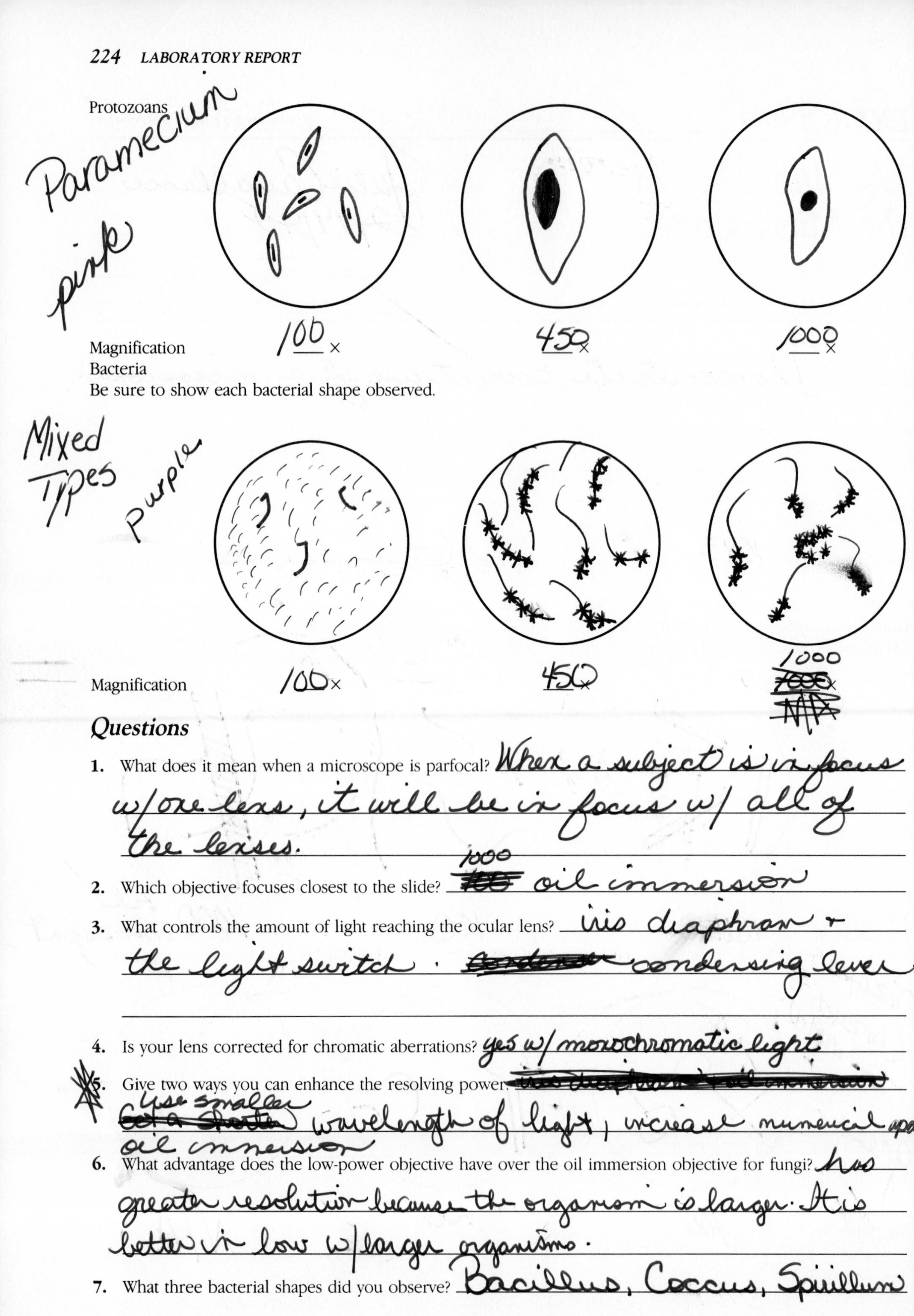

Protozoans

Magnification ___× ___× ___×

Bacteria

Be sure to show each bacterial shape observed.

Magnification ___× ___× ___×

Questions

1. What does it mean when a microscope is parfocal? When a subject is in focus w/one lens, it will be in focus w/ all of the lenses.

2. Which objective focuses closest to the slide? ~~100~~ 1000 oil immersion

3. What controls the amount of light reaching the ocular lens? iris diaphram + the light switch. ~~condenser~~ condensing lever

4. Is your lens corrected for chromatic aberrations? yes w/ monochromatic light

5. Give two ways you can enhance the resolving power. ~~iris diaphram + oil immersion~~ ~~Get a shorter~~ Use smaller wavelength of light, increase numerical aperture; oil immersion

6. What advantage does the low-power objective have over the oil immersion objective for fungi? has greater resolution because the organism is larger. It is better in low w/ larger organisms.

7. What three bacterial shapes did you observe? Bacillus, Coccus, Spirillum

EXERCISE 2 **LABORATORY REPORT**

Examination of Living Microorganisms

Name ______________________

Date ______________________

Lab. Section ______________________

Purpose ______________________

____

____

Data

Wet Mount Technique

Draw the types of protozoans, algae, and bacteria observed. Indicate their relative sizes and shapes. Record the magnification.

Hay infusion
Light

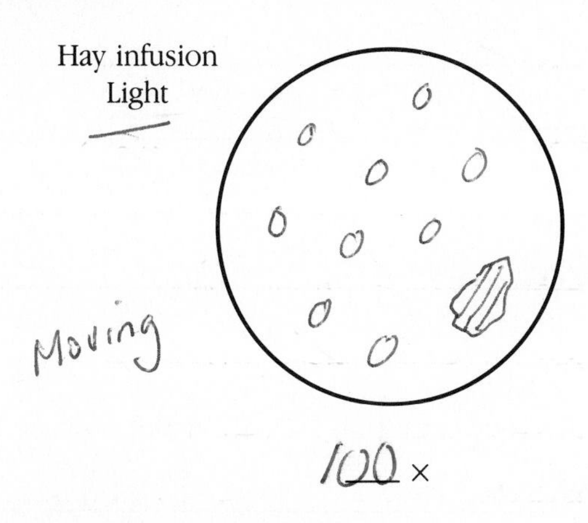

100 ×

Hay infusion
Dark

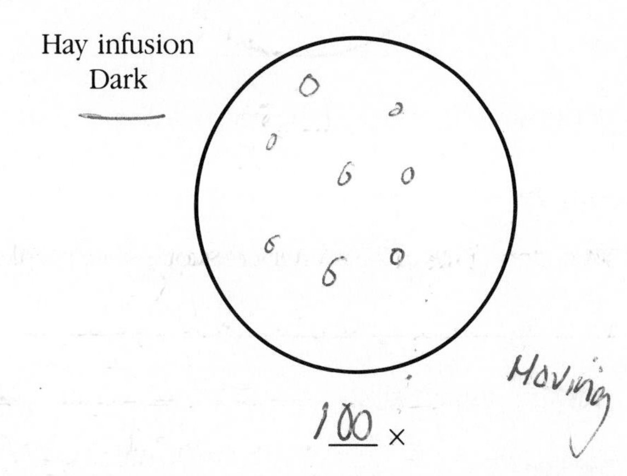

100 ×

Hanging-Drop Procedure

Draw the types of microorganisms seen under high-dry magnification. Indicate their relative sizes and shapes.

Peppercorn
Infusion

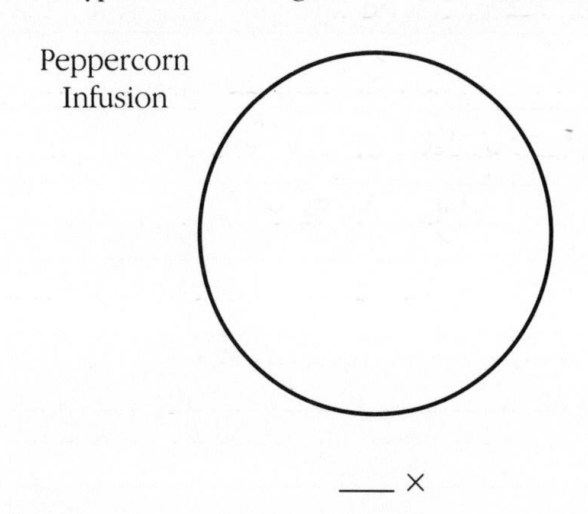

___ ×

Bacillus

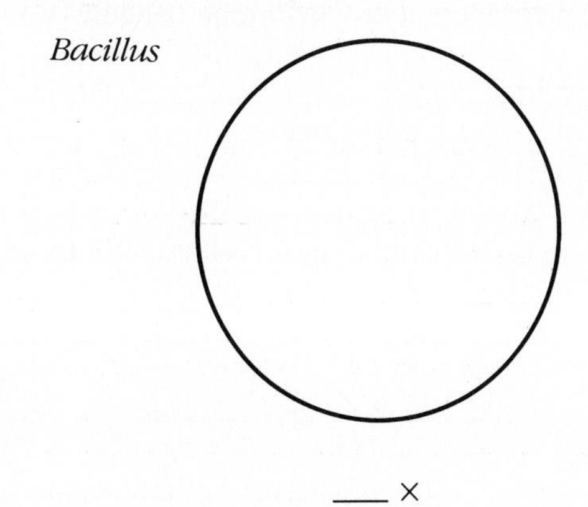

___ ×

Bacteria

On the next page record the relative numbers of each bacterial shape (Figure 1-6) observed. Record your data as 4+ (most abundant), 3+, 2+, +, − (not seen).

Culture	Shape		
	Bacilli	Cocci	Spirilla
Hay infusion, dark			
Hay infusion, light			
Peppercorn infusion			
Bacillus culture			

Questions

1. What, if any, practical value do these techniques have? ______________________

____

2. Did any of the bacteria exhibit true motility? Make sure that you can distinguish true motility from Brownian movement or motion of the fluid.

____

3. What is the advantage of the hanging-drop over the wet mount? ______________________

____

4. Why are microorganisms hard to see in wet preparations? ______________________

____

5. Can you distinguish the procaryotic organisms from the eucaryotic organisms (see the figure on page 87)?

Explain. ______________________

____

6. Why is petroleum jelly used in the hanging-drop? ______________________

____

____

7. Which infusion had the most bacteria? Briefly explain why. ______________________

____

____

8. From where did the organisms in the infusions come? ______________________

____

____

Specialized Microscopy

Name ______________________

Date ______________________

Lab. Section ______________________

Purpose ______________________

____

____

Observations

Carefully draw the organisms and their internal structure.

Pond water

40 ×

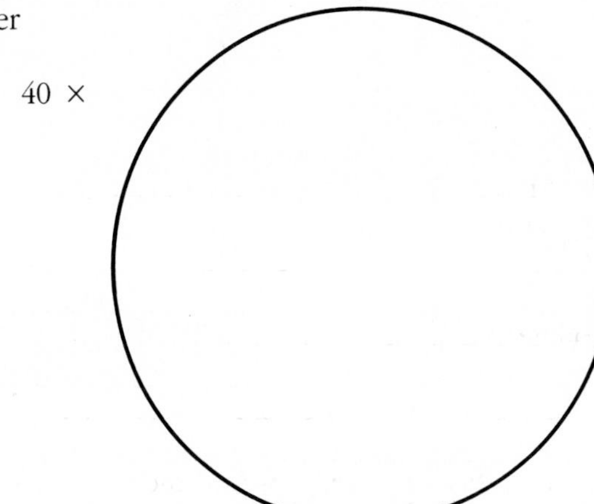

100 ×

Are any bacteria present? ______________________

Which organism did you study (*Euglena* or *Paramecium*)? ______________________

40 ×

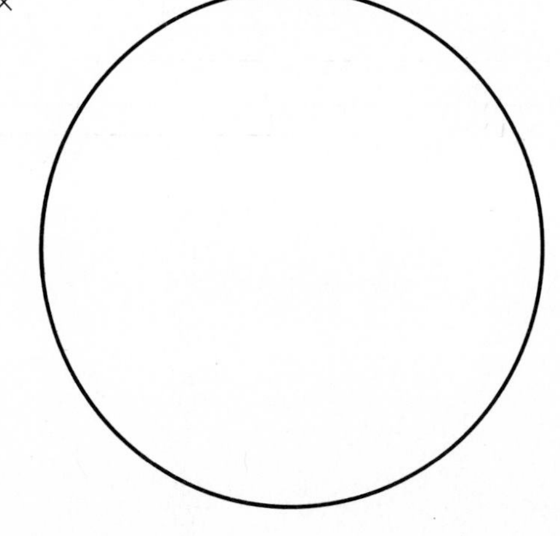

100 ×

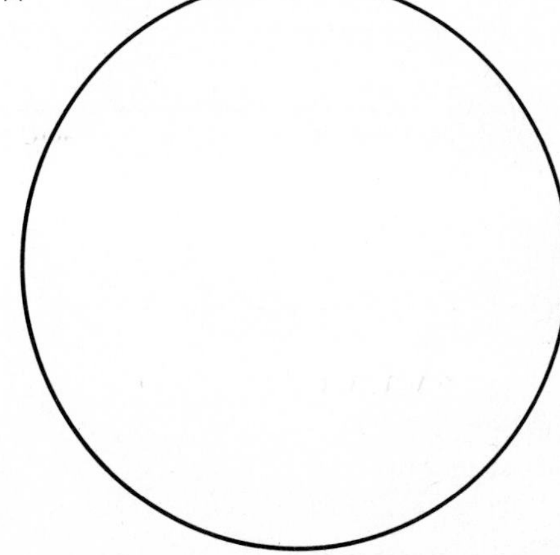

Are either of these motile? ______________________________

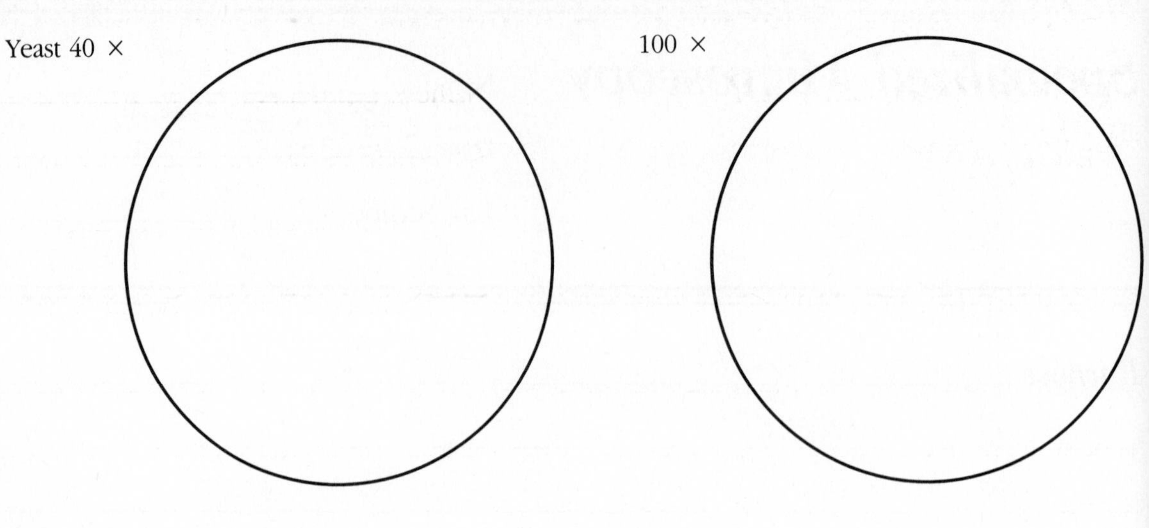

Questions

1. Compare the appearance of microorganisms in this exercise to those seen in Exercise 2. ______________________________

____

____

2. List one advantage of each of the following; that is, what would you use each technique for?

a. Phase-contrast microscopy ______________________________

b. Darkfield microscopy ______________________________

c. Fluorescent microscopy ______________________________

d. Electron microscopy ______________________________

3. Why does an electron microscope have a higher resolving power than a compound microscope? ______________________________

____

____

4. What are the "lenses" in an electron microscope? ______________________________

____

____

EXERCISE 4 — LABORATORY REPORT

Measurement of Microbes

Name Julie C.

Date 2/25/92 Tues.

Lab. Section ______

Purpose Estimate the size of organisms through a microscope. Measure cells w/ ocular micrometer.

Data

Microscope Number: 108 3

Estimating Size

Diagram your observations of the graph paper slide:

Low power ___ ×

High-dry ___ ×

Oil immersion ___ ×

Objective lens	Field of vision	
	mm	μm
Low power	10x10 = 100x	5
High-dry power	10x450 = 450X	1.25
Oil immersion	10x100 = 1000x	.5

Estimated size of:

Bacillus ______ μm at ___ × Yeast ______ μm at ___ ×

Accurate Measurements

Objective lens	Distance between ocular divisions	
	mm	μm
Low power		
High-dry power		
Oil immersion		

EXERCISE 4

Fill in the following table:

	Size (μm)
~~Length of *Streptococcus* chain~~ paramecium 45×5 =	225 μm
Diameter of *Streptococcus* cell	
Length of *Bacillus*	
Length of *Treponema*	
Diameter of yeast cell	
Dimensions of *Entamoeba histolytica*	
Dimensions of pinworm egg	
Diameter of human red blood cell	

Questions

1. What happens to the field of vision with increased magnification? the field of vision is a smaller field.

2. Determine the size of the following chain if each ocular division is 1.0 μm. .20

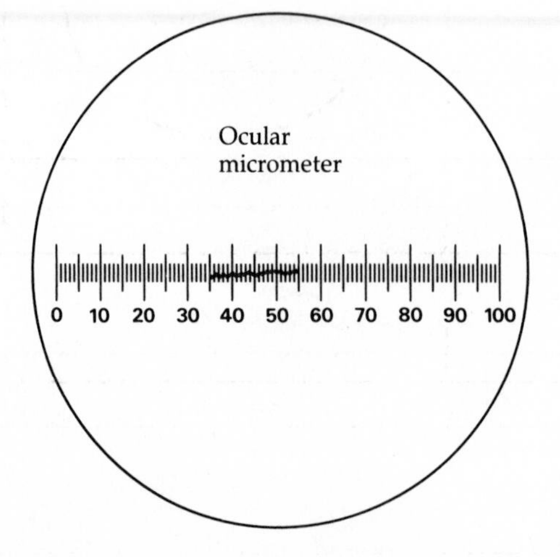

3. How are cells accurately measured through a microscope? _____

4. Why does the apparent size of the spaces in the ocular micrometer change with changing magnification? _____

5. Why is it necessary to calibrate the ocular micrometer for each set of lenses? because it changes.

EXERCISE 5 LABORATORY REPORT

Preparation of Smears and Simple Staining

Name ______________________

Date 2/18/92

Lab. Section ______________________

Purpose Make & heat fix a smear. Advantages of staining microorg. Basic mechanisms of staining. Perform a simple direct stain.

Data

Appearance	*Staphylococcus* ~~*epidermidis*~~ aureus	*Bacillus* ~~*megaterium*~~ subtilis
Sketch a few bacteria under 100× objective		
Morphology (shape)	circular, round	rod-like
Arrangement of cells relative to one another	grape-like structures	single or pairs

Questions

1. Of what value is a simple stain? When you only use one stain

2. Can dyes other than methylene blue be used for simple staining? Yes if the dye is on the pos. ion which is similar to methylene blue which is a basic stain

3. What would happen if no heat fixing were done? The bact. would not adhere to the slide which would prevent further use of the slide.

EXERCISE 5

Or too much heat applied Overheating will distort the shape of the cell.

EXERCISE 6 LABORATORY REPORT

Negative Staining

Name ______________________

Date ______________________

Lab. Section ______________________

Purpose ______________________

Data

Appearance	*Escherichia coli*	*Staphylococcus aureus*
Sketch		
Morphology and arrangement		

Questions

1. Which cell is a rod? ______________ How does its appearance differ from the rod used in Exercise 5? ______________

2. What microscopic technique gives a field similar in appearance to that seen in the negative stain? ______________

3. Why is the size more accurate in a negative stain than in a simple stain? ______________

EXERCISE 6

4. Could any dye be used in place of nigrosine for negative staining? ________________

What types of dyes are used for negative staining? ________________

EXERCISE 7 LABORATORY REPORT

Gram Staining

Name ______

Date 3/5/92

Lab. Section ______

Purpose Perform & interpret gram stains

Data

Appearance	*Staphylococcus epidermidis*	*Bacillus subtilis*	*Escherichia coli*
Sketch a few bacteria (100×)			
Morphology, arrangement, relative size	Clusters cocci 5mm 1.oc. unit	pairs} single} of rod shape	coccobacillus single
Color	purple	purple	pink
Gram reaction	(+)	(+)	(-)

Tooth and gum scraping sketch:

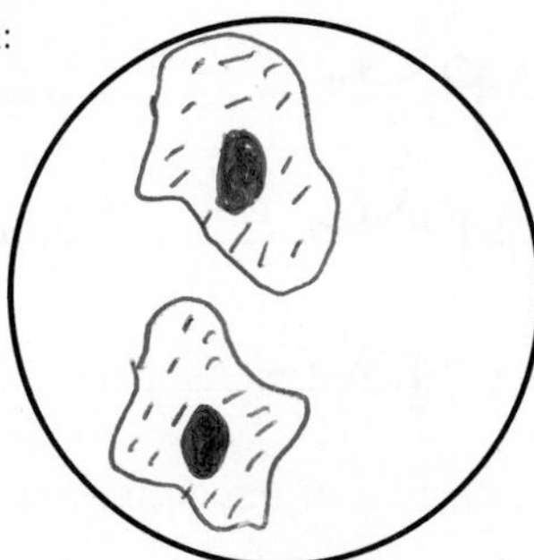

1100u

purple gram pos.

List shapes and Gram reactions of bacteria observed ______

Were skin cells observed? yes

EXERCISE 7

Questions

1. Can iodine be added before the primary stain in a Gram stain? No, it only intestifies ionic bond. no primary color

Gram-

2. If you Gram stained human cells, what would happen? They will stain pink because human cells don't have that cell wall to hold all of the purple.

3. Why is the age of the bacterial culture important in a Gram stain? Its only effective w/in the 1st 24 hours. Afterwords they will turn into Gram –.

4. Did your results agree with the information in your textbook? ______ If not, why not? Decolorization over or under; are the biggest mistakes

5. What other properties can be associated with the chemical differences of the cell wall? For example, can dye sensitivity be so associated?

Gram They can block antibiotics; crystal violet is sensitive to Gram (+), enzymes & lysozomes effect (+) much more.

6. List the steps of the Gram staining procedure in order (omit washings) and fill in the color of gram-positive cells and gram-negative cells after each step.

Step	Chemical	Appearance	
		Gram-positive cells	Gram-negative cells
1	crystal violet	purple	purple
2	Iodine	purple	purple
3	95% ETOH	purple	colorless
4	Sofranin	purple	pink

Which step can be omitted without affecting determination of the Gram reaction? the last step.

EXERCISE 8 LABORATORY REPORT

Acid-Fast Staining

Name ______________________

Date ______________________

Lab. Section ______________________

Purpose

____

Data

Appearance	Acid Fast red *Mycobacterium phlei*	Non-Acid Fast blue *Escherichia coli*	Demonstration slides #1	Demonstration slides #2
Sketch				
Morphology and color				
Acid-fast reaction				

EXERCISE 8

Questions

1. What are the large stained areas on the sputum slide? ______________________

OMIT

2. How do the acid-fast properties relate to the Gram stain? ______________________

Assuming you could stain any cell, would an acid-fast organism be gram-positive or gram-negative? Explain.

3. What diseases are diagnosed using the acid-fast procedure? Turberculosis, Leoparasy.

4. What is phenol (carbolic acid), and what is its *usual* application? aseptic surgery, disinfectant

EXERCISE 9 — LABORATORY REPORT

Structural Stains (Endospore, Capsule, and Flagella)

Name Julie Cavallucci

Date 2/27/92

Lab. Section ______

Purpose TO recognize the different types of flagellar arrangments & to prepare & interpret endospore, capsule & flagella stains.

Data

Endospores

Sketch your results and label each diagram as to color. Label the vegetative cells and endospores.

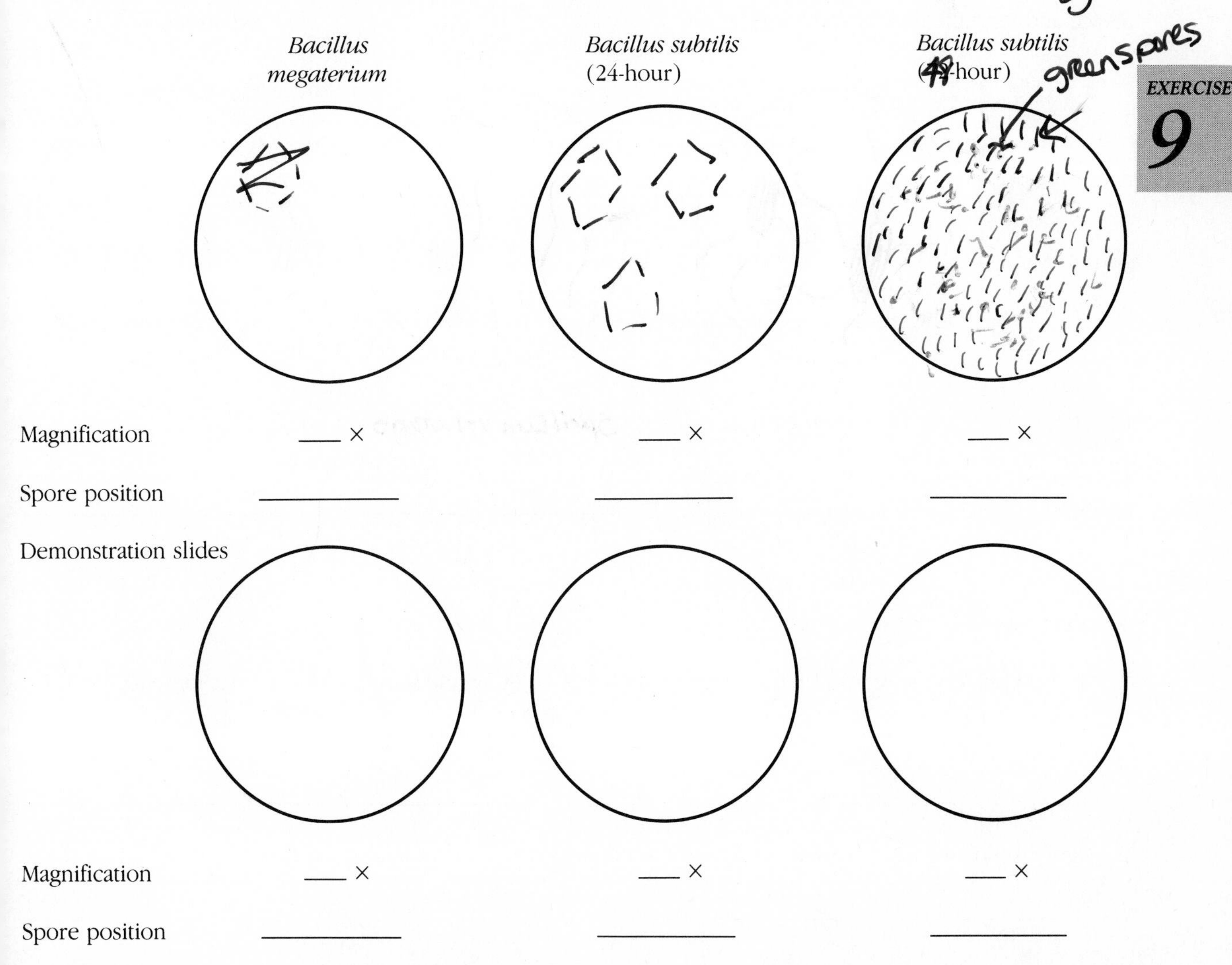

	Bacillus megaterium	*Bacillus subtilis* (24-hour)	*Bacillus subtilis* (48-hour)
Magnification	___ ×	___ ×	___ ×
Spore position	______	______	______

Demonstration slides

Magnification	___ ×	___ ×	___ ×
Spore position	______	______	______
Bacteria	______	______	______

Capsules

Sketch and label the capsules and bacterial cells.

Demonstration slide

Neg. Stain
Stain Background

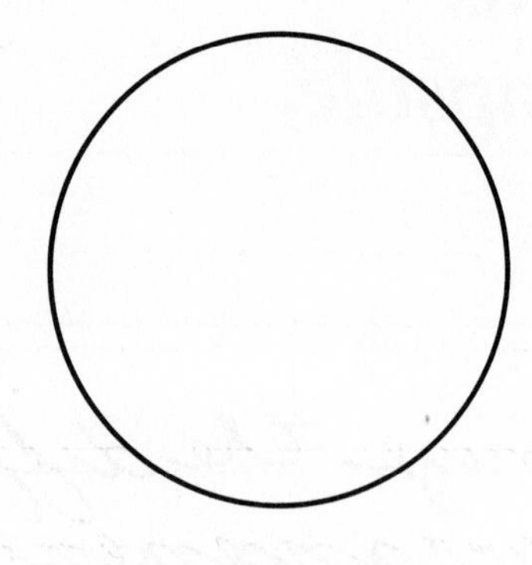

Bacteria ____________ ____________ ____________

Cell morphology ____________ ____________ ____________

Staining method ____________ ____________ ____________

Flagella

Sketch and label the flagella and bacteria.

Demonstration slides

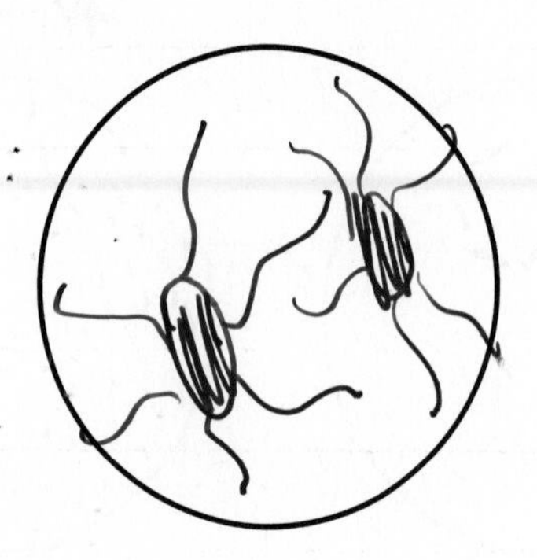

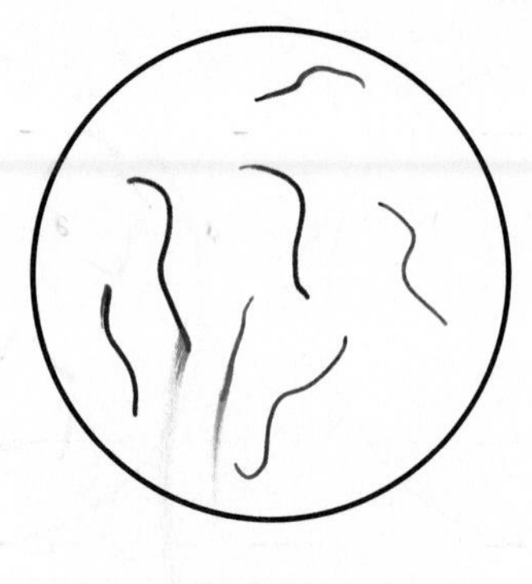

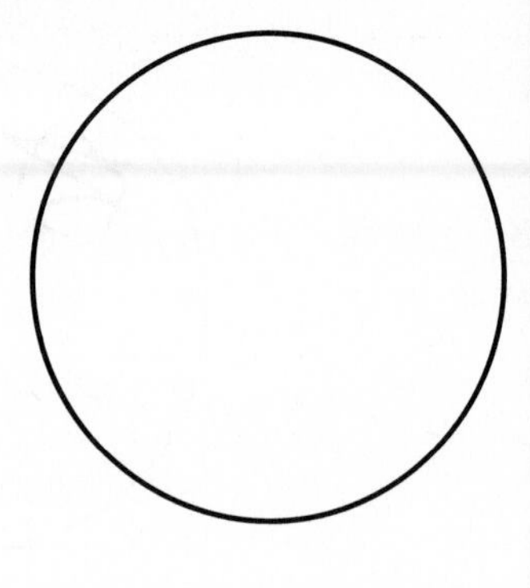

Bacteria ____ *Proteus vulgaris* ____ Spirillum volutans ____________

Flagella position ____________ ____________ ____________

Questions

1. You can see endospores by phase-contrast microscopy or by simple staining. Why not use these techniques?

____

____

2. How would an endospore stain of *Mycobacterium* appear? ____________

____

____

3. What are the Gram reactions of *Clostridium* and *Bacillus*? ______

______

4. What type of culture medium would increase the size of a bacterial capsule? ______

______

______

5. How might a capsule contribute to pathogenicity? ______

Flagella? ______

6. Of what morphology are most bacteria possessing flagella? ______

Which morphology usually does not have flagella? ______

7. What prevents the cell from appearing green in the finished endospore stain? ______

______

______

8. Of what advantage to *Clostridium* is an endospore? ______

______

______

9. Write a definition for each of the following flagellar arrangements:

a. Monotrichous ______

b. Lophotrichous ______

c. Amphitrichous ______

d. Peritrichous ______

Morphologic Unknown

Name ______________________

Date ______________________

Lab. Section ______________________

Results

Write *not done* by any category that does not apply.

Unknown # ______________________

Sketch of the Gram stain

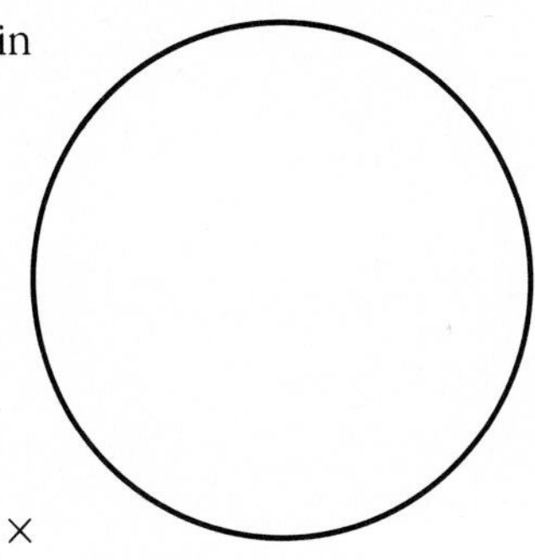

___ ×

Gram reaction ______________________

Morphology ______________________

Predominant arrangement ______________________

Endospores present? ______________________

Sketch if present

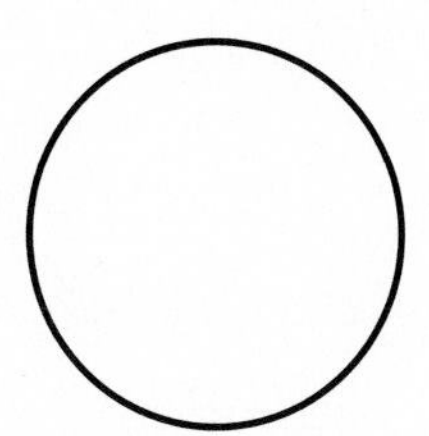

Capsules present? ______________________

Sketch if present

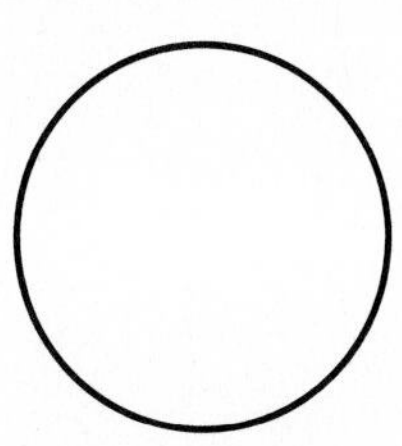

Motile? ______________________

Motility determined by ______________________

Acid-fast? ______________________

Culture Media Preparation

Name ______

Date ______

Lab. Section ______

Purpose

______

Data

Compare the appearance of the tube of nutrient broth with the flask or plates of nutrient agar after they have cooled to room temperature. ______

Questions

1. Which medium is chemically defined and which is complex? How can you tell? ______

2. What is (are) the carbon and nitrogen source(s) in nutrient agar? In the glucose-minimal salts broth? ______

3. What is the purpose of adding agar to culture media? ______

4. What physical and chemical properties of agar make it useful in culture media? ______

5. Why were the media autoclaved? ______________________________

____

____

6. Why are aseptic techniques employed in plate pouring? ______________________________

____

____

7. Why is the mouth of the flask heated before pouring the nutrient agar? ______________________________

____

____

EXERCISE 12 — **LABORATORY REPORT**

Microbes in the Environment

Name Julie Cavallucci

Date 2/6/92

Lab. Section

Purpose To explain the importance of aseptic technique, and describe colony morphology using accepted descriptive terms.

Data

Fill in the following table with descriptions of the bacterial colonies. Use a separate line for each different appearing colony.

	Colony Description					
	Diameter	Appearance	Margin	Elevation	Color	Number of this type
Area sampled	.4mm	circular	entire	convex	white	9
inner ear	.2mm	circular	entire	convex	white	24
Incubated at						
37 °C for						
48 hrs.						
Area sampled	.1mm	circular	entire	convex	white	1
inside mouth						
Incubated at						
37 °C for						
48 hrs.						
Area sampled						
window shade	NO Growth					
Incubated at						
37 °C for						
48 hrs.						

EXERCISE 12

	Colony Description					
	Diameter	Appearance	Margin	Elevation	Color	Number of This Type
Area sampled window ledge	3mm	circular	entire	convex	white	3
	4mm	circular	entire	convex	yellow	2
Incubated at 37 °C for 48 hrs.	1mm	circular	entire	convex	white	4

Description of the nutrient broth:

Area sampled window ledge. Incubated at 37 °C for 48 hrs.

Is it turbid? yes

Is a flocculant present? no

Is a sediment present? no

Is a pellicle present? no

Has the color changed? yes, its cloudy

Questions

1. How can you tell whether or not there is bacterial growth in nutrient broth? change in color, cloudiness, sediment at bottom.

2. What is the minimum number of different bacteria present on one of your plates? 1

   How do you know? by the color

3. Did all the organisms living in or on the environments sampled grow on your nutrient agar? No

4. Of what advantage is a solid medium over a liquid medium? easier to see diameter, appearance, margin, elevation, color & # of this type.

5. Why is aseptic technique important? prevents contamination from unwanted organism.

6. What is the value of Petri plates in microbiology? it allows bacteria & fungi to grow, see different colors, allows air in.

EXERCISE 13 LABORATORY REPORT

Transfer of Bacteria: Aseptic Technique

Name ______

Date 2/11/92

Lab. Section ______

Purpose To provide the rationale for aseptic technique, differentiate among broth, slant, + deep and aseptically transfer bacteria from one form of culture medium to another.

Data

Nutrient Broth

Describe the nutrient broth cultures.

Bacterium	Is It Turbid?	Is Surface Membrane or Pellicle Present?	Pigment?
Streptococcus lactis			
Pseudomonas aeruginosa	yes	yes	no
Proteus vulgaris	yes	no	no

EXERCISE 13

Nutrient Agar Slant

Sketch the appearance of each culture. Note any pigmentation.

Streptococcus lactis

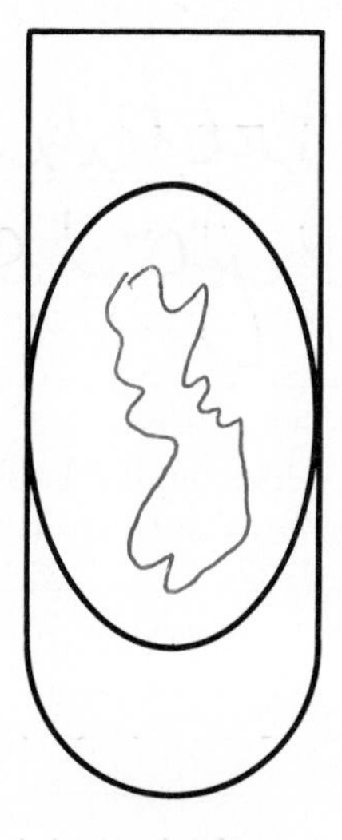

Pseudomonas aeruginosa

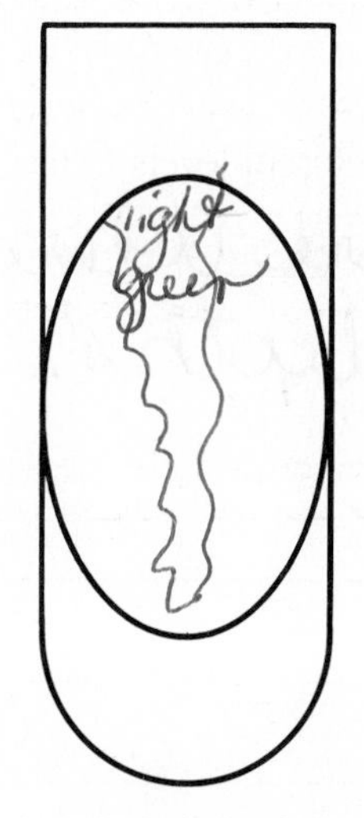

Proteus vulgaris

Type of growth:

N/A ______ ______ ______

Nutrient Agar Deep

Show the location of bacterial growth and note any pigment formation.

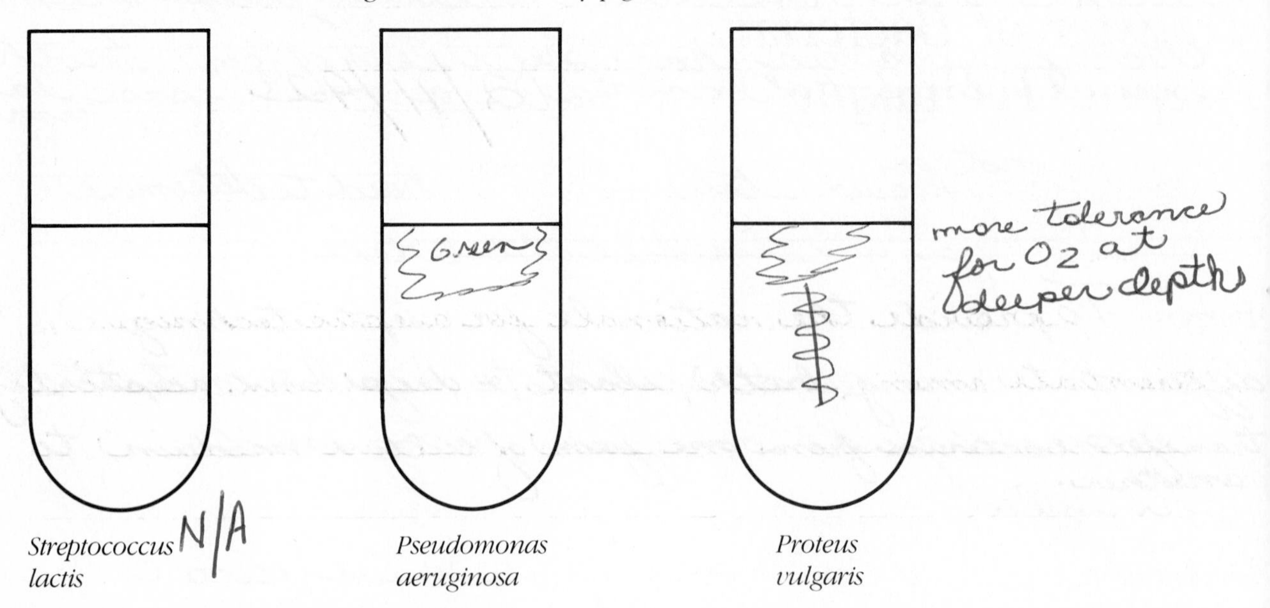

Comparison of Broth and Slant Cultures

	Streptococcus lactis	
	Original Broth Culture	Slant Culture
Gram Stain		
Morphology		
Arrangement		

Questions

1. Did growth occur at different levels in the agar deep? Proteins vulgaris had more tolerance for oxygen at deeper depths.

2. Were any of the bacteria growing in the semisolid agar deeps motile? Explain. yes

3. Why was the arrangement of *Streptococcus* from the broth culture provided in the first period different from the slant culture in the second period? N/A

4. What is the primary use of slants? Of deeps? Of broths? slants provide a solid surface which are easier to store. Deeps are motility checks for bacteria that prefer less oxygen. Broth provides large # of bacteria in a small space easily transportable

5. Can you determine whether a broth culture is pure (all one species) by visually inspecting it without a microscope? not in liquid state An agar deep culture? hard to determine An agar slant culture? yes, you should be able to tell.

6. When is a loop preferable for transferring bacteria? Use an illustration from your results to answer.

When is a needle preferable? needle-solid to liquid & motility loop - liquid to liquid, any media to solid.

7. What is the purpose of flaming the loop before use? After use? So we don't contaminate it w/ unwanted microorganisms.

8. Why must the loop be cool before touching it to a culture? Should you set it down to let it cool? How do you determine when it is cool? It should cool so you don't kill the bacteria you want to study. No unwanted microorg. will contaminate it. Until the redness goes away.

9. Define aseptic technique. The transferring technique when unwanted microorg. are not wanted.

10. How can you tell that the media provided for this exercise were sterile? They should be sterile before use & can be checked by the eye for unwanted bacterial growth.

EXERCISE 14 — LABORATORY REPORT

Isolation of Bacteria by Dilution Techniques

Name

Date 2/18/92

Lab. Section

Purpose Isolate bacteria by the streak plate + pour plate techniques. Prepare + maintain a pure culture.

Results

Streak Plate

Sketch the appearance of the streak plates.

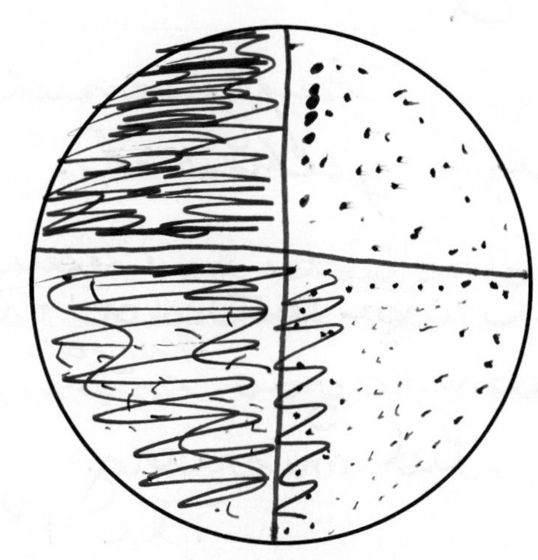

Mixed culture

orange

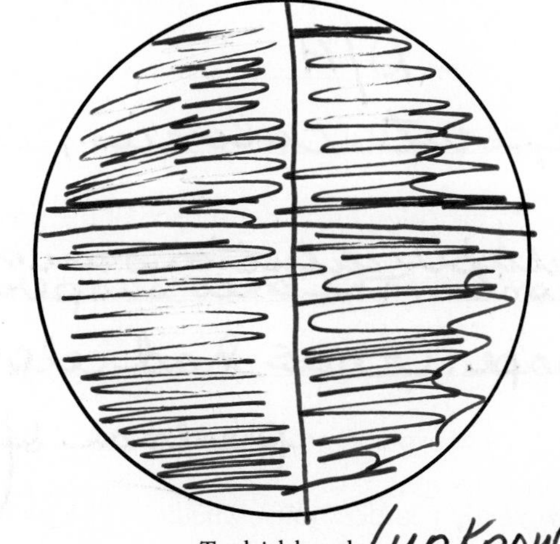

Turbid broth (UNKNOWN)

yellowish (light)

Fill in the following table using colonies from the most isolated streak areas.

Culture	Colony Description (Each different appearing colony should be described.)				
	Diameter	Appearance	Margin	Elevation	Color
Mixed culture	(1) .1 (2) .1 (3) .4 (4) .2	(1) circular (2) circular (3) circular (4) circular	entire entire entire entire	convex convex convex convex	orange orange orange orange
Turbid broth	(1) N/A (2) N/A (3) N/A (4) N/A	N/A N/A N/A N/A N/A	N/A N/A N/A N/A	flat flat flat flat	yellowish yellowish yellowish yellowish

EXERCISE 14

Pour Plate

Sketch the appearance of the pour plate having the fewest colonies. Which plate? Mixed culture 1

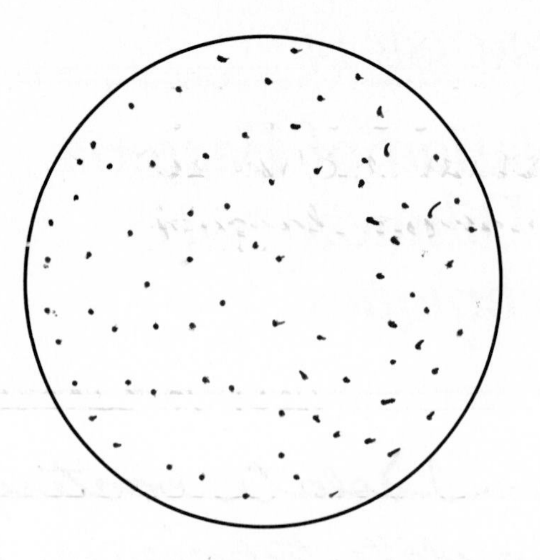

Subculture

Describe the growth on your slant. all the same bacteria on the entire plate. (orange) circular/convex

Questions

1. How many different bacteria were in the mixed culture? 2 How many in the turbid broth? N/A How can you tell? They formed seperate colonies, orange & yellowish.

2. How do the colonies on the surface of the pour plate differ from the colonies suspended in the agar? Multiple dilutions are made, some plates have seperated colonies. The ones suspended in agar lost color + abnormal shape. Others had color + room to grow. Aerobic + Anaerobic

3. What is a contaminant? presence of unwanted microorg.

4. How would you determine whether a colony was a contaminant on a streak plate? The 4th section should be isolated to 1 or 2 or no colonies contaminants look diff than orange

   On a pour plate? Much more difficult: Size + shape may be similar.

5. What would happen if the plates were incubated a week longer? Colonies would reach a max; A month? Same (they would not expand any further) drying out of bacteria

6. Could some bacteria grow on the streak plate and not be seen using the pour plate technique? yes

   Explain Bacteria that needs O_2 may not be present.

7. What is a disadvantage of the streak plate? Of the pour plate technique? Streak plate takes exactness. Pour plates => some bacteria won't show up + the difference in color from ones that have access to O_2.

EXERCISE 15 LABORATORY REPORT

Special Media for Isolating Bacteria

Name ______

Date ______

Lab. Section ______

Purpose

______

Data

Organism	Amount of Growth		Gram Stain Results
	24 hours	48-72 hours	

Questions

1. What do your results indicate? What difference did you see between 24 and 48 to 72 hours? ______

2. Is this a selective or differential medium? ______

3. How does this medium correlate with the Gram reaction? Explain. ______

4. Design another medium that would select for the same group of organisms. Explain how your medium is selective. ______________________________

____

____

5. Design an enrichment medium to isolate a DDT-detergent degrading bacterium that is found in the ocean and one that is found in soil. ______________________________

____

____

____

____

____

____

EXERCISE 16 LABORATORY REPORT

Bacterial Nutrition

Simmon's citrate agar
citric acid
Kreb's cycle
transport

Name

Date 3/10/92 Tues.

Lab. Section

e) + growth utilization of citrate - no growth (green) no utilization of citrate

Purpose indicators = chem. that change colors w/ different PH's
Brom thymol blue green = neutral PH
blue = basic PH

Data

Record the relative amount of growth in each medium and citrate test results.

Medium	Growth	
	Escherichia coli	*Pseudomonas aeruginosa*
Glucose		
Glucose, NaCl		
Glucose, NaCl, PO_4		
Glucose, NaCl, PO_4, $MgSO_4$		
Glucose, NaCl, PO_4, $MgSO_4$, Peptone		
Citrate agar: Color		
Positive or negative		

EXERCISE 16

Questions

1. Are the growth requirements of the two bacteria different?

In what way(s)? P. aurgernosa = blue use

How are they similar? Not

↑ see a color change ↑ only certain bact. grow ~~sele~~ selective medium

2. Is citrate medium a differential or a selective medium? differential How can you tell?

turns blue

3. Which of the media used in this exercise are chemically defined? Simmon's Citrate Ag

4. What is meant by "positive citrate test"? use citiate

EXERCISE 17 LABORATORY REPORT

Carbohydrate Catabolism

Name

Date 3/5/92

Lab. Section

Purpose Carbohydrate catabolism

①starch hydrolysis
exoenzyme = secreted by bact into starch (amylase or diastase)
starch → maltose (glucose-glucose)

Pour a thin layer of dilute Gram's iodine on plate.
starch + iodine = deep brown or blue
Maltose + iodine ⇒ clear

B. subtilis, P. aeruginosa, E. coli, unknown

+ is clear area around growth (starch hydrolysis)
unknown - dark (starch still present)

Data

Record your results in the following tables.

Starch Hydrolysis

Organism	Growth	Color of Medium Around Colonies After Addition of Iodine	Starch Hydrolysis: (+) = yes (−) = no
Bacillus subtilis		clear	
Pseudomonas aeruginosa		dark	—
Escherichia coli		dark	—

oxidizer
fermenter
use glucose in presence or absense of O2 - oxidizer / ferm.

OF-Glucose 4 tubes
stab into 4 tubes

[illegible]s glucose only when [illegible] is present (aero, resp only)

KNOWN (mineral oil) unknown

yellow yellow = fermenter
yellow green = oxidizer
green green = neither

Organism	Growth	Color		Gas Produced	Fermenter (F), Oxidizer (O), Neither (−)	Motile
		O2 Open Tube	Mineral oil Plugged Tube			
~~*Pseudomonas aeruginosa*~~ Acinetobacter calcoaceticus	yes	green	green	no	—	NO
Alcaligenes faecalis						
Escherichia coli						

glucose used = yellow (acid is produced)
" not used = green

blue = protein metabolism
gas = bubbles
Motility

Questions

1. Which organism(s) gave a positive test for starch hydrolysis? B. subtilis

How can you tell?

2. What would be found in the clear area that would not be found in the blue area of a starch agar plate after the addition of iodine? exoenzyme, maltose

3. How can you tell amylase is an exoenzyme and not an endoenzyme? added enzyme clear ring around starch. Amylase outside

4. How can you tell from OF-glucose medium whether an organism oxidizes glucose? 1 tube yellow = open tube 1 tube = green (air tube) no glucose

Ferments glucose? yellow - both

Doesn't use glucose? both - green

5. If an organism grows in the OF-glucose medium that is exposed to air, is the organism oxidizing or fermenting glucose? Explain. need to know color change in medium, what going on in clogged tube.

6. How can organisms that don't utilize starch grow on a starch agar plate? have to be other nutrients

7. If iodine were not available, how would you determine if starch hydrolysis had occurred? test for maltose

8. Locate the H^+ and OH^- ions from the water molecule that was split (hydrolyzed) in this reaction, showing esculin hydrolysis. omit

HO O CH_2OH H O O H OH H HO OH H OH + H_2O ⟶ CH_2OH H O H H OH H HO OH H OH + HO O O HO

Esculin

Glucose + Esculetin

37°C incubator

Fermentation of Carbohydrates

Name ______________________

Date ______________________

Lab. Section ______________________

Purpose

____

Data

Fermentation Tubes

Organism		Glucose				Lactose				Sucrose			
		Growth	Color	Acid	Gas	Growth	Color	Acid	Gas	Growth	Color	Acid	Gas
Escherichia coli	24 hrs.												
	48 hrs.												
Enterobacter aerogenes	24 hrs.	yes	yellow	yes	yes	yes	yellow	yes	yes	yes	yellow	yes	yes
	48 hrs.												
Alcaligenes faecalis	24 hrs.												
	48 hrs.												
Proteus vulgaris	24 hrs.												
	48 hrs.												

(Column groups: Carbohydrate — Glucose, Lactose, Sucrose)

MRVP Tests

Organism	Growth	MR		V–P	
		Color	+ or −	Color	+ or −
Escherichia coli	yes	red	+	yellow	−
Enterobacter aerogenes	yes	yellow	−	orange	+

Questions

1. Why are fermentation tubes evaluated at 24 and 48 hours? ______

   ______

   What would happen if an organism used up all the carbohydrate in a fermentation tube? ______

   ______ What would it use for energy? ______ What color would the indicator be then? ______

2. Could an organism be MR and V–P positive? Explain. ______

   ______

   ______

3. If an organism oxidatively metabolizes glucose, what result will occur in the fermentation tubes? ______

   ______

   ______

4. Were these media differential or selective? ______

5. How could you determine whether a bacterium fermented the following carbohydrates: mannitol, sorbitol, adonitol, or arabinose? ______

   ______

   ______

6. If a bacterium cannot ferment glucose, why not test its ability to ferment other carbohydrates? ______

   ______

   ______

   ______

Lipid Hydrolysis

Name ______________________

Date ______________________

Lab. Section ______________________

Purpose

____

Data

Sketch the tributyrin agar after incubation for ______________ hours.

Loop

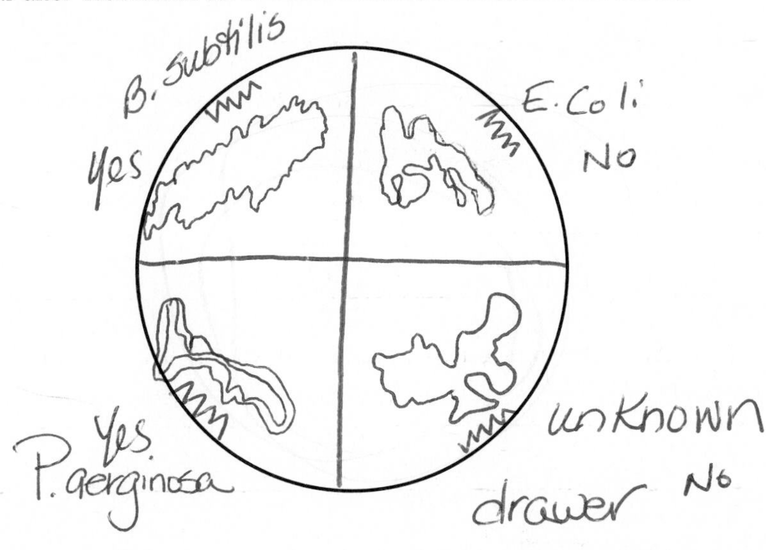

Questions

1. Which organisms were positive for lipid hydrolysis? ______________________

2. Besides clearing, how else could you determine if lipids had been hydrolyzed? ______________________

3. Write the chemical reaction for a positive tributyrin hydrolysis test.

____

Protein Catabolism, Part 1

Name ______________________

Date ______________________

Lab. Section ______________________

Purpose

____

____

Data

Complete the following table.

Test	Results	
	Pseudomonas aeruginosa	*Proteus vulgaris*
Gelatin hydrolysis at ______ days growth		
(+ or −)		
at ______ days growth		
(+ or −)		
Litmus milk Peptonization		
Acid		
Alkaline		
Coagulation		
Reduction		
No change		
Urea agar growth		
(+ or −)		

Diagram the appearance of the gelatin.

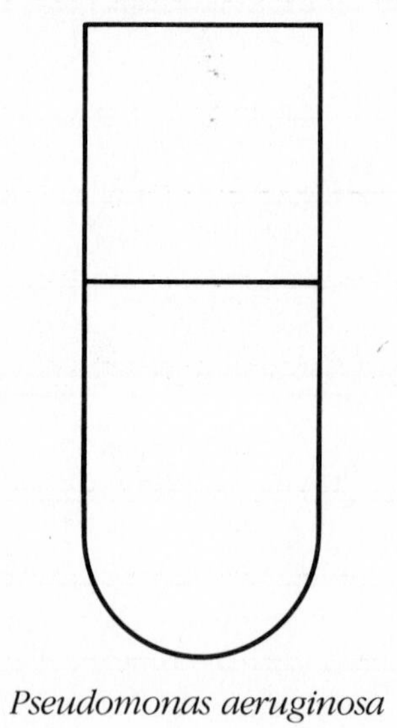

Pseudomonas aeruginosa

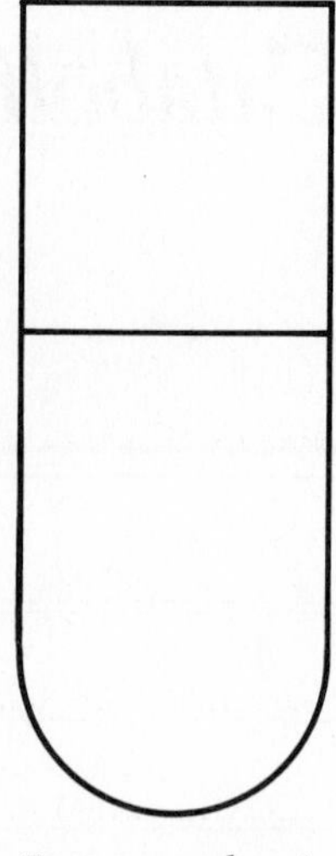

Proteus vulgaris

Questions

1. Nutrient gelatin can be incubated at 35°C. What would have to be done to determine hydrolysis after incubation at 35°C?

2. Why is agar used as a solidifying agent instead of gelatin?

3. What would you find in the liquid of hydrolyzed gelatin?

4. What is a disadvantage of litmus milk medium in diagnostic tests? not a very reliable test.

5. What happens to milk when the suspended (colloidal) proteins are hydrolyzed?

EXERCISE 21 — LABORATORY REPORT

Protein Catabolism, Part 2

Name ______________________

Date ______________________

Lab. Section ______________________

Purpose

____

Data

Fill in the following table.

Organism (unknown)	Test											
	Ornithine Decarboxylase			Phenylalanine Deaminase (4 or 5 Ferrac Chloride)			TSI Hydrogen Sulfide			Triptophane Indole (~~4 or 5~~ 10-15 Kovacs reagent)		
	Original color: purple			Original color: clear			Original color: ______			Original color: ______		
	Growth	Color	Reaction (+ or −)	Growth	Color	Reaction (+ or −)	Growth	Color	Reaction (+ or −)	Growth	Color	Reaction (+ or −)
Escherichia coli	Not tested											
Pseudomonas aeruginosa	Not tested						Not tested			Not tested		
Proteus vulgaris										Not tested		
Enterobacter aerogenes							Not tested					

Questions

1. Why look for black precipitate (FeS) in the butt instead of on the surface of an H_2S test? ______________________

2. When spoilage of canned foods occurs, what causes blackening of cans? ______________________

____

____

3. Show how lysine and arginine could be decarboxylated to give the end products indicated.

```
      H   H   H   H   H
      |   |   |   |   |
H2N - C - C - C - C - C - COOH  →                                    +
      |   |   |   |   |
      H   H   H   H   NH2
         Lysine                         Cadaverine                 ______
```

$H_2N-CH_2-CH_2-CH_2-CH_2-CH(NH_2)-COOH \rightarrow$ Cadaverine + ______

Lysine

```
HN    H   H   H   H   H
|     |   |   |   |   |
C  -  N - C - C - C - C - COOH  →                                    +
||            |   |   |   |
NH2           H   H   H   NH2
            Arginine                    Agmatine                   ______
```

Arginine

EXERCISE 22 — LABORATORY REPORT

Respiration

Name ____________________

Date ____________________

Lab. Section ____________________

Purpose

____

Data

Nitrate Test

Organism	Color After Nitrate Reagents A and B	Color After Zinc	NO_3 Reduction?
Bacillus megaterium			
Pseudomonas aeruginosa			
✓*Bacillus subtilis*	Red (+)	Red	

unknown : Red w/in 30sec.

Oxidase Test

Organism	Color After Reagent	Oxidase Reaction
Bacillus megaterium		
Pseudomonas aeruginosa		

Catalase Test

Organism	Appearance after H_2O_2	Catalase Reaction
Streptococcus lactis		
Bacillus subtilis		

EXERCISE 22

Questions

1. Define reduction. ______

______

2. Why does hydrogen peroxide bubble when it is poured on a skin cut? ______

______

______

3. Differentiate between aerobic respiration and anaerobic respiration. ______

______

______

4. Differentiate between fermentation and anaerobic respiration. ______

______

______

5. Would nitrate reduction occur more often in the presence or absence of molecular oxygen? Explain. ______

______

______

EXERCISE 23 — LABORATORY REPORT

Anabolic Activity

Name ______

Date ______

Lab. Section ______

Purpose ______

______

______

Data

Cellulose Pellicle Production

Describe the wet pellicle: ______

______

______

Wet strength: ______

Describe the dry pellicle: ______

______

______

Dry strength: ______

Motility Stabs

Sketch the growth patterns.

Proteus vulgaris

Motile? ______

Micrococcus luteus

______

Agar Slants

Describe the pigment.

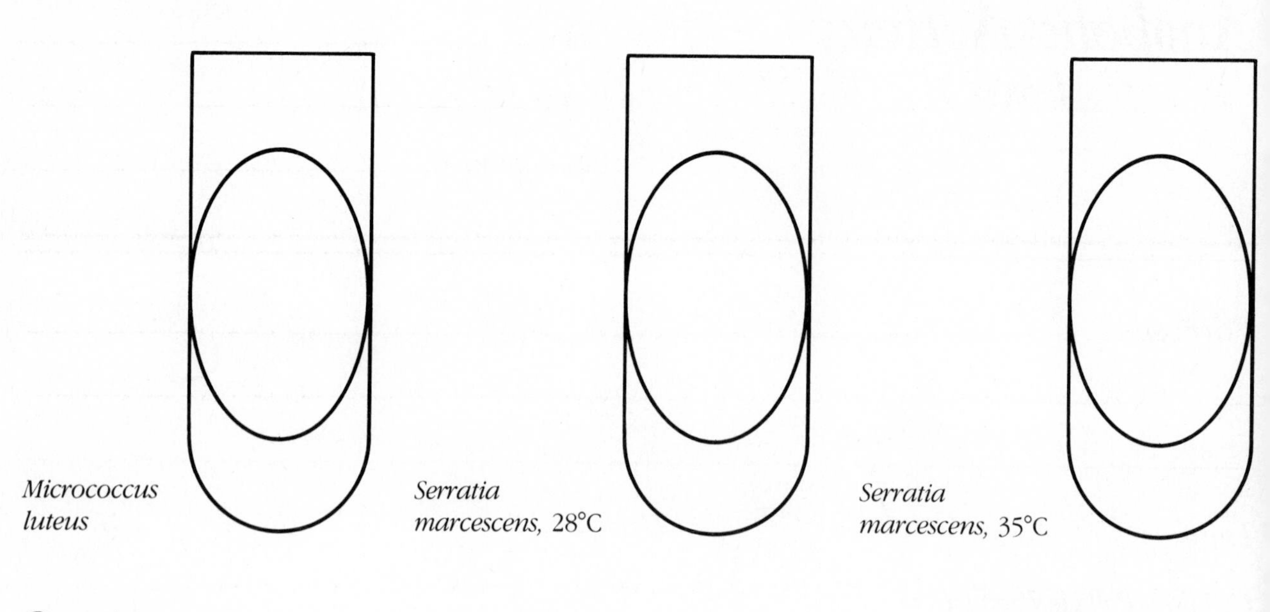

Questions

1. Could gelatin deeps be used to ascertain motility? Briefly explain. ______

2. List two other methods of determining motility. ______

3. What environmental conditions(s) influence(s) pigment production by *Serratia?* ______

4. Compare the slant culture of *Pseudomonas aeruginosa* grown in Exercise 13 with the *Serratia* slant culture grown in this exercise.

5. Is the cellulose produced by *Acetobacter* a secondary metabolite? How can you tell? ______

EXERCISE 24 — LABORATORY REPORT

Unknown Identification

Name ______________________

Date ______________________

Lab. Section ______________________

Purpose ______________________

____

Data

Write *not tested* next to tests that were not performed. Unknown # ______________

Morphological, staining, and cultural characteristics	Sketches—label and give magnification
The cell	
Shape ______	
Size ______	
Arrangement ______	
Endospores (position) ______	
Motility ______	
Determined by ______	
Staining characteristics	
Gram ______ Age ______	
Other ______ Age ______	
Colonies on trypticase soy agar	
Diameter ______	
Appearance ______	
Color ______	
Elevation ______	
Margin ______	
Consistency ______	
Agar slant Age ______	
Amount of growth ______	
Form ______	
Consistency ______	
Color ______	
Growth in broth Age ______	
Amount of growth ______	
Pellicle ______	
Flocculant ______	
Sediment ______	

Biochemical characteristics	Time (hr)						Results
	24	48	72	96	. .	. .	Temp. ____°C
Glucose							
Lactose							
Mannitol							
Catalase							
Gelatin							
Litmus milk							
Nitrate reduction							
Indole							
Methyl red							
V–P							
Citrate							
H_2S							
Oxidase							

LEGEND

A = Acid
G = Gas
a = slight acid
cd = curd
pep = peptonization
rd = reduction
alk = alkaline
+ = positive
− = negative
ng = no growth

Other characteristics, special media, etc. ______________________

____

____

EXERCISE 24

Questions

1. What organism was in unknown # ______ ? ________________________________

2. On a separate sheet of paper, neatly write your rationale for arriving at your conclusion.

3. Why is it necessary to complete the identification of a bacterium based on its physiology rather than its morphology?

____

____

____

4. Use the following key to identify a bacterium to Part. Using *Bergey's Manual* fill in the name of the group indicated by the key.

Key	Name of Part
I. Autotrophic	
A. Photoautotroph	______________
B. Chemoautotroph	
1. Methane not produced from carbon dioxide	
a. Gliding	______________
b. Does not glide	
(1) Ensheathed	______________
(2) Not sheathed	______________
2. Produces methane from carbon dioxide	______________
II. Heterotrophic	
A. Gliding	______________
B. Does not glide	
1. Ensheathed	______________
2. Not sheathed	
a. Budding or appendaged	______________
b. Not budding or appendaged	
(1) Cells helical or curved	
(a) Axial filament	______________
(b) Polar flagella	______________
(2) Cells rod or coccus	
(a) Gram-negative	
aa. Aerobic, nonfermenting	______________
bb. Facultative anaerobes	
i. Rods	______________
ii. Cocci	______________
cc. Anaerobic	
i. Rods	______________
ii. Cocci	______________
(b) Gram-positive	
aa. Endospores	______________
bb. Non–endospore-forming	
i. Rods	______________
ii. Cocci	______________
iii. Acid-fast or conidia	______________
(3) Cells pleomorphic, gram-negative	
(a) Intracellular parasites	
(b) Lacking cell wall	______________

EXERCISE 25 — LABORATORY REPORT

Rapid Identification Methods

Name ____________________

Date ____________________

Lab. Section ____________________

Purpose

____

Data

Unknown # ____________________

Appearance on nutrient agar: ____________________

Oxidase reaction: ____________________

EXERCISE 25

IMViC

Indicate positive (+) and negative (−) results for each test.

Indole ____________________

Methyl red ____________________

V–P ____________________

Citrate ____________________

Micro-ID

Indicate positive (+) and negative (−) results in the Test Results line. Then determine the five-digit code.

GENERAL DIAGNOSTICS — 941G930 *Trademark

MICRO-ID* ENCODING FORM

See Package Insert and Identification Manual for Detailed Instructions.

DATE ____________________

SPECIMEN IDENTIFICATION ____________________

TESTS	VP	N	PD	H_2S	I	OD	LD	M	U	E	ONPG	ARAB	ADON	INOS	SORB
NUMERICAL (Value of Positive Results)	4	2	1	4	2	1	4	2	1	4	2	1	4	2	1
TEST RESULTS															
SUM OF POSITIVE VALUES (in each group of three reactions)															

Enterotube®

Circle the number corresponding to each positive reaction below the appropriate compartment. Then determine the five-digit code.

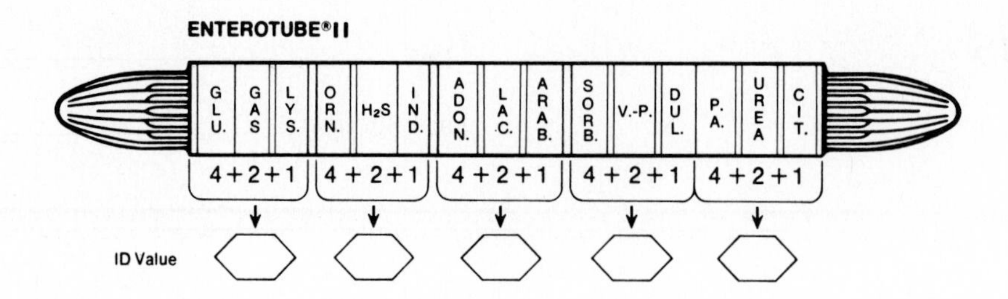

Questions

1. What species was identified in unknown # ______

by the IMViC tests? ______

by the Micro-ID®? ______

by the Enterotube®? ______

2. Did all three methods agree? ______ If not, explain any discrepancies. ______

______

3. Which method did you prefer? ______ Why? ______

______

4. Why are systems developed to identify enterics? ______

______

______

5. Use Table 25-1 to give an example of a limitation to the IMViC tests. ______

______

______

6. Why can one species have two or more numbers in the Enterotube® system? (For example, *E. coli* is number 41310 and 41340.)

______

______

7. Why should the first digit in the five-digit Enterotube® code number be equal to or greater than 4? ______

______

8. Why is an oxidase test performed on a culture before using Micro-ID® and Enterotube® to identify the culture?

______

Isolation of Bacterial Mutants

Name ______

Date ______

Lab. Section ______

Purpose

______

Data

Mark the location of colonies and note any changes in pigmentation. Organism used: ______

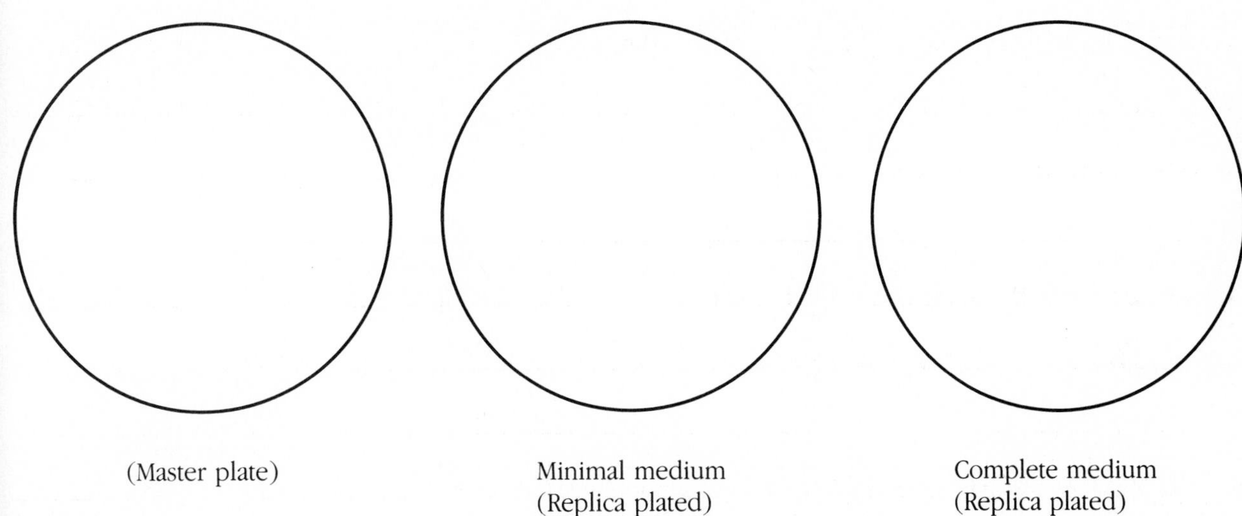

(Master plate)

Minimal medium (Replica plated)

Complete medium (Replica plated)

Questions

1. Circle the auxotrophs on the diagram of the complete medium.

2. Why was the complete medium, as well as the minimal medium, replica-plated? ______

3. How would you isolate a mutant from the replica plate? ______

Does your technique involve a catabolic or anabolic mutant? ______

4. How would you identify the growth factor(s) needed by an auxotroph? ______

______

______

______

5. How would you use the replica-plating technique to isolate a mutant that is resistant to an antibiotic? ______

______

______

______

Transformation

Name ____________________

Date ____________________

Lab. Section ____________________

Purpose

____

Data

Inoculum	Appearance on Citrate Agar		Gram Stain and Morphology
	Growth	Color	
E. coli control			
DNA control			
DNA/*E. coli* mixture			

Questions

1. What do the results indicate? ____________________

2. If there was no growth on the DNA/*E. coli* plate, what went wrong? ____________________

3. How would you prove that recombination had occurred? ______

______

______

______

4. How can you rule out mutation or contamination as an explanation for the growth of DNA/*E. coli* on citrate?

______

______

______

5. What is a protoplast? ______

______

A spheroplast? ______

______

6. What is the purpose of EDTA in this experiment? Of chloroform? ______

______

______

______

EXERCISE 28 — LABORATORY REPORT

Conjugation

Name ______________________

Date ______________________

Lab. Section ______________________

Purpose ______________________

____

____

Data

Time allowed for conjugation: ______________ minutes

Complete the following table.

Medium	Growth of			Genotype
	Hfr	F^-	Hfr/F^-	
Glucose, streptomycin, and threonine				
Glucose, streptomycin, and leucine				
Lactose, streptomycin, threonine, and leucine				

Questions

1. Did conjugation occur? ______________ Which genes? ______________

____

2. Why was streptomycin included in the media? ______________________

____

3. What control was exercised to detect back mutations? ______________________

____

4. Try to construct a rough genetic map. (Use data from your classmates to help.)

5. How could you map these genes with better accuracy? ______________________________

____

____

Transduction

Name ______

Date ______

Lab. Section ______

Purpose ______

Data

Auxotroph used: ______

Nutritional supplement required: ______

Record the appearance of each plate.

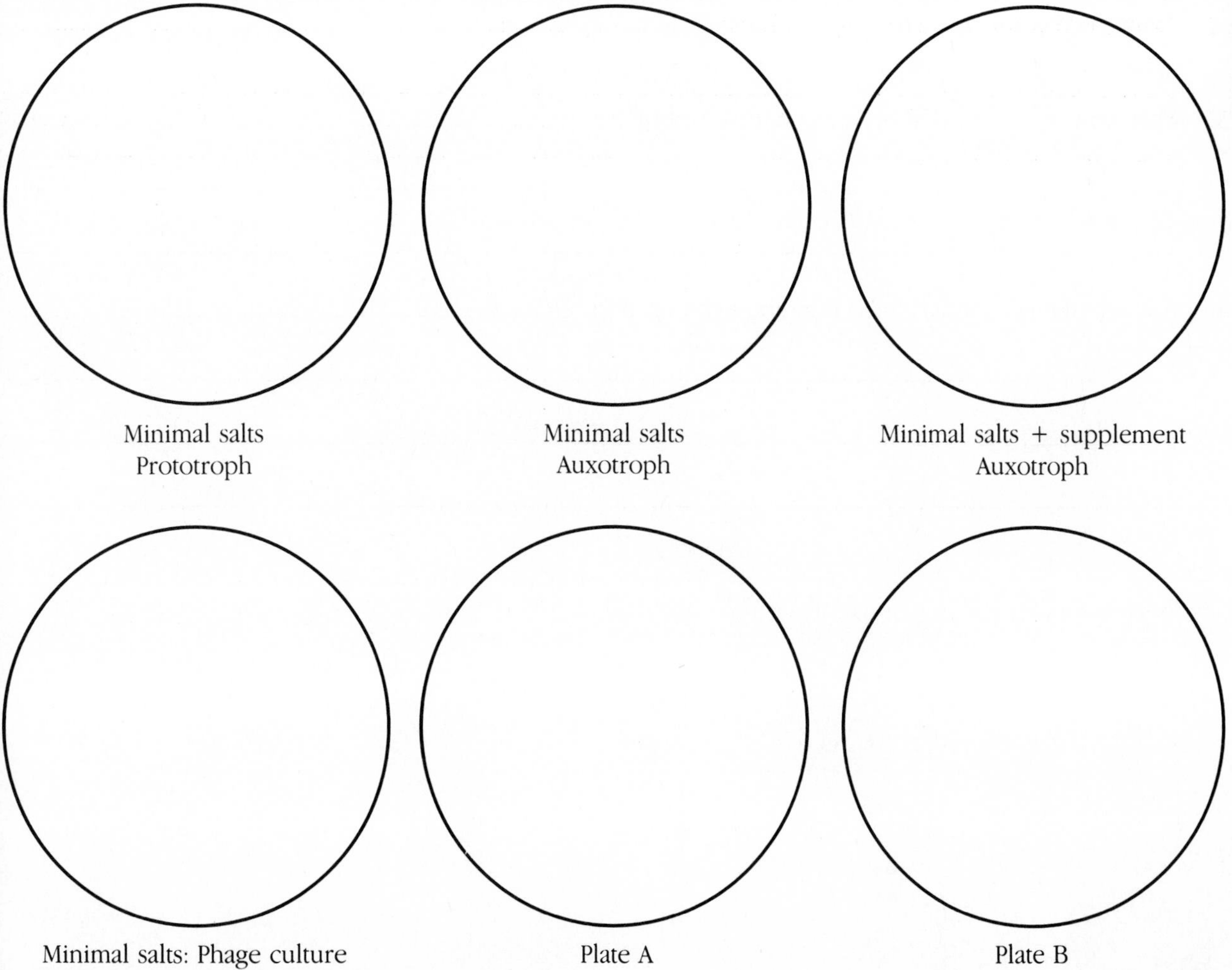

Conclusions

Questions

1. There were four control plates used in this exercise. What were they and what was the purpose of each?

2. How could you tell whether transduction occurred in this experiment?

3. What type of transduction occurred in this experiment?

4. Why are minute colonies formed in abortive transduction?

Ames Test for Detecting Possible Chemical Carcinogens

Name ______________________

Date ______________________

Lab. Section ______________________

Purpose ______________________

Data

Show the location of paper disks and bacterial growth on the plates. Number the disks to correspond to the following table.

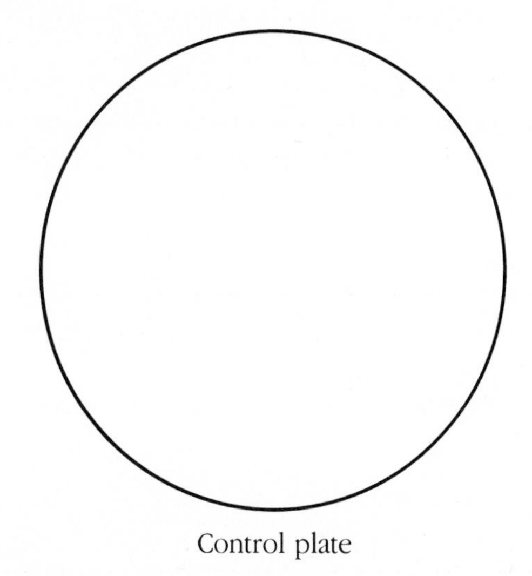

Control plate

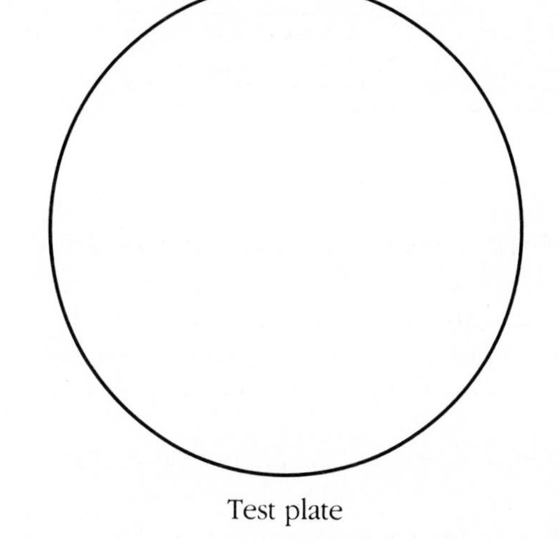

Test plate

Complete the following table:

Compound	Growth (large colonies)	Mutagenic
1		
2		
3		
4		
5		

Questions

1. What degree of spontaneous reversion occurred? ____________________

2. Why are mutants used as test organisms in the Ames Test? ____________________

3. Does this technique give a minimum or maximum mutagenic potential? ____________________

4. What is the advantage of this test over animal tests? Disadvantage? ____________________

EXERCISE 31 LABORATORY REPORT

Fungi: Yeasts

Name __________

Date __________

Lab. Section __________

Purpose __________

Data

Yeasts

Fermentation tubes

Organism	Glucose			Sucrose		
	Acid	Gas	Fermentation?	Acid	Gas	Fermentation?
Rhodotorula rubra						
Candida albicans						
Saccharomyces cerevisiae						
Baker's yeast						

Sabouraud agar plates

Organism	Draw a Typical Colony	Wet Mounts
Rhodotorula rubra Color ________ Smell ________		
Candida albicans Color ________ Smell ________		

EXERCISE 31

Organism	Draw a Typical Colony	Wet Mounts
Saccharomyces cerevisiae Color ______ Smell ______		
Baker's yeast Color ______ Smell ______		

Yeast Isolation

Plants used: ______

Describe the appearance of the glucose broth after ______ days incubation. ______

______

Was gas produced? ______

Sabouraud agar

Colony appearance	Color	Smell	Wet mounts
Mouth:			
Plants:			

Any bacteria seen? ______

If so, which colonies? ______

Questions

1. Could you identify the genus of baker's yeast? ______

2. Did you culture yeast from your mouth? ______ From the plants? ______ How do you know? ______

3. What was the purpose of the balloon on the glucose–yeast extract broth bottle? ______

4. Define the term *yeast.* ______

5. Compare and contrast yeast and bacteria regarding their appearance both on solid media and microscopically.

6. Name a genus of yeast that causes human disease. ______

7. Why are yeast colonies larger than bacterial colonies? ______

EXERCISE 32 LABORATORY REPORT

Fungi: Molds

Name ______________________

Date ______________________

Lab. Section ______________________

Purpose ______________________

Observations

Slide Cultures

Characteristics	Organism		
	Rhizopus stolonifer	*Aspergillus niger*	*Penicillium notatum*
Coenocytic?			
Type of spores			
Color of spores			
Diagram spore arrangement			

EXERCISE 32

Sexual spores

Zygomycete

Genus ______________________

______ X

Ascomycete

Genus ______________________

______ X

Basidiomycete

Genus ____________________

______ X

Plate Cultures

	Organism			
	Rhizopus stolonifer	*Aspergillus niger*	*Penicillium notatum*	Unknown from contaminated plate
Colony Appearance *Macroscopic* Hyphae color Spore color Underside				
Microscopic Diagram				

Wet mount and tape technique

Septate?				
Diagram spore arrangement.				

Mushroom Growing

Record your observations. ____

____

____

____

Questions

1. Can you identify the phylum of your unknown mold? ____________________ On what basis did you make your identification? ____________________

Or, what additional information do you need? ____________________

____

____

2. Complete the following table.

Organism	Asexual Spores	Sexual Spores	Phylum
Rhizopus			
Aspergillus			
Penicillium			

3. How can you make a selective medium for fungi that minimizes bacterial growth, other than by adjusting the pH and nutrients? ____________________

____

____

____

4. How do mold spores differ from bacterial endospores? ____________________

____

____

____

5. Why aren't molds *streaked* for isolation? ____________________

____

____

____

EXERCISE 33 — LABORATORY REPORT

Algae

Name ______

Date ______

Lab. Section ______

Purpose

______

Data

Name of Alga	Relative Abundance in Pond Water			
	Observed in Light	Observed in Dark	Observed with NO_3^- and PO_4^{3-}	Observed with $CuSO_4$
Drawings of other algae seen				

EXERCISE 33

Conclusions

1. What can you conclude regarding the effects of light on algal growth? ______

2. What can you conclude regarding the effects of the addition of nitrates and phosphates to pond water? ______

3. What can you conclude regarding the effects of the addition of copper sulfate to pond water?

Questions

1. Why can algae be considered indicators of productivity as well as of pollution?

2. How can algae be responsible for the production of more oxygen than land plants?

3. Why is it difficult to include all algae in the kingdom Plantae?

4. Describe one way in which algae and fungi differ.

   How are they similar?

EXERCISE 34 — LABORATORY REPORT

Protozoans

Name ______________________

Date ______________________

Lab. Section ______________________

Purpose ______________________

____

____

Observations

Use a series of diagrams to illustrate the following:
The movements of *Amoeba* across the field of vision. _______ ×

The ingestion of food, formation of a food vacuole, and movement of food vacuoles in *Paramecium.* Note any color changes in the food vacuoles. _______ ×

The movement of *Euglena* and its flagellum. ______ ×

Questions

1. How did *Euglena* respond to the acetic acid? ______

2. Which organism observed in this exercise would you bet on in a race? ______

3. Describe the arrangement of cilia on *Paramecium.* ______

4. What criteria are used to place an organism in the kingdom Protista? ______

5. How are protozoans classified as to phyla? ______

6. Why is *Euglena* often used to study algae (Exercise 33) and protozoans (Exercise 34)? ______

7. Why are green algae classified as plants and not protists? (*Hint:* Refer to Table 33-1.) ______

EXERCISE 35 LABORATORY REPORT

Isolation of Bacteriophages

Name ______

Date ______

Lab. Section ______

Purpose ______

Data

Draw what you observed.

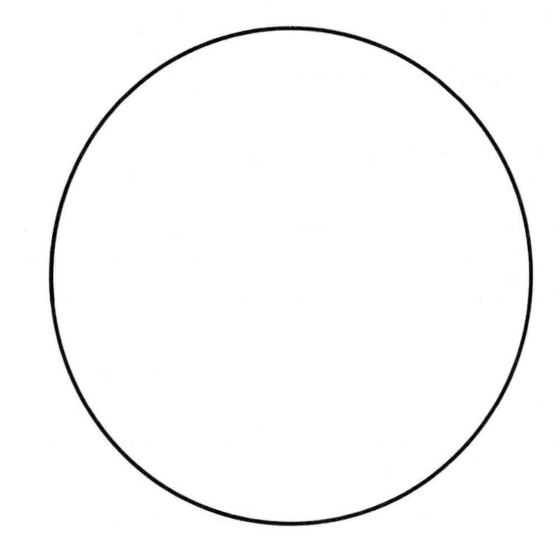

Questions

1. Are any contaminating bacteria present? How can you tell? ______
2. What caused the clearing? ______
3. Are there any differences in the size of the clearings? ______
4. To what would you attribute differences in plaque size? ______
5. Why did you add *Escherichia coli* to sewage, which is full of bacteria? ______

EXERCISE 35

6. How would you isolate a bacteriophage for a species of *Bacillus?* ____________

7. Do the areas of clearing represent a pure phage isolate? ____________

8. Why do flies contain *E. coli* phage? ____________

Titration of Bacteriophages

Name ____________________

Date ____________________

Lab. Section ____________________

Purpose ____________________

Data

Broth-Clearing Assay

Indicate whether each tube was turbid or clear.

Tube	Turbid or clear	Dilution
1		
2		
3		
4		
5		
6		

Plaque-Forming Assay

Choose one plate with 30 to 300 plaques:

Number of plaques = ____________________

Dilution used for that plate = ____________________

Interpretations

Broth-Clearing Assay

What was the end point? ____________________

What was the titer? ____________________

Plaque-Forming Assay

p.f.u./ml = ____________________

Show your calculations:

Questions

1. How would you explain turbidity in all of the tubes in a broth-clearing assay? ______________________

2. Are all the plaques the same size? Briefly explain why or why not. ______________________

3. Which of these assays is more accurate? Briefly explain. ______________________

4. What limits the size of the plaque? Why is an agar overlay used in plaque titration? ______________________

5. What does the end point represent in a broth-clearing assay? ______________________

6. How would you develop a pure culture of a phage? ______________________

7. If there were no plaques on your plates, offer an explanation. ______________________

Identification of Bacteria Using Phage: Phage Typing

Name ______________________

Date ______________________

Lab. Section ______________________

Purpose

____

Data

E. coli strain	Cleared area	Phage sensitive

Questions

1. Which *E. coli* strain(s) is (are) sensitive to phage T_2? ______________________

2. In phage typing, what is typed? ______________________

3. Is phage typing a method of classification or identification? Briefly explain your answer. ______________________

4. You have isolated a bacterium. Design a procedure to identify it by phage typing. ______________________

5. Why is it desirable to trace the origin of a nosocomial infection? ______________________

6. One of the criteria for determining the presence of an epidemic is to show that all cases of a specific infection are caused by the same organism. Explain how phage typing is used to determine whether an epidemic exists.

____

____

____

7. How does the phage-typing procedure done in a clinical laboratory differ from the procedure you have done?

Briefly explain. ____

____

____

EXERCISE 38 — LABORATORY REPORT

Plant Viruses

Name ______________________

Date ______________________

Lab. Section ______________________

Purpose

____

____

____

Data

Draw a plant showing locations of the labeled leaves and any signs of infection. Plant species: ____________

Tobacco Product Used	Appearance of Leaves							
	Lab Period							
	1	2	3	4	5	6	7	8
Control								
1.								

EXERCISE 38

Tobacco Product Used	Appearance of Leaves							
	Lab Period							
	1	2	3	4	5	6	7	8
2.								
3.								
4.								
5.								
6.								
7.								
8.								

Questions

. Did systemic disease occur? How can you tell? ____________________________

____

2. How do insects act as vectors of plant viruses? ______

3. Why can tobacco *or* tomato plants be used? Could any other plant be used to detect TMV? ______

4. If the control leaf was damaged, what happened? ______

5. Design an experiment to prove that damage to the test leaves was due to a viral infection. ______

6. A reservoir of infection is a continual source of that infection. What is a likely reservoir for TMV? ______

Determination of a Bacterial Growth Curve

Name ______________________

Date ______________________

Lab. Section ______________________

Purpose

____

____

____

Data

Optical Density Data

Time	O.D.	Time	O.D.

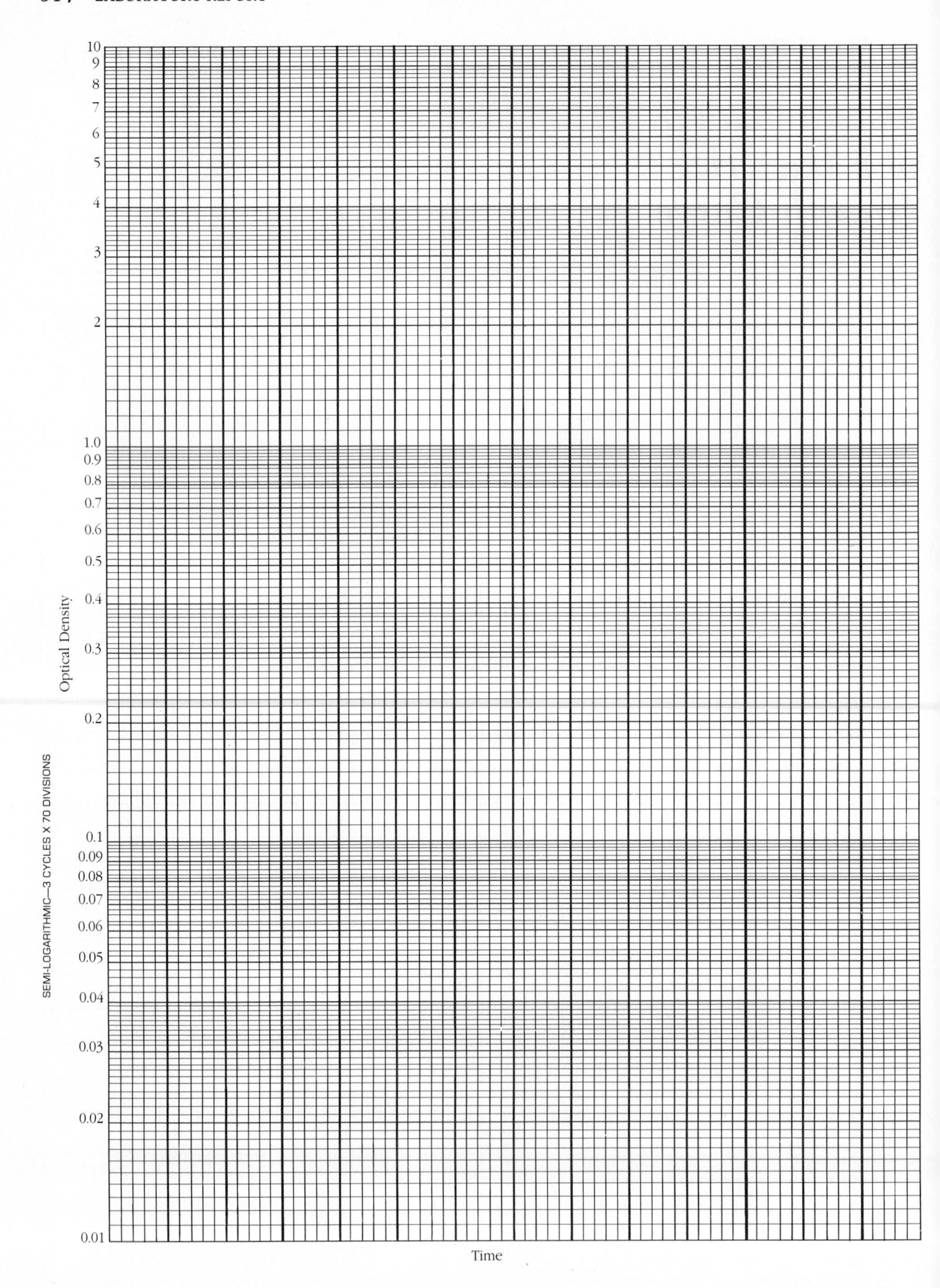
10
9
8
7
6
5
4
3
2
1.0
0.9
0.8
0.7
0.6
0.5
0.4
0.3
0.2
0.1
0.09
0.08
0.07
0.06
0.05
0.04
0.03
0.02
0.01
Optical Density
SEMI-LOGARITHMIC—3 CYCLES X 70 DIVISIONS
Time

Plot your data on the semi-log graph paper (Appendix D). Optical density is marked off on the *y*-axis (logarithmic scale) and time on the *x*-axis (linear scale). If other students used different incubation conditions, plot these data on the same graph using different lines or colors. Label the phases of growth.

Questions

1. Draw a bacterial growth curve showing the four typical phases of growth. Label each phase.

Log number of viable cells

Time

2. Why aren't you likely to get lag and death phases in this exercise? ______

______

______

3. How could you use turbidity to estimate the numbers of bacteria? ______

______

______

______

______

______

EXERCISE 40 ________ LABORATORY REPORT

Oxygen and the Growth of Bacteria

Name ____________

Date ____________

Lab. Section ____________

Purpose ____________

____

____

Data

Thioglycollate

Sketch the location of bacterial growth and label as to its oxygen requirements as done in Figure 40-1. Observe other students' thioglycollate tubes.

Colored indicator

Nutrient agar plates

Aerobic No. of different colonies ________		*Anaerobic* ________	
Colony description	Catalase reaction	Colony description	Catalase reaction

Does the "anaerobic" plate have an odor? ____________

Questions

1. Which of the colonies on the Petri plates could be anaerobic bacteria? ____________

____

EXERCISE 40

What would you need to do to determine whether they are anaerobes? ______________________

____

2. Could you have a catalase-positive colony on the plate incubated anaerobically? ______________

____

3. Why was the soil or feces heated? ______________________

What would be the Gram reaction and morphology of most of the organisms cultured? ______________

____

____

4. Why will obligate anaerobes grow in thioglycollate? ______________________

____

____

5. The catalase test is often used clinically to distinguish between two genera of gram-positive cocci:

______________ and ______________ .

It is also used to distinguish between two genera of gram-positive rods: ______________ and

______________ .

6. To what do you attribute the odors of anaerobic decomposition? ______________________

____

____

____

Role of Temperature in the Growth of Bacteria

Name ____________________

Date ____________________

Lab. Section ____________________

Purpose

____

Data

Indicate which temperature is the optimum or near the optimum for each organism tested. Classify each organism as psychrophile, mesophile, or thermophile.

Species	Amount of growth at				Optimum Temperature	Temperature Group
	15°C	Room temperature in °C:__	35°C	55°C		
S. marcescens						
P. fluorescens						
M. luteus						
B. sterothermophilus						

Questions

1. What is the optimum temperature for human pathogens? ____________________

2. Where would you most likely find a thermophile? ____________ A psychrophile? ____________

3. What is the effect of temperature on enzymes? Can you use any examples from your experiments? ____________________

Other Influences on Microbial Growth: Osmotic Pressure and pH

Name ______________________

Date ______________________

Lab. Section ______________________

Purpose

____

Data

Effect of Osmotic Pressure on Growth

Rate the relative amounts of growth on nutrient agar and nutrient agar-containing solutes (salt and sucrose): (−) = no growth; (+) = minimal growth; (2+) = moderate growth; (3+) = heavy growth; (4+) = very heavy (maximum) growth.

	Organism/Amount of Growth			
Medium	*S. aureus*	*E. coli*	Yeast	Mold
Nutrient agar				
+ 5% NaCl				
+10% NaCl				
+15% NaCl				
+10% sucrose				
+25% sucrose				
+50% sucrose				

Gradient plate.

Sketch your growth patterns and those of another group.

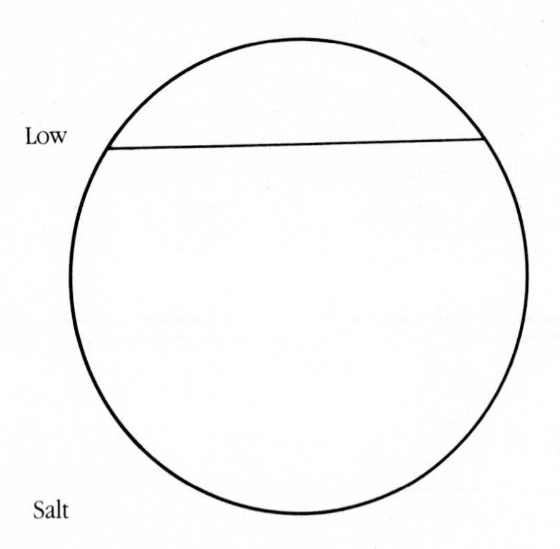

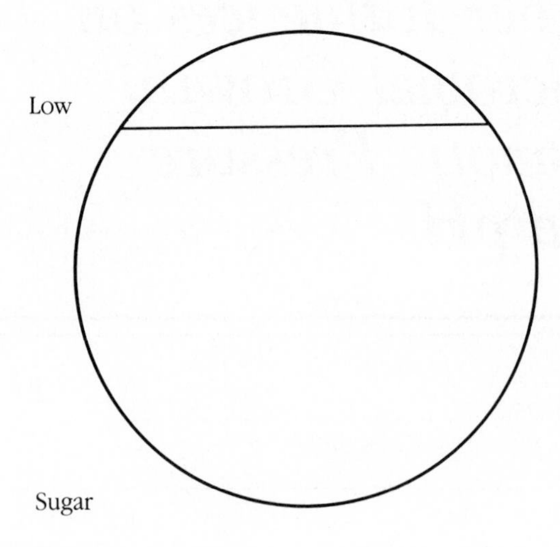

Effect of pH on Growth

Record the growth (+) or lack of growth (−) of each organism used.

	Growth of					
pH	*S. aureus*	*A. faecalis*	*E. coli*	*S. marcescens*	Mold	Yeast
2.5						
5.0						
7.0						
9.5						

Questions

1. Which organism tolerates high concentrations of salt best? __________

Of what advantage is this to the organism? __________

____

2. Which organism tolerates high concentrations of sucrose best? __________

Why is this advantageous to the organism? __________

____

3. Write a statement comparing the osmotic tolerance of bacteria to that of fungi. __________

____

4. Explain the growth patterns on the gradient plates. ______

______

______

______

______

5. What is the pH tolerance of bacteria compared to that of fungi? ______

______

6. Did any organisms grow in acidic conditions? ______ Why would you expect some organisms to grow in an acidic environment? ______

______

______

______

7. Why are bacteria more likely to survive in a hypotonic environment than in a hypertonic environment? ______

______

8. What is the principle of the gradient plate? ______

______

9. What is the purpose of the control plate? ______

______

10. Should "4+" growth occur only on the control plate? ______

Briefly explain. ______

______

11. Why were the media containing raisins and hamburger labeled "enrichments"? ______

______

12. What is meant by "neutral pH"? ______

______

EXERCISE 43 LABORATORY REPORT

Physical Methods of Control: Heat

Name ____________________

Date ____________________

Lab. Section ____________________

Purpose

____

____

____

Data

Record growth on a scale from (−) to (4+).

Organism	Temperature/Time									
	63°C					72°C				
	0	15 sec	2 min	5 min	15 min	0	15 sec	2 min	5 min	15 min
Old *Bacillus subtilis*										
Young *Bacillus subtilis*										
Staphylococcus epidermidis										
Escherichia coli										
Mold (*Penicillium*) spores										

Conclusions

____

____

____

____

Questions

1. Explain any unexpected results. ______________________________

____

EXERCISE 43

2. What is the heat sensitivity of fungal spores in comparison to bacterial endospores? ______________________

____

3. Give an application (use) of thermal death time. ____

____

4. In the exercise, was the thermal death time or thermal death point determined? ________________

____

5. Give an example of a nonlaboratory use of each of the following to control microbial growth: incineration, pasteurization, autoclaving. ____

____

____

____

____

6. What is fractional sterilization? ____

____

____

7. Define pasteurization. What is the purpose of pasteurization? ___________________________

____

____

____

Physical Methods of Control: Ultraviolet Radiation

Name ______

Date ______

Lab. Section ______

Purpose ______

Data

Sketch your results and those for the other culture used. Note any pigmentation.

Bacillus subtilis:

A B C

Serratia marcescens:

A B C

Conclusion ______

Questions

1. If the *Bacillus* had sporulated before exposure to radiation, would that affect the results? ______________________________

2. What are the two variables in ultraviolet radiation treatment? ______________________________

3. Can dark repair be a factor in this experiment? ______________________________

4. Ultraviolet radiation is used in hospital operating rooms; you've seen its limitations. What then is the object of using it? ______________________________

5. Why are there still some colonies growing in the areas exposed to ultraviolet light? ______________________________

EXERCISE 45 — LABORATORY REPORT

Chemical Methods of Control: Disinfectants and Antiseptics

Name ______________________

Date ______________________

Lab. Section ______________________

Purpose

____

Data

Time of exposure (minutes)	Chemical Control			Chemical Phenol			Chemical ______		
	Turbidity	Color	Gas	Turbidity	Color	Gas	Turbidity	Color	Gas
2.5									
5									
10									
20									

Conclusion

____

Questions

1. Was this a fair test? Is it representative of the effectiveness of the test substance? ______________________

2. How could the techniques be altered to measure bacteriostatic effects? ______________________

EXERCISE 45

3. Read the label of the preparation you tested. What is (are) the active ingredient(s)? ____________

____

Using your textbook or another reference, find the method of action of the test substance. ____________

____

____

____

4. What is the use-dilution method? ____

____

____

5. What does a phenol coefficient of 1 indicate? ____

____

____

Chemical Methods of Control: Antimicrobial Drugs

Name ______

Date ______

Lab. Section ______

Purpose

______

Data

Effect of garlic? ______ Of copper? ______

Antimicrobial Agent	Disk Code	*Staphylococcus aureus*		*Escherichia coli*		*Pseudomonas aeruginosa*	
		Zone size	S, I, or R*	Zone size	S, I, or R*	Zone size	S, I, or R*
1.							
2.							
3.							
4.							
5.							
6.							
7.							
8.							

*S = sensitive
I = intermediate
R = resistant

Questions

1. Which chemotherapeutic agent was most effective against each organism? ______

2. What other factors are considered before using the chemotherapeutic agent in vivo? ______

3. In which growth phase is an organism most sensitive to an antibiotic? ______

4. Is the agar diffusion technique measuring bacteriostatic or bactericidal activity? ____________

5. Why is the agar diffusion technique not a perfect indication of how the drug will perform in vivo? ____________

6. What effect would the presence of streptomycin in the body have on penicillin therapy? ____________

Effectiveness of Hand Scrubbing

Name ______________________

Date ______________________

Lab. Section ______________________

Purpose

____

Data

Indicate the relative amounts of growth in each quadrant:

Section	Water Alone	Soap
1. (No washing)		
2.		
3.		
4.		

Conclusion

____

Questions

1. What is a surgeon trying to accomplish with a 10-minute scrub with a brush followed by an antiseptic? ______________________

2. How do normal flora and transient flora differ? ______________________

3. Most of the normal flora aren't harmful. Then why must hands be scrubbed before surgery? ____________

____

____

Mannitol Salt Agar: selective; only Staphylococci + Micrococci grow
Differential: yellow (S. Aureus) zone around colony (ferm. of mannitol)
orange zone = no mannitol ferment.

EXERCISE 48 LABORATORY REPORT

Infection Control: Equipment Sampling

Name
Date
Lab. Section

Purpose

Data

E. Coli S. Aureus

Equipment sampled: glass rods A + B (contaminated bact.)

Colony Description on Trypticase Soy Agar	Gram Stain	
	Gram Reaction	Morphology

Colony Description on EMB or Mannitol Salt Agar*	Fermentation of Lactose or mannitol?	Gram Stain	
		Gram Reaction	Morphology

*Circle the medium that you used.

EMB Eosin methylene Blue Agar → light colonies (do not ferment lactose)
(selective + Differential) (inhibits Gram +)
(Dark colony (purple) (lactose fermented by bact.) ↔ green metallic sheen = E. Coli

Questions

1. What can you conclude about the cleanliness and safety of this piece of equipment?

2. How is this piece of equipment normally decontaminated?

3. Design an experiment to determine the effectiveness of the decontamination procedure noted in question 2.

4. Define nosocomial.

5. What are the sources of nosocomial infections?

6. How can a fomite transmit an infection?

7. Why is it necessary to have an infection control unit in a hospital?

Microbes in Water: Multiple-Tube Technique

Name ______

Date ______

Lab. Section ______

Purpose

______

Data

Water sample: ______

Presumptive Test

	Gas	+ or − test
1-ml		
10-ml		

Confirming Test

Sample: ______ Growth: ______

Appearance of colonies: ______

Are coliforms suspected? ______

Completed Test

Appearance of colony used: ______

Gas from lactose: ______

Gram reaction and cell morphology: ______

Questions

1. Were coliforms present in your water sample? ______

2. What is the MPN of the demonstration? ______ per ______ ml

3. Why isn't a pH indicator needed in the lactose broth fermentation tubes? ______

4. Why are coliforms used as indicator organisms if they are not usually pathogens? ______________________

____

5. Could the water have a high concentration of *Salmonella* and give negative results in the multiple-tube technique? Briefly explain. ____

____

6. Why don't we inoculate EMB agar directly and bypass lactose broth? ______________________________

____

____

Microbes in Water: Membrane Filter Technique

Name ____________________

Date ____________________

Lab. Section ____________________

Purpose

____

____

____

Data

Sample tested: ____________________

Describe the general appearance of the colonies on ______ medium. ____________________

____

____

____

Number of coliform colonies: ____________________

Number of coliforms/100 ml water: ____________________
Use the space below for your calculations:

Data from water samples tested by other students:

Sample	Coliforms/100 ml

Questions

1. Which water sample(s) is (are) potable? ____________________

2. Which water sample(s) is (are) contaminated? ____________________

3. Can you determine the source of contamination (for example, human, domestic animal, wild animal) from this test? ______

4. What basic assumption is made in this technique, if the number of bacteria is determined from the number of colonies? ______

______

5. Why can filtration be used to sterilize culture media and other liquids? ______

______

6. If you did Exercise 49, list one advantage and one disadvantage of each method of detecting coliforms. ______

______

______

______

______

______

7. Why is the membrane filter technique useful for a sanitarian working in the field? ______

______

8. Design an experiment to detect the presence and number of *Acetobacter* in wine. Why would you want to perform such a test? ______

______

______

EXERCISE 51 LABORATORY REPORT

Microbes in Soil: Quantification

Name ______________________

Date ______________________

Lab. Section ______________________

Purpose

____

____

____

Data

Soil source: ______________________

Soil type: ______________________

Media	Nutrient agar				Sabouraud agar			
Dilution	____	____	____	____	____	____	____	____
Number of colonies								
Bacteria	____	____	____	____	____	____	____	____
Actinomycetes	____	____	____	____	____	____	____	____
Fungi	____	____	____	____	____	____	____	____

Questions

1. Number of bacterial c.f.u. per gram of soil = ______________________
   Show your calculations below.

2. Number of actinomycetes per gram of soil = ______________________
   Show your calculations below.

3. Number of fungi per gram of soil = ______________________
   Show your calculations below.

EXERCISE 51

4. Why do these numbers represent the minimum number of microorganisms in the soil? ______

______

5. Which medium was selective for fungi? ______

6. Compare your data with your classmates' using different soil types. Which soil type has the most microorganisms? ______

Why? ______

______

7. Which type of microorganism predominates in each type of soil? ______

______

______

______

Microbes in Soil: Nitrogen Cycle

Name

Date

Lab. Section

Purpose

Data

Ammonification

Incubation	Growth	Nessler's Reagent		Ammonia Present?
		Color	Color of Control	
2 days				
7 days				

Denitrification

How did you test for denitrification?

Inoculum	Growth	Gas	Nitrate Reduction	Denitrification?
Soil				
Pseudomonas				

Nitrogen Fixation

Diagram of root nodule

Diagram microscopic appearance of cells in root nodule

Cell morphology ____________________

Diagram microscopic appearance of bacteria from culture

Demonstration slides
Label root, nodule, and bacteria

100 ×

Cell morphology ____________________

1000 ×

Describe colonies on the mannitol-yeast extract plate. ____________________

Questions

1. Write the equation for ammonification of any amino acid.

2. Why can't you smell ammonia in the peptone broth tubes? ______

3. Why is denitrification a problem for farmers? ______

4. What gas is in the gas tube of a positive denitrification test? ______

5. What genetic engineering project would you propose concerning nitrogen fixation? ______

6. Why do the bacteria in the root nodule look different from the bacteria cultured from the nodule? ______

7. Which step(s) in the nitrogen cycle require oxygen? ______

8. Why is the nitrogen cycle important to all forms of life? ______

Microbes in Soil: Photosynthetic Bacteria

Name ______________________

Date ______________________

Lab. Section ______________________

Purpose ______________________

____

____

Data

Diagram the appearance of your enrichment at weekly intervals. Label patches of photosynthetic bacteria. Light source: ______________________

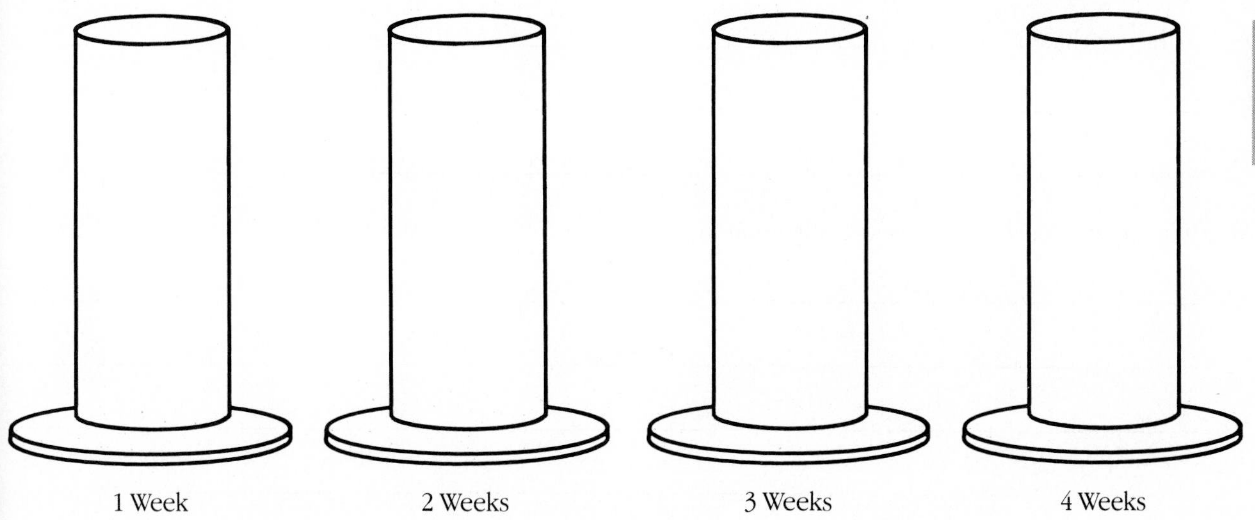

Microscopic observations

Source	Morphology	Motility	Sulfur Granules Present?

Questions

1. What was the purpose of each of the following chemicals in the enrichment?

$CaCO_3$ ________________ Na_2S ________________

Hay or paper ________________ KH_2PO_4 and K_2HPO_4 ________________

$CaSO_4$ ________________ NH_4Cl ________________

2. Indicate the aerobic and anaerobic regions on the diagrams of your enrichment. How can you tell? ________________

____

3. Is there evidence of any nonphotosynthetic growth in the enrichments? ________________

Explain. ________________________________

____

4. Why can't air be used as a source of carbon dioxide by photosynthetic bacteria? ________________

____

5. Design an experiment that would determine the effect of heat produced by the lights on the growth of bacteria in the enrichment. ________________________________

____

____

6. If other light sources were used, compare the appearance of the enrichments after 4 weeks. ________________

____

____

____

____

____

EXERCISE 54 — LABORATORY REPORT

Microbes in Food: Contamination

Name ____________________

Date ____________________

Lab. Section ____________________

Purpose

____

Data

Sample ____________________

Dilution	Colonies per Plate
1:	
1:	
1:	
1:	

Number of bacteria per ml (or gram) of original food: ____________________
Do your calculations in the space below.

Record data for other foods tested by other students.

Food	Bacteria/ml

EXERCISE 54

Questions

1. What could you do to ensure that the bacteria present in foods do not pose a health hazard?

2. Are bacteria identified in the standard plate count?

   Why is a standard plate count performed on food?

3. Why are plates with 30 to 300 colonies used for calculations?

4. In a quality control laboratory, each dilution is plated in duplicate or triplicate. Why would this increase the accuracy of a standard plate count?

5. There are other techniques for counting bacteria such as a direct microscopic count and turbidity (Exercise 39). Why is the standard plate count preferred for food?

6. Why is ground beef a better bacterial growth medium than a steak or roast?

7. List four foodborne diseases. Identify one that is primarily transmitted via milk.

8. Why does repeated freezing and thawing increase bacterial growth in meat?

Microbes Used in the Production of Foods (Yogurt, Root Beer, and Wine)

Name

Date

Lab. Section

Purpose

Observations

Yogurt

Characteristics:

Taste

Consistency

Odor

pH

Gram stain results:

Streak plate results:

Gram stain of isolated colonies:

Wine

Characteristics:

Color

Flavor

Odor

Sweetness ______

Carbonation ______

Final alcohol ______

Simple stain results: ______

Root Beer

Characteristics:

Color ______

Carbonation ______

Sediment ______

Taste ______

Streak plate results: ______

Simple stain of isolated colonies: ______

______

Questions

1. How can pathogens enter yogurt and how can this be prevented? ______

______

______

2. What product(s) would result from aeration of must? ______

______

3. What product(s) would result from aeration of wine? ______

______

4. What was the source of the bacteria and yeast originally used in dairy product fermentations, wine, and breads?

______

5. What could cause an inferior product in a microbial fermentation process? ______

______

6. How are microbial fermentations used to preserve foods? ______

______

7. Why isn't alcohol produced in root beer fermentation? ______

______

Epidemiology

Name __________

Date __________

Lab. Section __________

Purpose

____

Data

S. Marcescens: red growth)

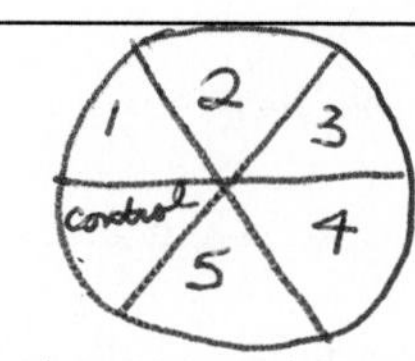

Results

Sector	Student's Name	Tootsie Roll® Number	Appearance of Colonies on Nutrient Agar
1.			
2.			
3.			
4.			
5.			

Conclusion

Who had the contaminated Tootsie Roll®? __________

What number was the Tootsie Roll®? __________ Explain how you arrived at your conclusion. __________

Questions

1. Could you be the "infected" individual and not have growth on your plate? Explain. ______

______

______

______

2. Do all people who contact an infected individual acquire the disease? ______

______

______

______

3. When does an epidemic stop? ______

______

______

______

4. Are any bacteria other than *Serratia* growing on the plates? How can you tell? ______

______

______

5. What was the method of transmission of the "disease" in this experiment? ______

______

______

Koch's Postulates

Name ______________________

Date ______________________

Lab. Section ______________________

Purpose ______________________

____

____

Vibrio—Goldfish

Gram Stains

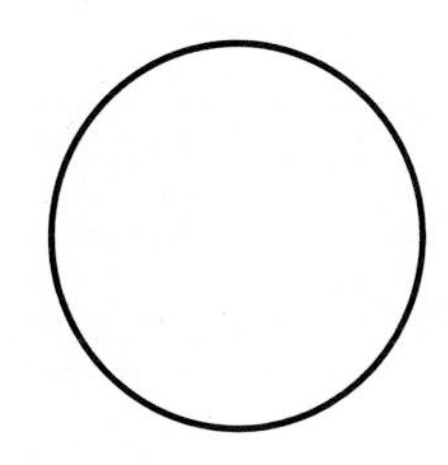

Vibrio anguillarium
(before inoculation)
Morphology ______________________

Gram reaction ______________________

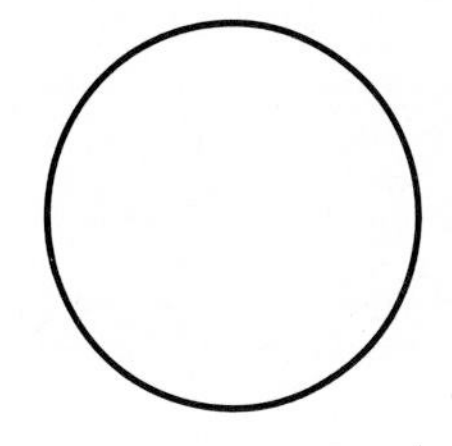

Peritoneum fluid from dead fish
Morphology ______________________

Gram reaction ______________________

Cultures

Describe the growth seen.

Source	Nutrient Agar	Gram Stain	Salt-Starch Agar	Starch Hydrolysis?
Saline				
Vibrio preinoculation				
Peritoneum fluid				

Necropsy observations:
Describe your findings. (Compare your fish to the dissected demonstration.) ______________________

____

____

Pus? ______ Hemorrhage? ______

Conclusions

______

______

______

Questions

1. Do you think this bacterium produces a toxin? ______

______

______

2. If your broth-inoculated fish died, what would this suggest about the method of disease transmission? ______

______

______

______

3. What evidence would indicate that the microbe obtained from the dead fish was *Vibrio anguillarium?* ______

______

______

______

4. Were Koch's postulates fulfilled? ______ Explain ______

______

______

Erwinia—Carrot Soft Rot

Gram Stains

Erwinia carotovora
Morphology ______

Gram reaction ______

Soft Rot
Morphology ______

Gram reaction ______

Cultures

Inoculum	Colony Description	Gram Stain
Erwinia		
Soft rot		

Describe the appearance of the carrots. ______________________________

____

____

____

Conclusions ______________________________

____

____

____

Questions

1. Why is the disease called soft rot? ______________________________

____

2. Why will the addition of water speed the disease process? ______________________________

____

____

3. Were Koch's postulates fulfilled? ______________________________

____

____

Agrobacterium—Crown Gall

Gram Stains

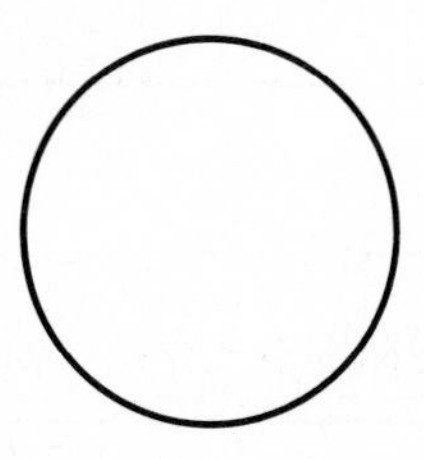

Agrobacterium tumefaciens
Morphology ______________________

Gram reaction ______________________

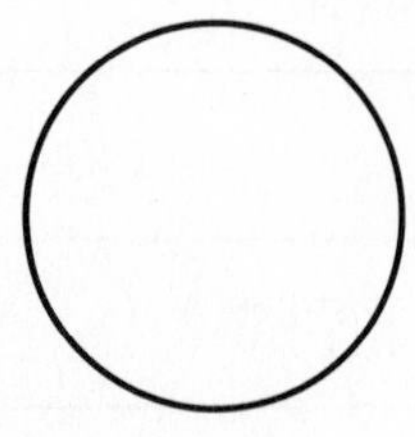

Carrot or tomato tumor
Morphology ______________________

Gram reaction ______________________

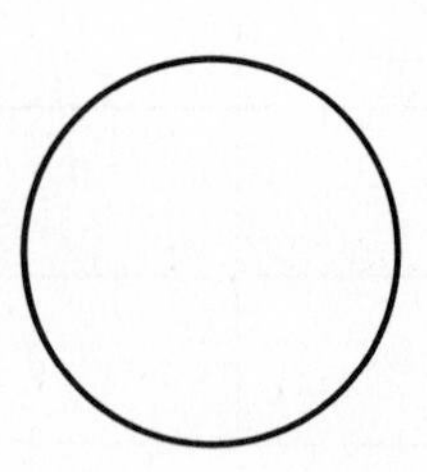

Demonstration
Morphology ______________________

Gram reaction ______________________

Cultures

Inoculum	Colony Description	Gram Stains
Agrobacterium		
Carrot		
Tomato		

Describe and sketch the galls.

Conclusions ____

____

____

____

Questions

1. Did you find bacteria in the carrot or tomato gall? ______________________
2. What did this experiment indicate about the host range of *Agrobacterium?* ______________________

   ____
3. Why aren't bacteria present in large quantities in galls? ______________________

   ____

   ____

4. Were Koch's postulates fulfilled? ______ Explain ______________________________

____

____

Summary Questions

1. Give two examples (other than those used in this exercise) of instances where Koch's postulates have not been fulfilled. ______________________________

____

2. What parts of Koch's postulates did these experiments fulfill? ______________________________

____

____

____

3. What is the normal method of transmission of the following:

a. *Vibrio anguillarium?* ______________________________

b. *Erwinia carotovora?* ______________________________

c. *Agrobacterium tumefaciens?* ______________________________

EXERCISE 58 — LABORATORY REPORT

Bacteria of the Skin

Name ______________________

Date ______________________

Lab. Section ______________________

Purpose

____

____

Data

Source of inoculum: ______________________

First mannitol salt plate

	Colony		
	1	2	3
Colony Description			
Pigment			
Mannitol fermentation			

Which colony was isolated? ______________________

Catalase reaction ______________________

Gram stain

Reaction ______________________

Cell morphology ______________________

Cell arrangement ______________________

Additional biochemical tests

____

____

____

Identification: ______________________

Questions

1. Why is mannitol salt agar used to select for normal skin flora?

2. After you have observed a gram-positive coccus, you need additional information before performing a coagulase test. What is this additional information?

3. List three identifying characteristics of *Staphylococcus aureus*.

4. List three factors that protect the skin from infection.

5. What is coagulase? How is it related to pathogenicity?

EXERCISE 59 — *LABORATORY REPORT*

Bacteria of the Respiratory Tract

Name ______

Date 5/21/92

Lab. Section DN1

Purpose

______

Data

S. Aureus
Beta Hemolysis Inhibition ~~no~~ catalase +

Throat Culture

	Colony			
	1	2	3	4
Appearance of colonies on blood agar	Clusters of little cells.			
Hemolysis	Beta			
Catalase reaction	+			

EXERCISE 59

Streptococcus

	Organism	
	S. pyogenes	*S. pneumoniae*
Gram stain Gram reaction		
Morphology		
Arrangement		
Blood agar plate Appearance of colonies		
Hemolysis		

Continued

	Organism	
	S. pyogenes	*S. pneumoniae*
Inhibition by Optochin		
Bacitracin		
Bile solubility		

Other Bacteria

Characteristics	*Mycobacterium smegmatis*	*Bordetella bronchiseptica*	*Cornyebacterium diphtheriae*	*Neisseria sicca*
Gram stain: Gram reaction				
Morphology				
Arrangement				
Acid-fast reaction				
Appearance of cells in simple stain	na			na
Oxidase reaction	na	na	na	
Appearance of colonies on Lowenstein Jensen medium		na	na	na
Bordet-Gengou medium	na		na	na
Tellurite agar	na	na		na
Brain-heart infusion agar	na	na	na	

na = not applicable

Questions

1. Why is blood agar a differential medium? ______________________________

____

2. If a Bacitracin disk is added to the area swabbed on a blood agar plate, a rapid identification technique (Exercise 25) results. Why? ______________________________

____

____

3. Is the Gram stain of significant importance in the identification of the organisms studied in this exercise?

_______ Explain. ______________________________

____

4. You have isolated a gram-positive coccus from a throat culture that you cannot identify as staphylococci or streptococci. A test for one enzyme can be used to distinguish *quickly* between these bacteria. What is the enzyme? ______________________________

5. Name one disease caused by each genus used in this exercise. Identify the species that causes that disease.

____

____

____

____

____

____

Bacteria of the Mouth

Name ______________________

Date ______________________

Lab. Section ______________________

Purpose

____

Data

Snyder test

	Positive or Negative?	Color
Results: at 24 hours		
at 48 hours		
at 72 hours		

Sucrose gelatin agar	Colony 1	Colony 2	Colony 3
Appearance of colonies			
Catalase production			

Questions

1. List three characteristics of streptococci found in the mouth. ______________________

2. *S. mutans* forms large mucoid colonies on sucrose gelatin agar. Would you expect the same type of colonies on glucose gelatin agar? ______________________ Briefly, explain your answer. __________

____

3. Studies have shown that both sucrose and bacteria are necessary for tooth decay. Why should this be true? ____

____

____

____

4. Did you observe any green dots in the Snyder agar? ________________________________

What caused them? ____

____

5. What is the value of having your teeth cleaned? Of brushing your teeth? __________________

____

____

Bacteria of the Gastrointestinal Tract

Name ______________________

Date ______________________

Lab. Section ______________________

Purpose

____

____

____

Data

EMB Agar

Source of inoculum ______________________

	Colony			
	1	2	3	4
Appearance of colonies				
Lactose fermentation?				

TSI

Colony picked ______________________

Appearance: Slant ______________________

Butt ______________________

Gas? ______________________

H_2S ______________________

Other TSI slants:
Description of colony picked ______________________ ______________________
Results of TSI slant:

Tomato Juice Agar
Description of colonies: ______________________

____

Gram reaction of colony picked: ______________________

Catalase test: ______________________

SF Broth

Growth ________ Acid ________________________

Gram stain: ______________________________

Questions

1. What genus do you think you have in your TSI slant? ________________________

What additional information do you need to identify it? ________________________

____

2. Did you culture *Lactobacillus* on the tomato juice agar? ________________________

How do you know? ____________________________________

3. What can you conclude from the SF broth? ____________________________

____

4. What tests, if any, would you need to perform to identify bacteria producing a green metallic sheen on EMB agar?

____

5. Differentiate between a coliform, an enteric that is not a coliform, and an enterococcus. Give an example of an organism in each group. ____________________________

____

____

6. Is EMB a selective medium? ________ A differential medium? ________________________

Bacteria of the Urogenital Tract

Name ______

Date ______

Lab. Section ______

Purpose

______

Data

Urine Culture

Blood agar plate:
Number of colonies: 69

Total count (bacteria/ml): 6.9×10^3

Hemolysis: ______

EMB plate:
Coliforms present: yes

If so, describe the colonies: ______

Noncoliforms: ______

Cystitis

unknown B =

	Organisms		
	E. coli (B)	P. vulgaris	P. aeruginosa
EMB agar			
Appearance of colonies	dark		
Lactose fermentation (+ dark, - light)	+		
~~Swarming~~	omit		
~~Gram stain~~	omit		
Pseudomonas agar P			
Appearance of colonies			
Pigmentation (+ blue-green, - NO green)	–		
Oxidase reaction	–		
OF-glucose (Fermentation or oxidation)	neither		
Urease production	+		

G. C. Smears

Diagram the appearance of a known G. C.-positive smear.

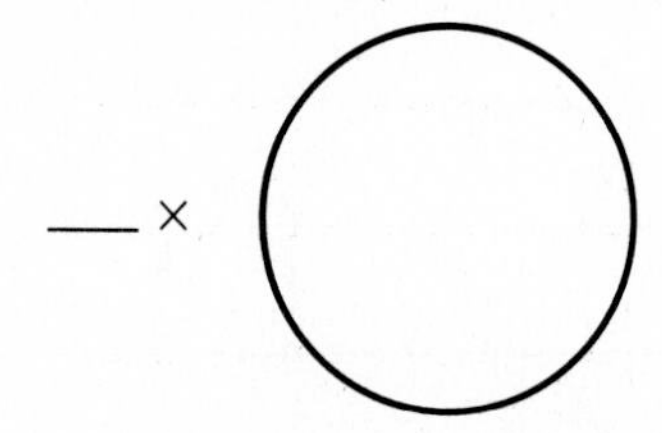

Diagram the appearance of a known G. C.-negative smear.

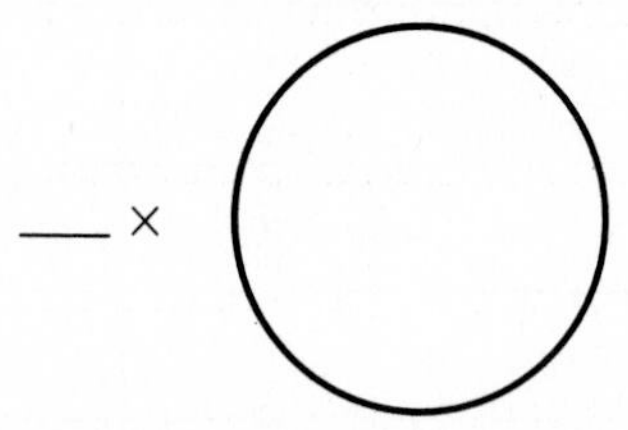

Unknown # ______________________

Diagram the appearance of your unknown smear.

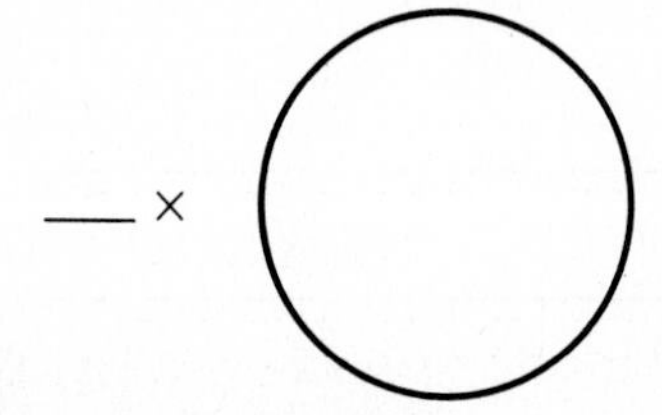

Could *N. gonorrhoeae* be present? ______________________

Questions

1. Why is EMB agar inoculated with a urine specimen? ______________________

____

2. Why wasn't it necessary to perform glucose fermentation and urease tests on *E. coli?* ______________________

____

3. Differentiate between the pigment of *P. aeruginosa* on Pseudomonas agar P and the "pigment" of *E. coli* on EMB agar without referring to the colors. ______________________

____

4. The enterics and pseudomonads look alike microscopically; how can you easily distinguish between these two groups of bacteria? ______________________

____

____

5. Why are females more prone to urinary tract infections than males? ______________________

____

6. What role does antibiotic treatment have in yeast infections of the urinary tract? ______________________

____

Identification of an Unknown from a Clinical Sample

Name ______________________

Date ______________________

Lab. Section ______________________

Purpose

____

____

____

Data

Unknown # ______________________

Source ______________________

Preliminary Gram stain results: ______________________

____

	Organism	
	A	B
Trypticase soy agar plate		
Appearance of colonies	______	______
Differential media		
(______________)		
Appearance of colonies	______	______
Additional information	______	______
Gram stain		
Morphology	______	______
Gram reaction	______	______
Trypticase soy agar slant		
Appearance of colonies	______	______
Other tests:		
______	______	______
______	______	______
______	______	______
______	______	______
______	______	______

	Organism	
	A	B
Antibiotic sensitivity to		
Ampicillin	______	______
Chloramphenicol	______	______
Erythromycin	______	______
Penicillin G	______	______
Streptomycin	______	______
Sulfonamide	______	______
Tetracycline	______	______

Questions

1. What two bacterial species were in unknown # ______________________ ? ______________

____

2. On a separate sheet of paper, neatly write your rationale for arriving at the conclusion.

3. Why would a clinical microbiology laboratory perform antimicrobial susceptibility tests as well as identify the unknown? ____

____

EXERCISE 63 — LABORATORY REPORT

Identification of an Unknown from a Clinical Sample

Name ______________________

Date ______________________

Lab. Section ______________________

Purpose

____

____

____

Data

Unknown # ______________________

Source ______________________

Preliminary Gram stain results: ______________________

____

	Organism	
	A	B
Trypticase soy agar plate		
Appearance of colonies	____________	____________
Differential media		
(______________)		
Appearance of colonies	____________	____________
Additional information	____________	____________
Gram stain		
Morphology	____________	____________
Gram reaction	____________	____________
Trypticase soy agar slant		
Appearance of colonies	____________	____________
Other tests:		
____________	____________	____________
____________	____________	____________
____________	____________	____________
____________	____________	____________
____________	____________	____________
____________	____________	____________

	Organism	
	A	B
Antibiotic sensitivity to		
Ampicillin	______	______
Chloramphenicol	______	______
Erythromycin	______	______
Penicillin G	______	______
Streptomycin	______	______
Sulfonamide	______	______
Tetracycline	______	______

Questions

1. What two bacterial species were in unknown # ______________ ? ______________

____

2. On a separate sheet of paper, neatly write your rationale for arriving at the conclusion.

3. Why would a clinical microbiology laboratory perform antimicrobial susceptibility tests as well as identify the unknown? ____

____

Animal Viruses

Name ______________________

Date ______________________

Lab. Section ______________________

Purpose ______________________

____

____

Data

Virus model with 32 capsomeres

Color of pentamers? ______________________

Color of vertices? ______________________

Color of hexamers? ______________________

Color of faces? ______________________

Questions

1. Capsids may have one of two shapes. The shapes are ______________________ and ______________________.
2. Since the geometry of an icosahedron cannot vary, how do icosahedral viruses vary in size? ______________________

   ____
3. Of what advantage is the capsid geometry? ______________________
4. In a real virus particle, the capsomeres are composed of ______________________ instead of paper.
5. What is (are) the function(s) of the capsid? ______________________

   ____
6. Measure the diameter of your icosahedral virus: ______________________ mm.

   How many capsomeres does it have? ______________________

   It is a model of ______________________ virus, which has a diameter of ______________________ nm

   How many real virus particles could fit inside the model? ______________________
7. How does your paper helical model differ from a helical virus? ______________________

   ____

8. What is the diameter of your paper helical virus? ____________________

9. Why are animal viruses difficult to culture in a laboratory? ____________________

____

Why are they dangerous to work with? ____________________

____

EXERCISE 65 LABORATORY REPORT

Introduction to Medical Mycology

Name ______________________

Date ______________________

Lab. Section ______________________

Purpose ______________________

____

____

Data

Slide Culture

Gymnoascus

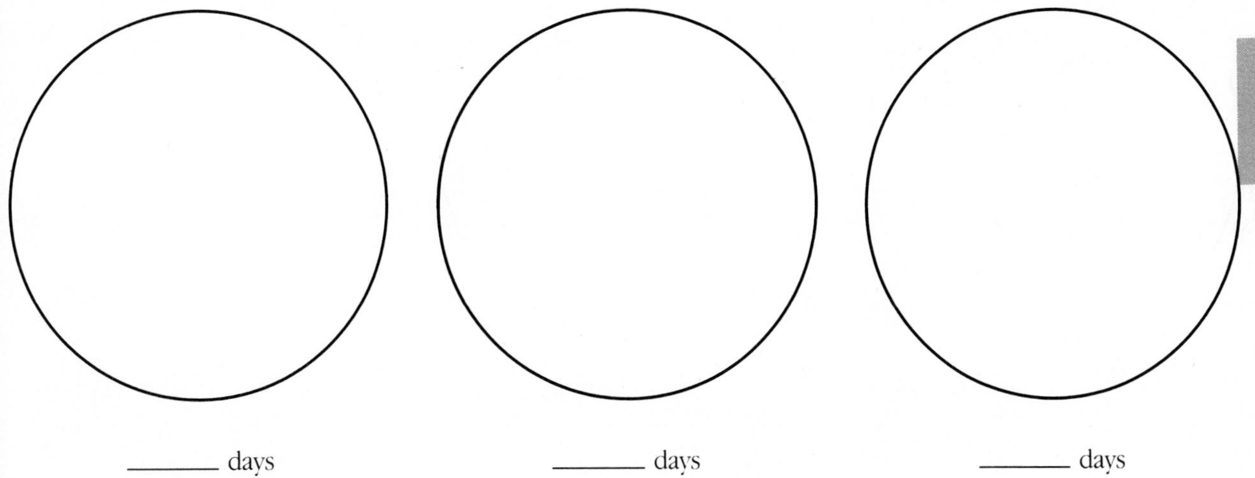

_______ days _______ days _______ days

Type of spores observed? ______________________

Sepedonium

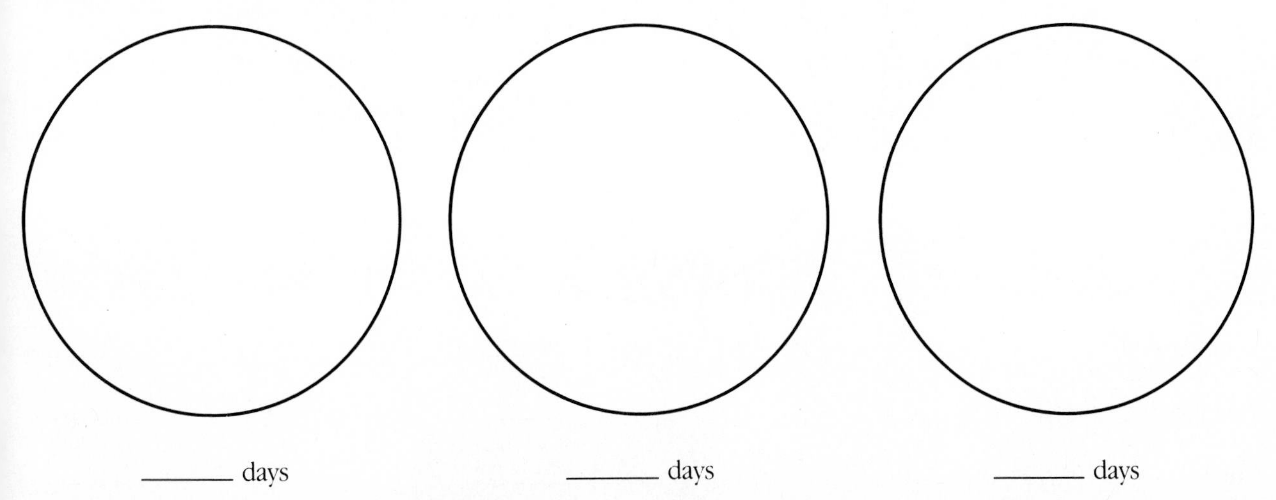

_______ days _______ days _______ days

EXERCISE 65

Type of spores observed? ____________________

Dimorphic Gradient

Top of agar:
(Draw the structures seen)

Magnification ___ × ___ × ___ ×

Middle

Magnification ___ × ___ × ___ ×

Bottom

Magnification ___ × ___ × ___ ×

Coenocytic or septate ______ ______ ______

Fungus ______ ______ ______

Questions

1. What determines the form of fungi seen in the dimorphic gradient?

2. Does dimorphism have any clinical importance?

3. Why might the Deuteromycota be considered a "holding category"?

4. Why do you suppose most pathogenic fungi are in the Deuteromycota?

5. Why do media used to culture fungi contain sugars?

6. Why are antibiotics frequently added to Sabouraud agar for isolation of fungi from clinical samples?

7. How can fungi cause respiratory tract infections?

8. Why weren't pathogenic fungi used in this exercise?

Introduction to Medical Protozoology

Name ______

Date ______

Lab. Section ______

Purpose ______

______

______

Observations

Trichomonas vaginalis

Identify the undulating membrane, nucleus, and flagella. How many flagella did you observe on one cell? ______

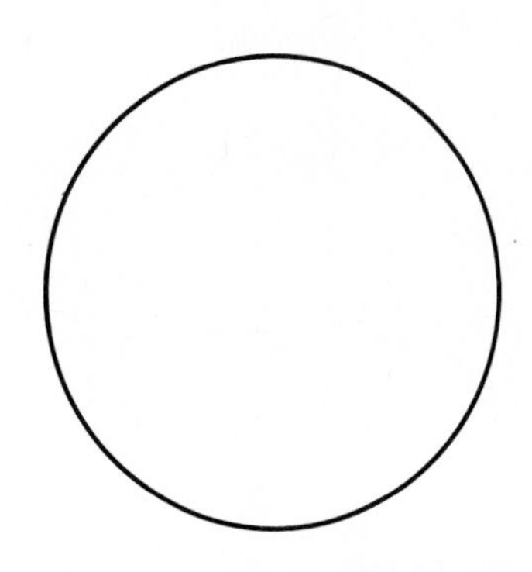

___ ×

Trypanosoma gambiense

Identify the undulating membrane, nucleus, and flagella. Show a red blood cell. How many flagella did you observe on one cell? ______

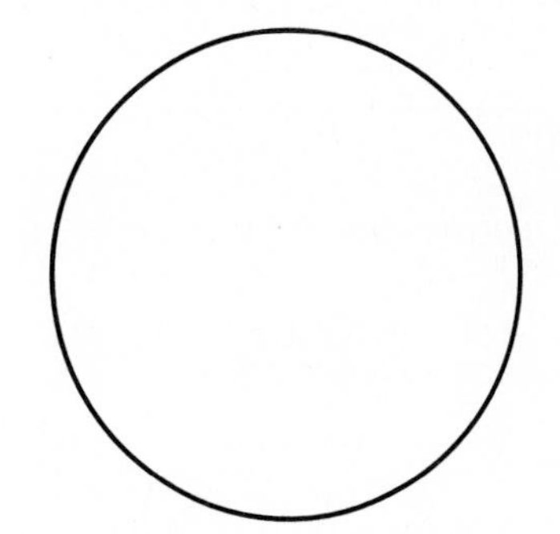

___ ×

Balantidium coli

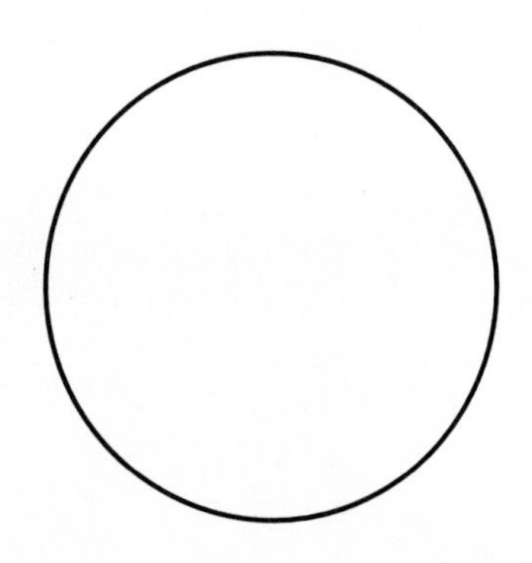

Trophozoite

___ ×

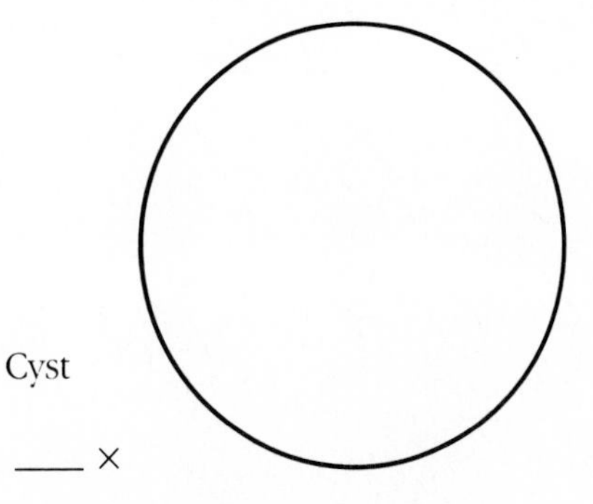

Cyst

___ ×

Identify the nucleus, cilia, contractile vacuole, and oral groove in the trophozoite. Locate the nucleus and refractile body in the cyst. Does the cyst have cilia? ____________

Toxoplasma gondii

Can you find any method of locomotion on these cells? ____________

___ ×

___ ×

Entamoeba histolytica

Identify the granular endoplasm and the clearer ectoplasm.

How does this cell move? ____________

How can you tell? ____________

____

___ ×

Plasmodium falciparum, ring stage

Identify a red blood cell. What will happen to this young trophozoite? ____________

____

____

____

Questions

1. Complete the following table.

Phylum	Method of Motility	Example Observed in This Exercise
Sarcodina		
Ciliata		
Mastigophora		
Sporozoa		

2. Why would a cyst be advantageous to a parasitic species? ____________________

____

3. What is the mode of transmission of *Entamoeba histolytica?* ____________________

Of *Plasmodium falciparum?* ____________________

Of *Trichomonas vaginalis?* ____________________

EXERCISE 67 — LABORATORY REPORT

Introduction to Medical Helminthology

Name ______________________

Date ______________________

Lab. Section ______________________

Purpose

____

Observations

Free-living

Using words and/or diagrams, describe the movement of planaria (a free-living flatworm). ______________________

Can you see the cilia on its ventral surface? ______________________

Using words and/or diagrams, describe the movement of vinegar eels (free-living roundworms). ______________________

Prepared slides

Fasciola hepatica. Carefully draw each stage of the sheep liver fluke *(Fasciola).* Identify the host or habitat of each stage and label each stage.

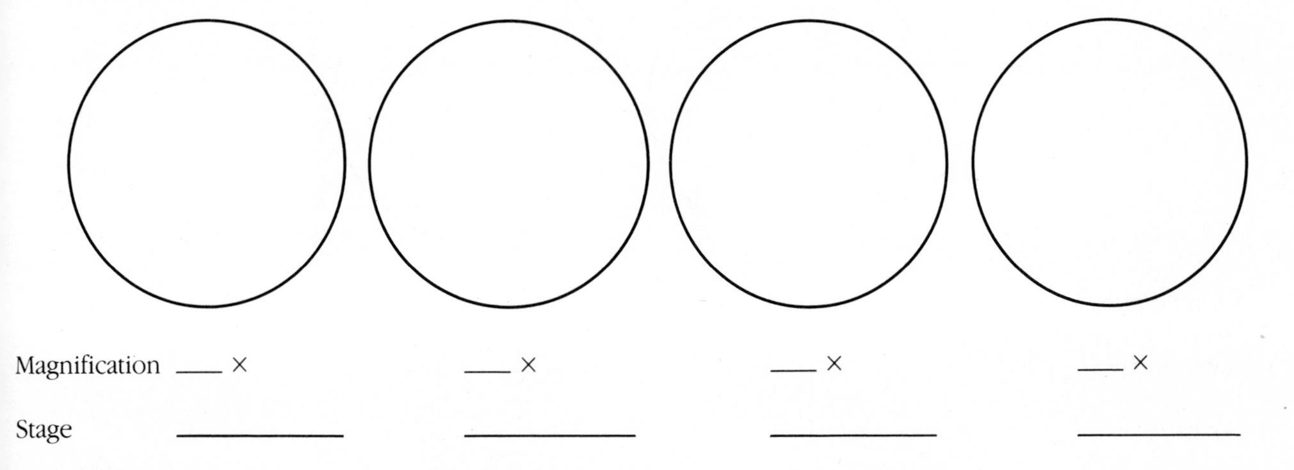

Magnification	___ ×	___ ×	___ ×	___ ×
Stage				
Where found?				

Adult human liver fluke *(Clonorchis sinensis).* Identify the oral and ventral suckers on your diagram. Point out the features that indicate this animal has an incomplete digestive system.

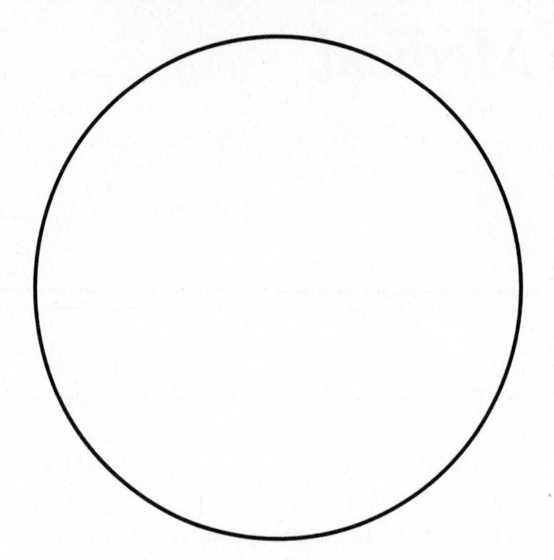

Magnification____ ×

Schistosoma. Indicate features that distinguish the male from the female.

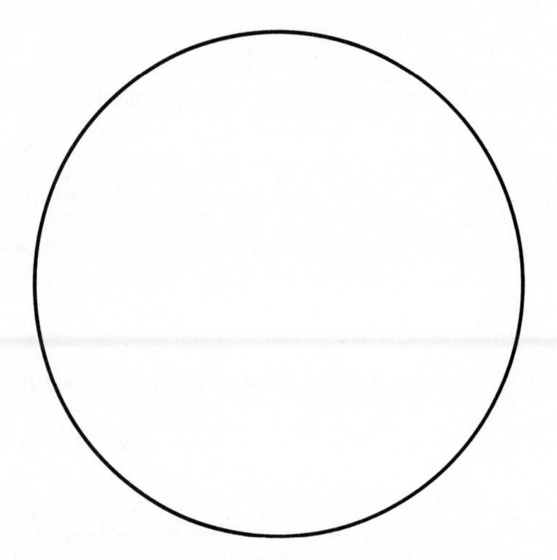

Magnification____ ×

Hookworm. Indicate the habitat/host of each stage.

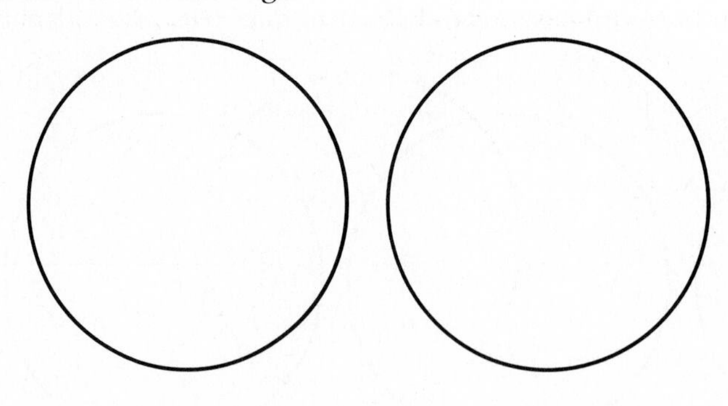

Magnification	___ ×	___ ×
Stage	____________	____________
Habitat/host	____________	____________

Pinworm.

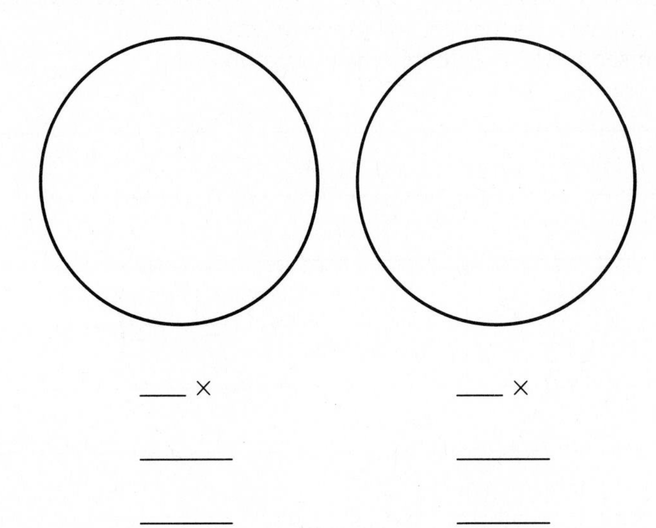

Magnification	___ ×	___ ×
Stage	_______	_______
Where found?	_______	_______

Tapeworm. Identify the suckers and hooks on the scolex. Show the scolex and proglottids in the correct sequence of an adult tapeworm.

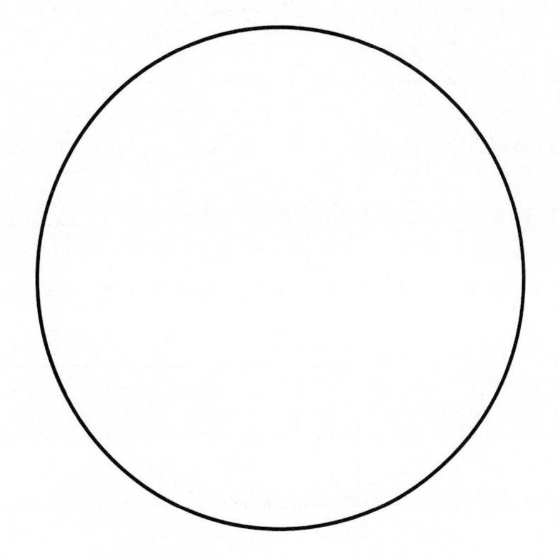

Magnification___ ×

Trichinella spiralis. In what animal(s) is this stage found? ___________________________

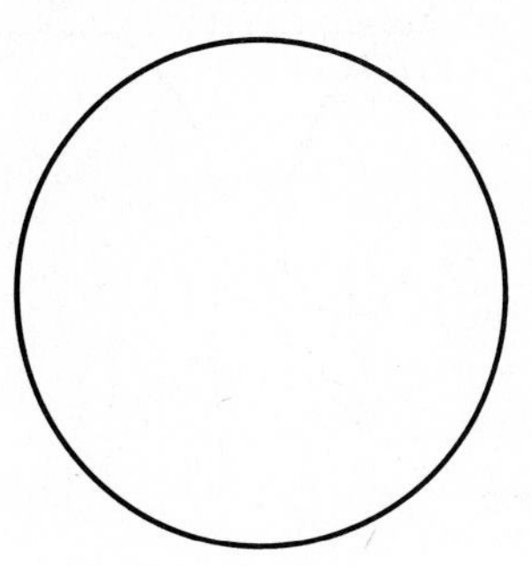

Magnification___ ×

Questions

1. What similarities did you see in each of the platyhelminths observed? ______

   ______

   What differences? ______

   ______

2. What similarities did you observe in each of the nematodes? ______

   ______

   What differences? ______

   ______

3. Of what advantage is hermaphroditism to a parasite? ______

   ______

   Disadvantages? ______

4. Which helminth studied in this exercise employs humans as an intermediate host? ______

5. What is (are) the best method(s) to prevent helminthic infections? ______

   ______

   ______

   ______

6. Humans are considered a dead-end for *Trichinella.* Why? ______

   ______

   ______

Nonspecific Resistance

Name ______________________

Date ______________________

Lab. Section ______________________

Purpose

____

____

____

Data

Lysozyme Activity

Optical Density			
Time	Saliva	Tears	Egg White Lysozyme
30 sec			
60 sec			
120 sec			
180 sec			
240 sec			
5 min			
10 min			
15 min			
20 min			

Plot your data for lysozyme activity on the graph paper on the next page. Optical density is marked on the y-axis and time on the x-axis. Make one line for tears or saliva and another line for egg white lysozyme.

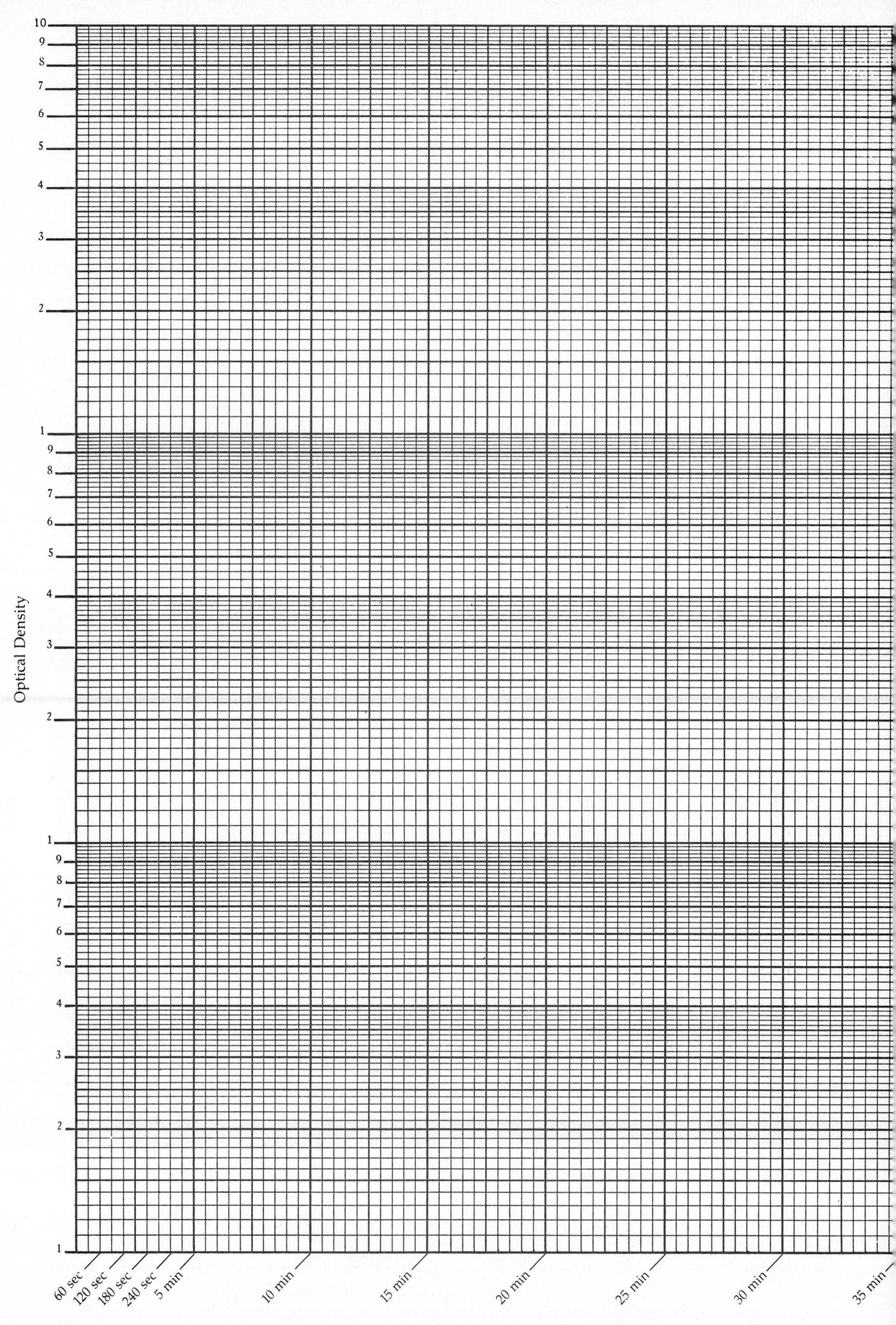
Optical Density
10
9
8
7
6
5
4
3
2
1
9
8
7
6
5
4
3
2
1
9
8
7
6
5
4
3
2
1
60 sec
120 sec
180 sec
240 sec
5 min
10 min
15 min
20 min
25 min
30 min
35 min
Time

Normal Serum Bactericidins

	Number of Colonies	% Increase or Decrease
$1{:}10^2$ Unheated serum		
Heated serum		
Control		
$1{:}10^4$ Unheated serum		
Heated serum		
Control		
$1{:}10^6$ Unheated serum		
Heated serum		
Control		

Questions

1. What effect did lysozyme have on the bacterial suspension? ______

______

2. Was lysozyme present in your tears or saliva? ______

How can you tell? ______

3. Did you test tears or saliva? ______ Compare your data with someone who tested the other. Which secretion has more lysozyme activity? ______ How can you tell? ______

______

4. Does normal serum have bactericidal properties? ______

5. What can you conclude regarding the effect of heat on the bactericidal properties of normal serum? ______

______

______

6. What factors might account for the bactericidal properties of normal serum? ______________________

____

Which part of your data helps to substantiate this? ______________________________________

____

Inflammation

Name ______

Date ______

Lab. Section ______

Differential Count

Purpose ______

______

______

Data

Number per 100 leukocytes counted

Lymphocytes ______

Monocytes ______

Neutrophils ______

Eosinophils ______

Basophils ______

Any platelets present? ______

Skin Window

Purpose ______

______

______

Data

	Time (hours)	Cells	Amount	Comment
1.				
2.				
3.				
4.				
5.				

Continued

	Time (hours)	Cells	Amount	Comment
6.				
7.				
8.				
9.				
10.				

Conclusions ____

____

____

____

Compare your results with someone who did a different time sequence.

Questions

1. What might cause a rise in the percentage of neutrophils? Of lymphocytes? ______________________

____

____

2. What is the life span of a neutrophil? A lymphocyte? ______________________

____

____

3. What is a macrophage? A histiocyte? ______________________

____

____

4. How could the skin window technique be made quantitative? ______________________

____

Phagocytosis

Name ______

Date ______

Lab. Section ______

Purpose ______

______

______

Data

India Ink
25 neutrophils: Number with India ink = ______
Calculate the % viable (alive) phagocytes, or viability index.

Viability index $= \frac{\text{Cells with ink}}{25} \times 100$

Viability index = ______

Yeast
Record the number of yeast in each of 25 neutrophils

1. ______	6. ______	11. ______	16. ______	21. ______
2. ______	7. ______	12. ______	17. ______	22. ______
3. ______	8. ______	13. ______	18. ______	23. ______
4. ______	9. ______	14. ______	19. ______	24. ______
5. ______	10. ______	15. ______	20. ______	25. ______
				Total Yeast ______

Calculate the phagocytic index.

Phagocytic Index $= \frac{\text{Number yeast in 25 cells}}{25}$

Phagocytic index = ______

Questions

1. Does the phagocytic index measure the same function as the viability index? ______

______

2. How could you measure the phagocyte-killing index? ______

______

______

EXERCISE 71 LABORATORY REPORT

Precipitation Reactions

Name ______________________

Date ______________________

Lab. Section ______________________

Purpose ______________________

____

____

Data

Precipitin Ring Test

Diagram the appearance of the precipitin tubes.

Unknown # ______________________

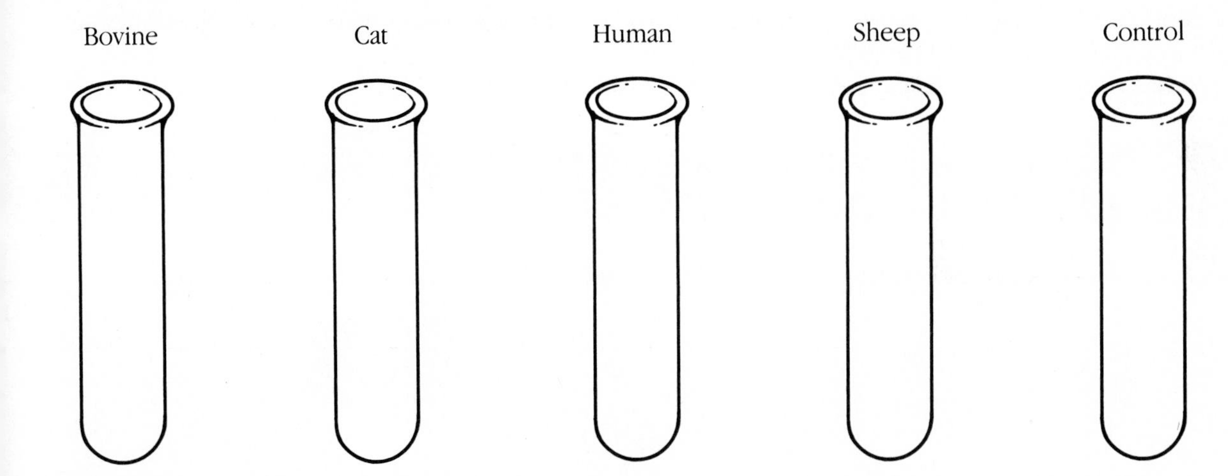

Ouchterlony Test

Diagram the appearance of the Ouchterlony test. Label the wells.

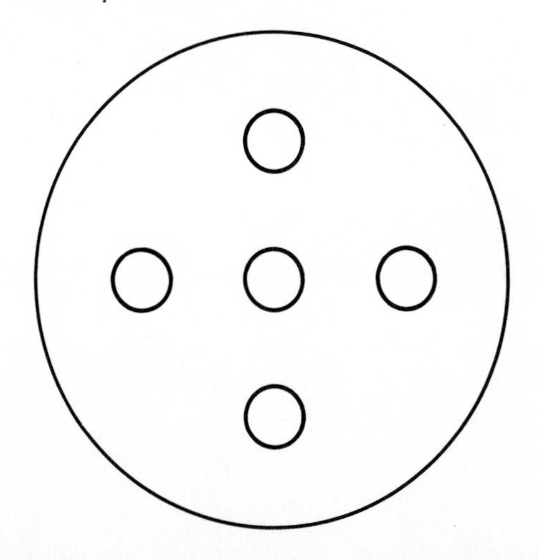

Questions

1. What is the antigen in unknown #__________? ____

   How can you tell? ____

2. What is the purpose of the control tube in the precipitin ring test? ____________________________

3. Label the lines of identity, partial identity, and nonidentity in your diagram of the Ouchterlony test.

4. From the Ouchterlony test, which antigens were related? ____________________________________

   Which were entirely different? ____

5. What is a precipitation reaction? ____

6. How could you detect the presence of horse meat in a commercial product labeled "ground beef"? __________

7. For what is a precipitation reaction used in a clinical laboratory? _______________________________

Immunoelectrophoresis

Name ______________________

Date ______________________

Lab. Section ______________________

Purpose

____

Data

Draw the pattern seen on the slide and try to identify each line. Mark which end is the cathode and which is the anode.

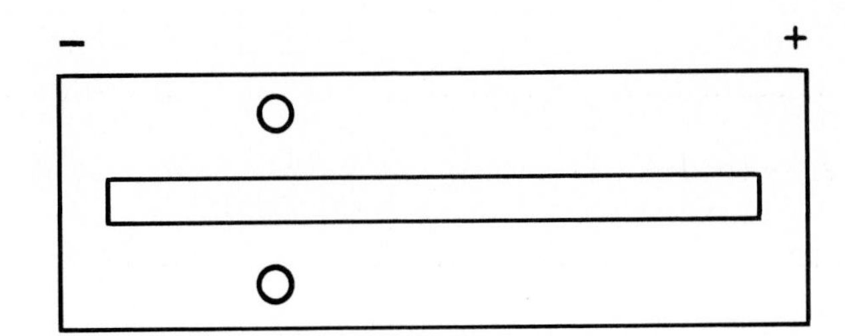

Questions

1. Why is this technique used clinically? ______________________

2. Why not use plain agar? ______________________

3. Could polysaccharides be electrophoresed? Explain briefly. ______________________

Blood Group Determination: Slide Agglutination

Name ______

Date ______

Lab. Section ______

Purpose ______

______

______

Data

Antiserum	Hemagglutination
Anti-A	
Anti-B	
Anti-D	

Questions

1. What is your blood type? ______

2. What is the blood type of the person on your left? ______

Is your blood compatible? ______ Explain briefly. ______

3. What is the blood type of the person on your right? ______

Is your blood compatible? ______ Explain briefly. ______

4. What is hemolytic disease of the newborn? How does Rhogam® prevent it? ______

______

______

5. How can blood type O be considered the "universal donor"? ______

______

Agglutination Reactions: Tube Agglutination

Name ____________

Date ____________

Lab. Section ____________

Purpose

____

Data

Tube #	Final Dilution	Agglutination
1		
2		
3		
4		
5		
6		
7		
8		
9		
10		

Diagram the appearance of the bottom of a positive and negative tube:

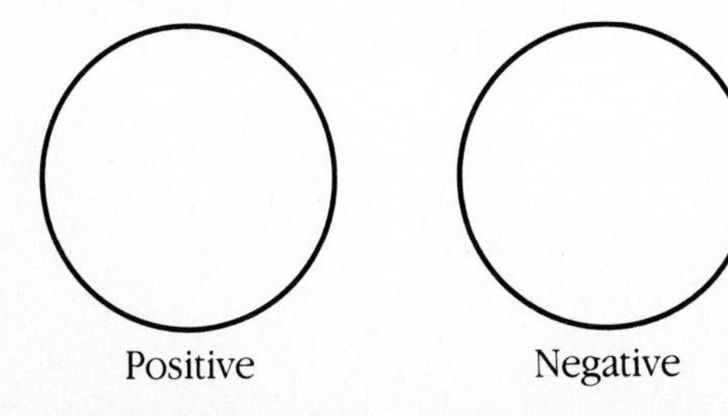

Questions

1. How did you determine the presence of agglutination? ______________________

____

2. What was the end point? ______________________

3. What was the antibody titer? ______________________

4. What was the antiserum used in this experiment? ______________________

5. How would agglutination reactions help in locating the source of an epidemic? ______________________

____

6. The following antibody titers were obtained for three patients:

	Antibody Titer		
Patient	Day 1	Day 5	Day 12
A	1:128	1:128	1:128
B	1:128	1:256	1:512
C	0	0	0

What can you conclude about each of these patients? ______________________

____

____

Fluorescent-Antibody Technique

Name ______

Date ______

Lab. Section ______

Purpose

______

Data

Gram stain results:

Culture# ______

Culture# ______

F.A. test results:

Culture	Microscopic Appearance	
	Brightfield	Fluorescent
# ____		
# ____		

Questions

1. Which culture was *E. coli?* ______

2. Briefly outline the procedure you would have to follow to identify the *E. coli* culture without fluorescent microscopy. ______

3. Was the test performed in this exercise a direct or an indirect F.A. test? ______

4. Compare and contrast direct and indirect F.A. tests. ______________________

____

____

____

____

5. Outline a procedure using fluorescent-antibody testing for identifying the other culture used in this exercise.

____

____

Figure Acknowledgments

Figures: 1-3, 1-4, 1-5, 1-6, 2-1, 2-2, 3-1, 3-2, 3-3, 3-4, 4-1, 4-2, 4-3, 4-4, Part 2, 5-1, 6-1, 6-2, 7-1, 9-1b, 9-2, 9-3, 9-4, 11-1, 12-1, 13-1, 13-2, 13-3, 13-4, 13-5, 13-6, 14-1, 14-2, 14-3, 14-4, 15-1, Part 4, 17-2, 18-1, 18-3, 23-1b, Part 5, 26-1, 26-2, 26-3, 27-1, 29-1, 31-1, 32-1, 32-3, 33-1, 34-1, 34-2, 34-3, 35-2, 36-1, 37-1, 37-2, 38-1, 39-2, 40-2, 41-1, 41-2, 42-1, 42-2, 42-3, 44-1, 45-2, 46-1, 48-1, 50-1, 51-1, 51-3, 52-2, 52-3, 53-1, 53-2, 54-1, 54-2b, 54-3, 55-1, 57-1, 57-2, 57-3, 59-1, 59-2, 61-1, 64-2, 64-3, 64-4, unnumbered figure in Exercise 64, 65-1c, 65-2, 65-3, 67-2, 68-1, 69-1, 69-2, 69-3, 69-4, 69-5, 71-3, 71-4, 72-1, 72-2, 72-3, 74-1 (insets), 74-2, 75-1, A-1, A-2, A-3, A-4, Problem 1, Problem 2, C-1, D-1, E-1, F-1, Figures in Laboratory Reports 39 and 68.

Figures: 1-1 (top), 1-2, 9-1a, 30-1, Part 6, Part 7, 35-1, Part 8, 39-1, 40-1, Part 9, 49-1, 51-2, 52-1, 65-1a and b, 66-1, 66-2, 67-1, 67-3, 67-4, 67-5, 71-1, 71-2, 73-1, 75-1, 75-2, B-1.

Other Figure Sources

Part 1: Biological Photo Service. 1-1 (bottom): AO Scientific Instruments. Part 3: a, b, and d from Center for Disease Control, Public Health Service, U.S. Department of Health and Human Services, Atlanta, GA; c from R. J. Hawley, Ph.D., Department of Microbiology, University of Maryland, College Park, Maryland, and T. I. Imaeda, M.D., Ph.D., Department of Microbiology, University of Medicine & Dentistry of New Jersey, Newark, NJ. 25-1: General Diagnostics, Division of Warner-Lambert Co., Morris Plains, NJ. 25-2: Roche Diagnostics, Division of Hoffman-La Roche, Inc., Nutley, NJ. 49-1a: Thomas D. Brock, Katherine M. Brock, BASIC MICROBIOLOGY: With Applications, (c) 1973, p. 276. Reprinted by permission of Prentice-Hall, Inc., Englewood Cliffs, NJ. Part 11: Dwayne C. Savage, University of Illinois at Urbana-Champaign. 64-1: Alyne K. Harrison, Viral Pathology Branch, Center for Disease Control, Atlanta, GA. 65-1d: Center for Disease Control, Atlanta, GA. Part 12: J. G. Hadley, Batehe-Pacific Northwest Laboratories/Biological Photo Service. 74-1: Biological Photo Service.

Index

Bold numbers represent procedure descriptions.